Theodor J. Stewart
Robin C. van den Honert (Eds.)

Trends in Multicriteria Decision Making

Proceedings of the 13th International Conference
on Multiple Criteria Decision Making
Cape Town, South Africa, January 1997

Springer

Editors

Theodor J. Stewart
Robin C. van den Honert
University of Cape Town
Department of Statistical Sciences
Rondebosch 7701
South Africa

Cataloging-in-Publication Data applied for

Die Deutsche Bibliothek - CIP-Einheitsaufnahme

Trends in multicriteria decision making : proceedings of the 13th
International Conference on Multiple Criteria Decision Making, Cape
Town, South Africa, January 1997 / Theodor J. Stewart ; Robin C.
van den Honert (ed.). - Berlin ; Heidelberg ; New York ; Barcelona ;
Budapest ; Hong Kong ; London ; Milan ; Paris ; Singapore ; Tokyo :
Springer, 1998
  (Lecture notes in economics and mathematical systems ; Vol. 465)
  ISBN 3-540-64741-4

ISSN 0075-8442
ISBN 3-540-64741-4 Springer-Verlag Berlin Heidelberg New York

Typesetting: Camera ready by author
SPIN: 10546692   42/3143-543210 - Printed on acid-free paper

Preface

This volume contains a collection of papers presented at the 13th International Conference on Multiple Criteria Decision Making held in Cape Town during 6-10 January 1997. This was one of the regular conferences of the International Society on Multiple Criteria Decision Making, which are held at approximately two-yearly intervals. The 13th conference was attended by 143 delegates from 35 countries, and 123 papers were presented. Some of the papers which were presented are included in another volume published by Springer, namely the "festschrift" for Professor Stanley Zionts of Buffalo, New York, entitled *Essays in Decision making: A Volume in Honour of Stanley Zionts* (edited by Mark H Karwan, Jaap Spronk and Jyrki Wallenius).

The present volume provides a record of the technical part of the proceedings of the conference. In reflecting back on January 1997, however, memories return of the social occasions, and the camaraderie which developed amongst the MCDM community. Early January, at the height of the southern summer, is carnival time in Cape Town, and the mood is infectious, and was enhanced amongst those from the northern hemisphere (the majority of delegates) who overnight went from winter chills (as low as -30°C) to over 30°C. This spirit was in no way dampened when the "sunset cruise" on Table Bay, after three days of blazing sunshine, was so fogged in that the boats had to return to port inside an hour!

The holiday atmosphere did not detract from the high quality of the technical presentations, however. Within the general theme of *"where the currents meet"* (motivated by Cape Town's geographical position), a wide range of topics in multiple criteria decision making (or decision analysis, or decision aid) were covered, ranging from fundamental theoretical developments to many areas of application. In preparing these proceedings, the editors have attempted to put together a collection of papers which represent both the quality and the range of topics which characterized the conference.

The papers have been grouped into seven sections. The first contains the opening and closing plenary papers which were presented by Po Lung Yu and Stanley Zionts respectively. The remaining sections represent a classification of the remaining papers as follows:

- Preference modeling

- Methodologies of MCDA

- Negotiation and group decision support

- Systems and philosophical issues

- MCDA in environmental and natural resources management

- General applications of MCDA

Finally, the editors wish to take this opportunity to express their thanks to the referees who put so much effort into reviewing the papers. We greatly appreciate their assistance, without which this volume would not have been possible. Their names are listed at the end of the volume.

Theo Stewart and Rob van den Honert
Cape Town, May 1998

Contents

Part 6 MCDA in Environmental and Natural Resource Management   283

Part 7 General Applications of MCDA   345

Part 1
Plenary Addresses

Habitual Domains: Stealth Software for Effective Decision Making[1]

Herbert Moskowitz[2] and Po-Lung Yu[3]

[2] Lewis B. Cultman Distinguished Professor of Manufacturing Management
Director of the Center for the Management of Manufacturing Enterprises (CMME)
Krannert Graduate School of Management, Purdue University, West Lafayette, IN 47907, USA

[3] Carl A. Scupin Distinguished Professor
Director of the Center for Habitual Domains
School of Business, University of Kansas, Lawrence, KS 66045, USA

Abstract. Each human is endowed with a priceless supercomputer - the brain. Unfortunately, due to its limited software (habitual concept and thinking), the supercomputer could not fully utilize its potential. To be a great MCDM scholar, we need to expand and enrich our domains of thinking, which usually can lead us to strikingly better decisions than ever conceived possible. Using habitual domain (HD) theory, we provide easy but effective human software programs to expand and enrich our thinking for re-engineering effective decision making.

Keywords. Habitual domains, Charges, Competence sets, Decision traps, Decision elements, Decision environments, Decision blinds.

1 Introduction

Consider that each human is endowed with a priceless supercomputer of virtually unlimited capacity - a brain which consists of 100 billion neurons, which is virtually of limitless capacity. Each neural cell can be lit up or remain dark in processing information. To get a sense of the magnitude we are dealing with, consider the following numbers:

$$2^8 = 256$$
$$2^{100} = 1\ 267\ 650\ 600\ 228\ 329\ 401\ 496\ 703\ 205\ 376$$
$$2^{100\ 000\ 000\ 000} = \text{limitless!}$$

[1] This research is partially supported by Chiang Ching-Kuo Foundation for International Scholarly Exchange, Taipei, Taiwan, to the second author.
The authors thank Thomas J. Rzeszotarski of the University of Kansas in his assistance in composing this article.

Thus potentially we all have the capacity to be a super MCDM scholar:

> (1) to have a knowledge/skill base which is unlimited in breadth and depth;
> (2) to be infinitely liquid, flexible and well organized; and
> (3) to instantaneously understand all events and solve all problems.

We also have the capability to be very kind and loving, relieving our pains and frustrations and those of others, thus creating happiness in our lives and in the workplace. Unfortunately, because of our limiting habits, only a small portion of this potential has been realized - much less than 10%! Imagine the power and effect on an organization if each MCDM scholar could increase the utilization of his/her brain, i.e. improve his/her thought process by just 10%.

	% of Utilization	Neurons Utilized
Potential Realization	100%	100 000 000 000
Current Realization	10%	10 000 000 000
Increase Utilization by 10%	11%	11 000 000 000

The added brain power (as measured by the utilization ratio) for *each* individual would be:

$$\frac{2^{11\,000\,000\,000}}{2^{10\,000\,000\,000}} = 2^{1\,000\,000\,000}$$

Each concept and thought process in our brain - our supercomputer - may be viewed as human software. The totality of these programs and memory, together with their formation, organization, and activation we call a *habitual domain* (HD). Without efficient and effective software (human software), supercomputers (our brain) can become ineffective. The concepts (software) we use most often, psychologists tell us, are those we learned when we were young. Can we expect efficient or effective computing (thinking) if we are using software (concepts) that we learned 10, 20 or 30 years ago?

In the following, we shall introduce the concept of habitual domains (HD) as a tool to expand and enrich our thinking (Yu, 1985, 1990, 1995). We shall begin with HD analysis and core concepts underlying habitual domains in Section 2, and then examine decision traps and decision blinds in Section 3. We shall solve the decision traps by incorporating methods for effective decision making in Section 4. In Section 5, we discuss methods which will ultimately lead to a continuously expanding habitual domain. Some conclusions are offered in Section 6.

2 Habitual Domains: Constraints, Variables, and Concepts

Consider the following example:

Example 1: Child and Net Profits
Once there was a Chief Executive Officer (CEO) who was trying extremely hard with his wife to have a baby. Although the couple worked very hard, using many methods at having a child, the wife would not get pregnant. So they decided to adopt a child instead. The couple became very comfortable and happy with the child they had; the stress to have a child was now over. Unexpectedly, within the next six months the wife became pregnant. Why? Both felt large amounts of pressure, charges if you will, to have a child, which prevented them from having a child. When this charge was removed, the couple could have a child joyfully and biologically. This incident made the CEO think about the management of his enterprise. He used to focus on the "bottom line", net profit. The effort did not, however, give him much profit. He now changed his focus to "happiness and satisfaction of his employees and customers". Unexpectedly, just like getting his child, he reaped large profits. There is an old saying "happy cows produce better and more milk".

In terms of MCDM with respect to business, the CEO wanted to have profits and satisfaction of his employees (2 criteria). With respect to his family, he wanted a baby and a joyful relationship with his wife. When the focus was misplaced and high levels of charge were created, the CEO could not get what he wanted.

The above example suggests that by being willing to look at MCDM problems differently from one's own habitual ways, that person can see a better structure and solution to the problem. Indeed, by breaking loose the habitual constraints of the problems, we can expand and enrich our domain of thinking and discover a better solution than ever existed, greatly improving the quality of our decision making. Furthermore, releasing the high charges that we experience in everyday life will enable us to consider different perspectives and reach our goals, much like the CEO and his wife trying to have the child and profits.

Although different people or an individual at different times may behave very differently, it has been recognized that each person has habitual ways of responding to stimuli, sometimes called conditioned or programmed behaviour patterns. For example, each of us has habitual ways of eating, dressing, speaking and thinking.

An habitual domain consists of four elements:

(i)   The *potential domain* (PD) - the collection of ideas and actions that can potentially be activated.

(ii)   The *actual domain* (AD) - the set of ideas and actions that are actually activated.

(iii)   The *activation probabilities* (AP) - the probabilities that idea and actions in the PD also belong to the AD; the *core* of an HD is the collection of those strong ideas and actions that have a very high probability to be activated at any moment in time.

(iv)   The *reachable domain* (RD) - the set of ideas and actions that can be attained from a given set in an HD.

The relationship between the AD and the PD is similar to that between the realized value and the sampling space of a random variable. The concept of the RD arises from the fact that ideas can generate other ideas and that actions can prompt other actions. Within one's RD are the goals that can be achieved through efficient planning. The concept of a RD plays an important role in human cognition and the recognition of suggestions.

Since an HD can change over time, it is dynamic, as are its four elements: PD, AD, AP, and RD. But it will stabilize most of the time except during some transient periods. This arises because (i) the more one learns the less the probability that an arriving message or experience is new; (ii) as existing memory is utilized to interpret arriving messages and experiences, there is a tendency towards consistency; and (iii) the environment may tend towards a regular rhythm and thus yield regularity in its input of information and experience. Thus in the absence of extraordinary events, an individual PD and AP may stabilize. For a mathematical formulation and the proof of the stability of the HD, the reader is referred to Chan and Yu (1985) or Yu (1985, Chapter 9).

The stabilization, on the one hand, makes us more efficient in dealing with routine problems. On the other hand it may restrict our new thinking and creativity to occur and prevent us from discovering better solutions.

Ideally, we want our Potential Domain (PD) so large that we could identify with every human and every living thing and understand them. We want our Actual Domain (AD) to be so liquid that proper concepts can be instantaneously activated so as to release the pains and frustrations of ourselves and others. Note AD is only a small part of our PD.

Example 2: Hospital's Lesson
Once there was a young Vice President of Finance who was on the fast track. She worked long hours and often didn't leave the office until 7 or 8 at night. Even then she usually was carrying work to be completed at home. The fast-track Vice President was under considerable pressure. Unfortunately, when she needed to vent, the target usually was her husband. She spoke rudely to him and often criticized him. He didn't appreciate this and reciprocated in kind. The marriage soon deteriorated into an uneasy and frustrating relationship marked by frequent arguments. One day the husband visited a friend in a hospital's intensive care unit. There he saw patients who were so sick and weak that they couldn't breathe without the help of a respirator. It suddenly struck the husband how fortunate he was to have a healthy wife.

That night when the wife came home and began to unload on him, the husband just kept smiling at her as if he was appreciating and discovering something wonderful. She was puzzled (the behaviour was beyond the comprehension of her habitual domain), but on this evening the usual argument did not develop. "What's going on?" she finally asked her husband. He described the plight of the patients in intensive care, his gratefulness that his wife had avoided such a fate, and how much he loved her. The wife was moved by this and admitted that she loved her husband as well. The pair became best friends and the marriage thrived. It happened because the husband was willing to change his actual domain from the usual angry reciprocation to one of gratitude. This, in turn, led to a profound change in his wife's behaviour. This example illustrates that if our actual domain could be liquid in activating effective and powerful thoughts we could solve problems effectively and create happiness and wisdom.

Example 3: Long Hairs

A President of a company often, because of his business, needed to socialize with his customers at night at the social clubs or bars. The President would return home very late in the evening, sometimes drunk. His astute wife was not so receptive to his midnight excursions. When she found long hairs on his clothing that were not hers, she would argue and even fight with the President. After a long while, the wife could not find any long hairs. One day when the President came home, she broke out in tears. Being asked, the wife said "I could not find any long hairs for a long while". The President remarked, "Isn't that what you wanted?" Much to his surprise, the wife stared at him with cold eyes. "I cannot imagine that you have downgraded yourself so much as to fool around with women with no hair!" This example illustrates that we tend to see what we want to see and our actual domain (AD) is usually not very liquid to change.

3 Decision Blinds and Decision Traps

For each decision there is a *competence set* (CS) consisting of skill, knowledge, information. When the decision maker (DM) believes that he has the needed CS, he/she would be confident and comfortable in making decisions; otherwise the DM would be hesitant in making a decision. Unfortunately, the truly needed competence set CS, and that *perceived* by the DM, denoted by CS^* could be different. Note that CS^* is a part of a DM's HD, a reachable domain (RD). When CS^* contains only a small part of CS, i.e. $CS\backslash CS^*$ is large, the DM has large *decision blinds*, measured by $CS\backslash CS^*$.

Decision blinds result from the narrowness of reachable domains in thinking processes. The more narrow one's thinking process is, the more susceptible one will be to overlooking the solution to the problem. Every day everyone has many things that can create high stress or charge. If you can find a way to lower the charge down to an acceptable level, you can stop thinking about the problem and switch your attention to other things. As a result, the thinking domains (RD) are usually not broad and deep enough.

Example 4: Habitual Domains (HD) and Total Quality Management (TQM)
After World War II, the quality gurus in the U.S., such as Deming, Juran, and others, began espousing new ideas about quality and quality practices. The new ideas of the quality gurus were shunned in the U.S. However, they were imported to and accepted readily in Japan, during their post WW II reconstruction and recovery. This led to their transformation from a defeated nation, with a reputation for manufacturing "junk", into a feared and revered quality leader in the world marketplace. Meanwhile, the HD of Corporate America had become ossified over time, largely due to its past economic success and dominance. A triggering event -*charge* - was needed to deossify and expand the HD of Corporate America. That charge was the shocking realization that economic survival was at stake, as many of the most respected U.S. corporations (e.g. Hewlett Packard, Motorola, Xerox) suddenly found themselves on the brink of insolvency due to Japanese product superiority and cost!

U.S. Corporations expanded and adjusted their thinking (HD) about quality to respond to the new market environment in the following ways:

1. There was a shift in quality perspective from the manufacturing function to the corporate boardroom. Quality thus became defined from the customer's perspective, and quality now required inclusion in the strategic planning process.
2. Quality became viewed as an aggressive competitive weapon from both the market (demand) as well as cost (supply) sides. Defect-free production (the goal of statistical quality control and quality assurance) became too limited a perspective; quality was now viewed as a strategic and competitive weapon.

In this process of evolution in quality thinking, long-held beliefs about quality (old HD) now turn out to be myths (decision traps), replaced with more enlightening concepts (expanded HD). We state some of these myths below.

Myth 1: Inspection (product control) - defect detection - is the primary means for controlling quality
Truth 1: Defect prevention (process control) is the means for controlling quality

Myth 2: Higher quality yields lower productivity
Truth 2: Higher quality yields higher productivity

Myth 3: Achieving shorter cycle time and higher quality yields higher costs
Truth 3: Achieving shorter cycle time and higher quality yields lower costs

Myth 4: The operators in manufacturing are responsible for quality
Truth 4: Quality is everyone's responsibility

Myth 5: Quality is defined as conformance to specifications
Truth 5: Quality is defined from the customer's perspective

The above example illustrates that decision traps resulting from rigid and narrowly defined habitual domains can be fatal. Until we can jump out of the traps and transform our perceptions to powerful ones, we may not survive in competitions.

On perception, judgement, information collection, contingency calculation, strategic decision making, we usually have *frequently used* patterns of behaviour allowing us make quick conclusions and judgements and believe that our conclusions and judgement will not be wrong. As a result, our AD cannot move or transform, and our RD will become trapped. A trapped AD and RD will inevitably lead to decision traps, unable to see the full picture of decision problems.

Russo and Schoemaker (1989) have made an extensive study of decision traps. Let us discuss two of their ten decision traps. The first one is called *plunging in*. This decision trap occurs when one begins to gather information and reach conclusions without first taking a few minutes to think about the crux of the issue one is facing; this also includes the failure to think about how one believes decisions should be made. Such decision traps are common to most of us. How many times have you attempted to solve an MCDM problem by plunging in, formulating a utility function or finding non-dominated solutions? In Examples 1-3, do you see that each participant has some plunging in behaviours? Behaviour tendencies such as this will invariably lead to rigid habitual domains. More importantly, the more one incorporates these behaviours in his/her everyday thinking, the more one can expect to fall into such traps. Effectiveness is minimized if we are not mindful to prevent ourselves from plunging in.

The next decision trap is called *frame blindness*. In frame blindness, one sets out to solve the wrong problem because that person has created a mental framework for a decision, with little thought, that causes one to overlook the best options and lose sight of important objectives. Have you ever tried to find a linear solution with the end in mind? By doing so, you have fallen prey to frame blindness. In Example 1-3, do you see that each participant committed some kind of frame blindness? When we engage in frame blindness, we anticipate what the outcome will be. When we subjectively choose the variables, we are also selectively choosing the answers.

4 Avoiding Decision Traps: Effective Decision Making

Consider the following example:
Example 5: During the 1960s, commitments that were made by city authorities to the Woodlawn ghetto organization of Chicago, Illinois were not being met. The organization was powerless. Alinsky, a great social movement leader, came up with a unique solution. He would mobilize a large number of supporters to legally line up and occupy all the restroom facilities of the busy O'Hare Airport. Imagine the chaotic situation of disruption that would occur if thousands of passengers who were hydraulically loaded (very high level of charge) rushed for restrooms but could not find the facility to relieve their charge. They must find somewhere to

release their charge eventually. This would create great excitement, through media, around the world and Chicago would instantaneously become notorious. The supporters were extremely enthusiastic about the project, sensing the sweetness of revenge against the City. The threat of this tactic was leaked to the administration. Within forty-eight hours, the Woodlawn Organization was meeting with the city authorities. Alinsky was successful by going down deep into human physiology to obtain such a creative, yet effective strategy.

Ask yourself how Alinsky avoided decision blinds and traps in this instance. He was extremely mindful to avoid plunging in. Rather, he focused on striking the core of the HDs of leading political officials. Imagine if the largest airport in the world did not have restrooms readily available! Alinsky also did not limit himself through frame blindness as the protesters did in the beginning. Civil groups thought that picketing would solve the problem. The solution decided the action which led to frame blindness. Considering all perspectives allowed Alinsky to see the problem and solve it effectively.

To avoid or minimize the decision blinds and traps, we could let our AD and RD move and expand systematically according to *decision elements* and *decision environments* as described below. Decisions are composed of five elements: (1) *Alternatives* - the available alternatives that may solve the problem; (2) *Criteria* - which are used to judge the decision quality; (3) *Outcomes* - predicted outcomes measured by the criteria; (4) *Preference structure* - preferences over the outcomes; and (5) *External information* - information that arrives with or without solicitation. Every decision that is made, in one form or another, overlaps with each of these decision elements. Consider Example 5. How many alternatives do you think Alinsky entertained? What were the variables involved? What was the desired outcome? How did Alinsky's preference structure come into play? What role did external information play in the outcome? Upon careful analysis, we see that each decision element is an integral part of the process.

Comprehension and understanding of decision environments are also essential in understanding the decision problem. The four decision environments are:

(i)   *Part of the Behaviour Mechanism* - the decision process is only a small part of our behaviour mechanism
(ii)   *Stages of the decision process*
(iii)   *Players* - those who have a stake in the decision process
(iv)   *Unknowns* - the factors that are unknown to us in the decision process.

These unknowns can typically lead to decision blinds and decision traps. In Example 5, because of being able to greatly expand the thinking of decision environments, Alinsky was able to derive an extraordinary winning strategy, which was far beyond the HD of most people. Whenever we are confronted with a difficult decision, analysing the parameters of decision environments will maximize our efficiency in forming winning strategies. Analysis of these factors can greatly decrease the probability that one will fall into a decision trap or decision blind.

5 Expansion and Enrichment of HD

Each one of us has a large but different PD (potential domain). Due to its organization and activation probability (AP), the actual domain (AD) and the reachable domain (RD) may be bounded. For instance, in Examples 1-5, we see that most people may be bounded by their view and formation of decision problems, which may be due to the bounding of AD and RD and possible PD.

By expanding and enriching HD, we mean (i) expanding our AD and moving our RD so that we could pay attention to most PD to avoid blind points or traps; and (ii) deepening our thinking and absorbing new ideas so that our AD could indeed exceed the existing PD (potential domain), a real breakthrough of the existing HD.

When our HD is expanded and enriched, we tend to be more effective, happier, and wiser. There are many sources such as proverbs, common sense, quotations from the Bible, Koran and Taoism that can help us expand and enrich our HD. These expansion and enriching ideas are classified into 3 tool boxes (Yu, 1990 1995): eight methods for expansion of HDs, nine principles of deep knowledge, and seven empowering operators for expanding and enriching our HD (see Tables 1-3).

Table 1. Eight Basic Methods for Expanding and Enriching Habitual Domains

1. Learning Actively
2. Take the Higher Position
3. Active Association
4. Changing the Relative Parameter
5. Changing the Environment
6. Brainstorming
7. Retreat in order to Advance
8. Praying or Meditation

Table 2. Nine Principles for Deep Knowledge

1. Deep and Down Principle
2. Alternating Principle
3. Contrasting and Complementing Principle
4. Revolving and Cycling Principle
5. Inner Connection Principle
6. Changing and Transforming Principle
7. Contradiction Principle
8. Cracking and Ripping Principle
9. Void Principle

Table 3. Seven Self-Perpetuating Operators: Reaching Toward the Ideal Habitual Domain

1.   Everyone is a Living Priceless Entity
2.   Clear, Specific, and Challenging Goals Produce Energy for our Lives
3.   There are Reasons for Everything that Occurs.
4.   Every Task is Part of My Living Domains
5.   I am the Owner of My Living Domain
6.   Be Appreciative and Grateful and Don't Forget to Give Back to Society
7.   Our Remaining Lifetime is our Most Valuable Asset.

Each of these may be regarded as a human software program. Do not worry, they are timeless concepts that are continually upgradable. Being limited by space, in the following we share with the reader two methods for HD expansion, two principles, and two operators. The readers are encouraged to read the rest or discover new ones.

5.1 Projecting from a Higher or Lower Position (from Eight Basic Methods for Expanding and Enriching Habitual Domains)

In the abstract, each person or organization may be regarded as a living system within a hierarchical structure. As our role or position changes in the hierarchical structure, we tend to change AD and RD, our views and perceptions. Therefore, if we mentally project ourselves to a higher position in the hierarchy, our views will become broader and our HDs (AD, RD and possibly PD) will likely be expanded. For instance, if we are the Vice President of Research and Development and mentally project ourselves to the President's position, we tend to think of issues of not only the research and technical details, but also marketing, production, finance, public relations, etc. Our HD will then become broader and larger, which may pave the way for our next promotion.

Similarly, in Operations Research we tend to find optimal solutions to a given system, which is one of many available systems. By projecting the problem to a higher dimension, we will begin to think of the problem of selecting or designing the optimal system, instead of just finding the optimal solution to a given system. A technical analysis of such problems is discussed in Lee, Shi and Yu (1990) and Shi, Yu and Zhang (1996).

5.2 Active Learning (from Eight Basic Methods for Expanding and Enriching Habitual Domains)

By active learning we mean obtaining those concepts, ideas, and thoughts from various channels including consultations with experts, reading relevant books and following the radio, television, journals, etc. Libraries are filled with literature which contain human wisdom, experiences and thoughts. As long as we are willing to study, we certainly can learn from this abundant literature contributed by our ancestors and current scholars.

To understand why positive learning is so important in expanding our habitual domains, we must realize that through positive learning we will not only acquire new concepts, ideas and operators, but we will also restructure our memory and integrate the relevant concepts and ideas to make them easily retrievable and applicable. Furthermore, through our newly acquired concepts and ideas by means of restructuring, integrating and self-suggestion, we may be able to generate more new concepts and ideas.

There are four fundamental steps for success to this active learning. First, define your goal. Be sure to free yourself from your insecurities, shortcomings and fears -you are master of your own domain. Second, identify those persons - past or present - who have achieved the goal you are striving for. Third, study these role models. Learn from the thought patterns of mathematical geniuses such as Albert Einstein or Sir Isaac Newton. As you study them, their HD becomes part of your PD (potential domains). Finally, use the model examples as a means to form your decisions. Based upon successful completion of step three, ask yourself what your role model would think in this situation; the answer will activate your role model's thoughts (thus retrieve them from your PD and make them part of your AD) and help you solve your decision problems.

5.3 The Deep and Down Principle (from the Nine Principles of Deep Knowledge)

In our daily lives, we are continuously presented with routine - and occasionally urgent - events and problems, and our attention is thus forced to shift to them as they come to our attention. Against this background, it is easy to imagine that some good ideas stored in our potential domains may not catch our attention, and thus may not contribute to enriching our actual domains and reachable domains.

To overcome this neglect, and to prepare our HDs for better responses to non-trivial problems, we need to let our minds be peaceful and open - that is, go deep and down - to allow ideas in the potential to catch our attention. This can be accomplished by periodically setting a period of time during which we empty our unfinished goals, tasks or decisions. It is a commonly reported experience that good ideas often surface during the common routines of life, such as taking a shower and a walk, or during times set aside for quiet thought.

The other inference that should be drawn from this method is that it is useful to assume a low and humble posture, thus making it easy for other people to offer ideas and suggestions which can change and expand our AD, RD and PD. Such an open environment stimulates new and creative ideas. Lao Tze, a Chinese philosopher, put it this way: "Why is the ocean king of all rivers? It is because the ocean assumes a lower position." Thus it allows all the water to flow into it easily.

5.4 The Alternating Principle (from Nine Principle of Deep Knowledge)

A door which is always closed or always open will lose its usage as a door. Likewise, an assumption which is always imposed or left out will lose its usage as an assumption. We all carry, no matter how invisible, a set of assumptions for seeing problems and events (HD). Being aware of the set of assumptions people use, and alternating the combinations of assumptions (by adding or removing subsets of assumptions), we will realize the usage of the assumptions and clarify the understanding of the events and the depth of HD involved. By alternating 0 and 1, we can reconstruct the numerical systems; by alternating the three primary colours (red, yellow, blue), we can create infinitely many pictures, and by alternating the seven basic tones (do, ra, mi, fa,...), we can compose infinitely many beautiful musical compositions. Therefore, by alternating the combinations of assumptions, man should be able to conceive an infinite number of good ideas. To counteract the common tendency of always making the same assumptions as part of our HDs, the alternating principle urges that they be changed from time to time to see what new ideas can emerge.

A linear program typically has a single criterion and a fixed set of feasible solutions defined by linear inequalities. But the single criterion can be replaced by multiple criteria (MC), and an MC-simplex method can be developed. Similarly, the fixed sets of feasible solutions can be replaced by a flexible set defined by the multiple constraints (also denoted by MC) of several levels of inequalities. These two changes lead to the development of an MC^2-simplex method (Yu, 1985, pp. 219-263).

As another example, the numerical system is a great human invention and we tend to think (as assumption) that everything could be numerically ranked or ordered. Thus, in mathematical programs we impose upon ourselves to consider objective and constraint functions which are based on numerical order. In MCDM, there are scholars continuously focusing on representing preference in terms of numerical order (an HD). If we alternate the assumption so as to say that decision problems may not be formulated in numerical order, then we will have to force ourselves to consider other kinds of formulation and re-examine the process of human decision making. Then we will discover that the human being is a complex and dynamic system. He needs to reach and maintain an ideal or equilibrium state. To do so, effective decisions which emphasize expanding competence sets and avoiding blinds may be more natural and important than optimal decisions. In front of fuzzy or challenging problems such as illustrated by the previous examples, he would like to expand and enrich his HD or competence set.

This leads to new dimensions of research in decision making emphasizing effectiveness instead of optimality, and focusing on set expanding and covering concepts on analysing challenging decision problems instead of just numerical ordering. Common sense that has been accumulated in human history becomes alive and important in this new look at problems (See Chapters 7-13 of Yu, 1990; Yu, 1995; and Yu and Zhang, 1990).

5.5 Everyone is a Priceless Living Entity (from Seven Empowering Operators)

We are all unique creations who carry the spark of the divine. If you are an atheist or an agnostic, you may be in protest. However, do not pass judgement yet based solely on words such as "divine". Rather, let your judgements go from your mind. Let's talk a moment about the idea that everyone is a priceless living entity. We do not have to talk in religious terms; instead, let us talk in business terms. Suppose you were offered $1 million for your eyes. Would you take it? What if someone wanted to buy your limbs? Or your kidneys? What would they be worth to you? Being the wise MCDM scholars that you are, I would expect that you would probably not sell vital parts of your body. They are priceless to you. A motivational speaker from the United States, Zig Ziglar, has asked, "If you have a race horse worth $1 million, "would you let it drink alcohol, smoke, stay awake all night or be poisoned by drugs?" Is not your priceless body worth the care you would lavish on a race horse?

It is not just our bodies that require nurturing. Our minds (and souls, if you will) require it as well. In fact, we must be wary of establishing destructive behaviour patterns (human software programs). If reading the biography of a great person can result in our incorporating some of that person's habitual domain into our own, then one must assume that hours devoted to violent entertainment, destructive gossip or pornography will result in those negative forces forming behaviour patterns as well. Ziglar notes that if someone dumped garbage in our living rooms, we would be furious. Ironically, we unwittingly pollute our minds with intellectual garbage. Do you monitor the food for thought you incorporate into your habitual domain? Do you allow time each day to clean the garbage out of your mind?

Take care of yourself, because you are invaluable. You are blessed with a priceless gift: your body. Treat it well. This operator works not only for personal development but also for the development of organizational culture. It can make you well liked by your colleagues. If everyone in your organization could subscribe to this operator, how wonderful would be your work environment. Eventually this would lead to higher productivity, profitability and satisfaction.

5.6 Learn to be Appreciative and Grateful: Give Back to Society (from the Seven Empowering Operators)

This world is full of beautiful living things, scenery and events. At a simple level, we must never fail to appreciate the world around us that reminds us every day how precious life is. Appreciation, gratification and making contributions are all modes of behaviour which can be cultivated until they become second nature. Such behaviour patterns benefit us first, because these patterns make us feel good. But they also benefit others. Through reciprocation, by making a contribution and giving back some of what we have gained we will assure that our behaviour patterns create an upward spiral of joy and satisfaction that affects not only ourselves but those around us.

Do you think that sounds too good to believe? As true as we are writing this to you right now, it works. The sharp rise of volunteering in America is due to the fact that people receive immeasurable satisfaction from helping others. On a simple level, when you receive compliments, make sure you reciprocate to show your appreciation. It is important to say "Thank you". When others complain about something you have done, your first response may be resentment. Do not fall into that trap. As hard as it may be, try to appreciate an opportunity to improve yourself. View their criticism as a form of encouragement. Thank them for sharing their views, and your HD will automatically expand and be enriched.

7 Conclusion

Continuous expansion and enrichment of our HD is, as discussed above, crucial for transforming and empowering ourselves to become great MCDM scholars or operations researchers. In order to avoid being puzzled and not knowing what to do, we need to continuously understand and absorb the HD of other people, disciplines, and entities. We will reach the real void state in which nothing is unknown to us. In such a state, we can understand all events and solve all problems quickly and become a great operations researcher, if not the greatest.

References

Alinsky, S.D. *Rules for Radicals*, Vintage Books, New York, NY, 1972.

Chan, S.J. and Yu, P.L. "Stable Habitual Domains: Existence and Implications," *Journal of Mathematical Analysis and Applications*, Vol. 110, No.2, 469-482, 1985.

Garvin, D.A. *Managing Quality*, Free Press, 1988.

Lee, Y.R., Shi, Y. and Yu, P.L. "Linear Optimal Designs and Optimal Contingency Plans", *Management Science*, Vol.36, No.9, 1106-1119, 1990.

Li, H.L. and Yu, P.L., "Optimal Competence Set Expansion Using Deduction Graph", *Journal of Optimization Theory and Applications*, Vol. 80, No.1, 75-91, 1994.

Russo, E. and Schoemaker, P. *Decision Traps*, Simon & Schuster, New York, 1989.

Shi, D.S. and Yu, P.L. "Optimal Expansion and Design of Competence Sets with Asymmetric Acquiring Costs", *Journal of Optimization Theory and Applications*, Vol.88, No.3, 643-658, 1996.

Shi, Y., Yu, P.L. and Zhang, D. "Generalized Optimal Designs and Contingency Plans in Linear Systems", *European Journal of Operational Research*, Vol 89, 618-641, 1996.

Yu, P.L. *Multiple Criteria Decision Making: Techniques and Extensions*, Plenum,

New York, 1985.

Yu, P.L. *Forming Winning Strategies, An Integrated Theory of Habitual Domains*, Springer Verlag, Berlin, Heidelberg, London, New York, 1990.

Yu, P.L. "Habitual Domains", *Operations Research*, December 1991.

Yu, P.L. *Habitual Domains: Freeing Yourself from the Limits of Your Life*, Highwater Editions, Kansas, 1995.

Yu, P.L. and Moskowitz, H., "Habitual Domains: Stealth Software for Human Re-engineering", *International Journal of Operations and Quantitative Management*, Vol. 2, No.1, 3-24, April 1996.

Yu, P.L. and Zhang, D. "A Foundation for Competence Set Analysis", *Mathematical Social Sciences*, Vol.20, 251-299, 1990.

Ziglar, Z. (cassette tape) *How to be a Winner*, Nightingale Conant Corporation, Chicago, IL.

Bringing MCDM to the Forefront of Management Practice

Stanley Zionts

School of Management, State University of New York, Buffalo, NY 14260-4000, USA

ABSTRACT

A great deal of effort has been expended in Multiple Criteria Decision Making in developing methods, and preparing computer programs implementing the methods. Though many articles have been published in the academic literature, and most management science and operations research texts devote space to multiple criteria decision making, there is a dearth of <u>real</u> applications and successes of multiple criteria decision making. Though it takes time to come up with and implement real applications, the success of our field ultimately depends on such applications. In this presentation, I reflect on what's been done, and explore ways in which participants in the field can try to further our application successes.

1 Introduction

I was asked to talk about the history and the future of MCDM. The field is really too young to talk about its history. I will say a few words about the history, however. I will then take stock of where our field is today, make some comments about the future, and then draw on what I have said to make some comments about what we should be doing and where we should be going.

The first recorded information I have seen regarding multiple criteria problems is over two hundred years old, attributed to Ben Franklin, the famous American statesman and is well documented. No doubt we could find references to multiple criteria problems in the Bible, but I haven't looked there. The modern roots of the field are probably in the work of Charnes and Cooper on goal programming (See, e. g., Charnes and Cooper, 1961). Many authors have written on multicriteria problems and methods since the early work. However, our past is prologue. The important thing to do is to take stock of where we are, and decide where we're going.

2 Where Are We? - The Present

What is the purpose of multiple criteria decision making? Multiple criteria decision making is making decisions in the face of multiple conflicting objectives. I would say the purpose of MCDM is to help decision makers learn about their problems, generate alternatives and make decisions, thereby solving their problems. What do we do in MCDM? We develop methods for solving MCDM problems, but, with few exceptions, we don't help people solve problems.

To evaluate where we are, I plan to use the SWOT technique of management strategy, to analyze the Strengths, Weaknesses, Opportunities, and Threats that I see for our field. The SWOT technique, for those not familiar with it, looks at the positives and negatives both inside and outside an organization. The positives and negatives inside the organization are referred to as the strengths and weaknesses, respectively, and the positives and negatives outside the organization are referred to as the opportunities and threats, respectively. By listing and analyzing the strengths, weaknesses, opportunities, and threats, an organization can try to capitalize on its positive factors, and compensate for or fortify itself against the negative factors.

We now take stock of our field. Though many articles have been written on MCDM (see below), and many methods have been developed, I would say that there are a handful of methods that have been successful. I'll indicate which ones I think are successful. Please excuse me if your list disagrees with mine. My definition of success is informal. It is based on my experience, and draws on what I have heard from others. Successful implementation of a method generally goes hand in hand with an appropriate computer program. Accordingly, a current summary of computer packages may be found in Buede (1996), who lists thirty packages together with sources for obtaining the methods. An earlier summary by Buede (1992) had roughly the same number of (and roughly the same) computer packages.

The essential point is that out of the myriad of approaches, only a handful of methods have achieved any success. I would include the Analytic Hierarchy Process (AHP) (Saaty, 1980) and its computer implementation, Expert Choice, the Pareto-race approaches, VIG, VIMDA (Korhonen and Wallenius, 1988), the methods Electre (see, e. g., Roy and D. Bouyssou, 1993), as well as the Multiattribute Utility Theory (MAUT) approaches of Keeney (1992) and its descendants. As an example of some indication of success in another field, marketing, Green and Krieger (1996) cites the MAUT and AHP methods as approaches for conjoint analysis, a specialized analytic marketing tool "close" to MCDM.

2.1 Strengths

Within the field of operations research and management science (OR/MS), in 1993, (Committee ..., 1993) MCDM was ranked in terms of its importance in an OR/MS graduate curriculum, based on a survey of over 200 professionals as the fifth and seventh most important topics of fifteen, respectively, for recent graduates of OR/MS programs and for seasoned professional practitioners of OR/MS. The related topic decision analysis was ranked fourth and third, respectively. Perhaps if the two had been combined, the combined ranking might have been higher.

We have many publication outlets for our research work. There are two dedicated journals: the *Journal of Multi-Criteria Decision Analysis*, the *Group Decision and Negotiation Journal*. But there are many others in which articles have been and continue to be published. During the six years from 1987 through 1992, 1216 refereed journal articles on MCDM were published in 153 journals, with authors from over fifty countries (Steuer et al., 1996). That's over two hundred articles per year, and the number of articles published per year was relatively evenly distributed over the period. The top ten journals, in terms of number of articles published, are, in descending order, European Journal of Operational Research, Journal of the Operational Research Society, Computers and Operations Research, Mathematical and Computer Modeling, Journal of Optimization Theory and Applications, Decision Sciences, Fuzzy Sets and Systems, Management Science, Central European journal for Operations Research and Economics, and Naval Research Logistics. The definition of an MCDM journal article there includes "... articles whose titles convey a strong multiple criteria message and whose abstracts corroborate a clear multiple criteria content. ... (W)e may also look at an article's list of references...".

Though I'm not familiar with all the publication outlets cited, I do know many of them. They all appear to be academic journals. In addition to the journal articles (which appeared in regular issues of the journals as well as 31 special issues from 1980 through 1993), the authors of the study also identified 217 books and proceedings volumes on MCDM. There were 143 conferences, workshops, and summer schools on MCDM from 1972 through 1994. Though Steuer et al. (1996) don't give a count on the number of authors of the papers, the latest MCDM WorldScan (Vol. 10, No. 1, April, 1996 - published by the International Society on Multiple Criteria Decision Making by Ralph Steuer, U. of Georgia, Athens, Georgia), lists 1,208 members in 82 countries.

Summarizing the strengths of the field, we have the following:

1. A field that is highly rated within the MS/OR field.
2. Many publications including some outstanding research, and many publication outlets including two "dedicated" journals.
3. A large number of intelligent people, all over the world, working on MCDM;
4. The International Society on Multiple Criteria Decision Making, a vibrant society that has great meetings, colleagues, and research, and some application work;
5. Multiple societies (in addition to our society there are the European Working group, Society for Judgment and Decision Making, the INFORMS special group on decision analysis and other groups).

2.2 Weaknesses

Our weaknesses all stem from our being in the academy. We now consider them further:

1. There is little or no source of funding for our work, other than the usual research sources available to academics.

2. As seen above, our outlets for publication are academic journals. We tend to publish very little, if at all, outside of academic journals. As a result, we tend to "talk a lot" to others in the field, and not to people outside the field.

3. In spite of MCDM being an important part of management science, management science is a declining field of business education. At many universities, including my own, management science is being de-emphasized, and required management science courses are being eliminated.

2.3 Opportunities

There are many opportunities, some of which may be derived from the strengths and weaknesses presented above:

1. There are many important multiple criteria management and societal problems to be attacked. Almost all real problems involve multiple objectives. By attacking and helping to solve such problems, we may be able to help legitimize our field and make a contribution at the same time.

2. By trying to solve practical problems, we may win some support for our work, and perhaps be able to generate funding.

3. Current technological innovations provide many opportunities for us to implement our ideas and do so in user-friendly ways. These include computer developments, such as high-speed reliable desk-top (or lap-top) computers with high-quality graphical interfaces, as well as the telephone and other communications ties to network computers, and the internet.

2.4 Threats

Currently, the threats to our field, based on my findings, are rather limited. There is the possibility of competition from others outside our field, though as of now, I believe the threat is minimal. However, unless we find some payoffs outside our field, there may be a problem sustaining the momentum that our society and others in the field have created.

2.5 Summary

We may summarize the SWOT analysis as follows:

Strengths
1. A highly-rated field within management science/operations research.

2. Many publications including some outstanding research, and many publication outlets including two "dedicated" journals.

3. A large number of enthusiastic and intelligent people, all over the world, working on MCDM;

4. Our society and others with great meetings, colleagues, and research.

Weaknesses

1. Little or no funding.

2. Our outlets for publication are academic journals. Little publication outside of academic journals. We "talk a lot" to others in the field, and not to people outside the field.

3. Though MCDM is an important part of management science, management science is currently a declining field of business education.

Opportunities

1. We can use MCDM approaches to attack many important management and societal problems.

2. Solving practical problems may help win support for our work, and perhaps be a source for funding.

3. Technological innovations provide many opportunities for our approaches.

Threats

Very limited.

To summarize, we have many excellent people all over the world working on interesting approaches in a highly regarded field that is part of a declining field. We tend to work very much in our own little fraternity, and have limited resources for our work. On the other hand, we have virtually no threats to the field, and the opportunity to do "REAL" work that may generate both financial and other support for our field.

3 A Digression - A Celebrity

During the conference presentation, I presented a photograph of a man and asked whether or not participants recognized him. Unfortunately, I could not reproduce the photograph here, but fewer than ten per cent of the participants recognized the mystery person. I know that I didn't from the photograph. The picture was of Boris Berezovsky, who was recently quoted by Forbes magazine (December 30, 1996, pp. 90ff.), an important American business magazine, as being the "most powerful man in Russia". The article states that Berezovsky is a "wealthy Russian car dealer, ... a powerful gangland boss, and the prime suspect in Russia's most famous murder investigation. ... (Berezovsky) is certainly one of the country's first dollar billionaires." His interests include Russia's biggest television network, one of Russia's most respected newspapers, an 80% interest in a major oil company, and considerable real estate in Moscow and St. Petersburg. Berezovsky is also accused of taking money from Aeroflot, the Russian airline. Berezovsky is also a confidant of Boris Yeltsin, and is said to be the "godfather of Russia's godfathers". Just who is Berezovsky then? The same article says, "In appearance and in background, Berezovsky is no thug. Boasting a Ph. D. in applied mathematics, the 50-year-old Berezovsky says he spent 25 years doing research on decision-making theory at the Russian Academy of Sciences." Ah! The plot thickens. Until recently, Berezovsky's name appeared on the list of members of the International Society on Multiple Criteria Decision Making, and he knew and was known by members of our society. After reading the article, I tried to contact Berezovsky, through Forbes magazine, through the TASS news agency, and through Russian and other colleagues. I wanted him to make some comments on the practice on MCDM, from the perspective of a practicing businessman.

I can only speculate on what he might have said. I can hear him say, "The real world is completely different. We don't have time for mathematical models. We make decisions and go on. Any analysis must, of necessity, be short and sweet. If you want me to use a model, it must be something that I can easily use and find useful."

Unfortunately, I haven't heard from him, and various Russians in Capetown cautioned me that it might be dangerous for me to speak about him there, because there may be links back to him. However, I have stated nothing except what I read in the media. As much as I would have liked for him to give us the benefit of his expertise, I reflected and then thought that we have many other members of our society who have become successful businessmen and academic administrators (chairmen, deans, and rectors). Why don't we ask them about their use of MCDM models in their work? I have yet to do that, but I am sure that we could gain from their wisdom.

4 The Future

It is always fun to speculate about the future. We can think of ourselves as fortune tellers! There are some things that we can be confident of predicting. Perhaps the major thing is that computing and communication will become faster, more reliable, and cheaper. As this happens, we can have more and more powerful computers capable of long-range network interaction.

As mentioned earlier, we currently have a backlash against management science. Management science is in decline in many, if not most, business schools, and there is a de-emphasis of quantitative methods in management. Though this is of course not the case in engineering, mathematics, and statistics, most of the expected users of models for important decisions are likely to come from the management area. Whether this will continue or not is hard to say though it does appear to be a trend. At my university, the State University of New York at Buffalo, we had several focus groups consisting of business people from small and large businesses, as well as recent alumni. They studied our MBA curricula. Based on their recommendations, our school changed the MBA curriculum. Though the groups were not unanimous on any issue, the most decisive issue was their unhappiness with quantitative methods. As a result, our basic MBA (required) management science course was dropped from the curriculum. The de-emphasis of management science is going on in most schools. It is likely to continue. So MCDM is an important area in a declining field. On the positive side, the use of quantitative methods has permeated most aspects of business education.

5 What Can (and Should) We Do?

We must *DO* more with MCDM. Publishing technical articles that get read by our peers is just not good enough; we need to peddle and sell our wares better to the management community. We must promote and sell what we do. Let's learn from the peddlers, even if they only *talk* a good game.

Before management will buy our wares, there are two things they must do. First, they must understand what we have and can do. Understanding does not mean that they must understand the mathematical intricacies or theories. When driving a car, a driver doesn't need to understand the workings of an internal combustion engine. All he needs to know is how to drive a car. Second, a user must find an approach useful, and must want to use it. In the same way, a driver must find driving the car attractive, and he must trust the car. He must have confidence that a car will take him where he wants to go, safe and comfortably, without breaking down. When Jake Burton developed the snowboard about twenty years ago, he was unable to sell it. So he had to build demand for it. Building demand for the snowboard, introducing it to potential users, not developing it, brought about its success.

Related to the selling of our wares, it is easier for us to publish in the usual places rather than in general management publications. Some time ago, Jyrki Wallenius, Pekka Korhonen, and I wrote a paper that was designed to informally present the concepts of MCDM to managers. The object was to present several simple concepts, including dominance, and the problems of using simple weights. We also included several examples of how easy it is to make a poor decision, using weights or ignoring dominance considerations. Though our paper has made the rounds of several highly visible management publications, it has been rejected for publication every time. Each time it is rejected, we rewrite the paper, and try again. We feel confident that we could easily publish it in an Operations Research outlet. Even though that might be relatively easy to do so, we continue to rewrite the paper so that we can publish it in a general management outlet. We are still working on it. It would be easy for us to publish the paper in our usual outlets. Yet the returns come not from publishing in operations research outlets, but from publishing in general management outlets!

6 What Can We Learn and Do Differently? Some Ideas

What can we learn from our SWOT analysis and the above vignettes? First, as just illustrated, it is probably easier for an established academic to publish articles in academic journals than to publish in practitioner or management outlets. It may be easier to take the path of least resistance, and to publish where we can, especially considering that we get more academic 'points' by publishing in academic journals. The message is that we should persevere and try to publish more of our works in the general management field. Below are some other possible lessons.

6.1 Encourage Simpler Approaches

We need simple approaches. Vahid Lotfi, Theo Stewart, and I intended the Aspiration-Level Interactive Method for MCDM (AIM) (Lotfi, et al., 1992), to be a really simple approach. The idea was that people tend to think in terms of aspiration levels, and that we could use the aspiration levels and feedback to help in decision making. We thought that by the use of aspiration levels, we could do two things:

1. Identify alternatives that satisfied aspiration levels;
2. Rank them, so that if no alternatives satisfy the aspiration levels, we would have them arranged in the order of 'most closely satisfying the specified aspiration levels.

Our working title for the approach was SIMPLE. We have yet to achieve what was intended in our work in terms of implementation. We continue to work on it.

6.2 Spreadsheet Solvers

I have proposed several times at these meetings the idea of a spread sheet solver for MCDM. The idea is that in the context of an electronic spreadsheet, we can include a method for solving problems using one or more MCDM methods, much as today's spreadsheets such as EXCEL include solvers for management science methods such as linear programming. The user of a spreadsheet could use the solver to help think through a decision. Theo Stewart's early implementation of our AIM approach was in fact a spreadsheet approach, relying upon spreadsheet macro instructions. As such, it was as close to a spread sheet solver of MCDM problems as I have seen. Steuer, in a paper presented at this meeting, (Steuer, 1997) presented another implementation of an MCDM approach in a spreadsheet.

6.3 Niches and Opportunities Using MCDM Methods

There may be special opportunities for MCDM methods to solve certain business problems. For example, our AIM approach (Lotfi et al., 1992) can be used to identify favorable business niches for new product development. Though we have yet to do it, we can categorize products using attributes (or objectives) in AIM. Then we can identify for every possible set of aspiration levels (a vector) a most preferred product for that vector. The products that are preferred most often are candidates for similar new product development. This is just an idea, and must be further developed.

6.4 Use MCDM to help construct new alternatives

One of the most important things to be done in decision making is to invent new alternatives. One of our problems in MCDM is that we think of the set of alternatives as fixed. In decision making and negotiation, managers generate new alternatives. It may be possible to use MCDM methods in a manner similar to that just proposed to help in the construction of new alternatives.

6.5 Use Technology

The role of technology in our lives has been fantastic. Communication and computing have developed in ways we could not have imagined only a few years ago. It is incumbent on us to use technology to develop and promote our work.

7 Conclusion

In this paper I have taken a brief look at some current aspects of MCDM, and speculated somewhat on the future. There is much more to do. We must communicate with the management community, and listen to what they want, develop wares they can and will use, and then sell our wares. We have a great community, and a fine though short history. Let's use the momentum to keep multiple criteria decision making a viable and important field.

8 References

Buede, D., "Software Review - Overview of the MCDA Software Market", Journal of Multi-Criteria Decision Analysis, July, 1992, Vol. 1, No. 1, pp. 59-61.

Buede, D., "Software Review - Second Overview of the MCDA Software Market", Journal of Multi-Criteria Decision Analysis, December 1996, Vol. 5, No. 4, pp. 312-316.

Charnes, A., and W. W. Cooper, Management Models and Industrial Applications of Linear Programming, John Wiley and Sons, New York, 1961.

Committee for a Review of the OR/MS Master's Degree Curriculum, "Suggestions for an MS/OR Master's Degree Curriculum", <u>OR/MS Today</u>, February, 1993, pp. 16-31.

Chen, A.H.Y., Jen, F. C. and Zionts, S., "Portfolio Models with Stochastic Cash Demands," <u>Management Science</u>, November, 1972, Vol. 19, No. 3, pp. 319-332.

Green, P. E., and A. M. Krieger, "Individualized Hybrid Models for Conjoint Analysis", <u>Management Science</u>, 42, 6, 1996 pp. 850-867.

Keeney, R. L., *Value Focused Thinking: A Path to Creative Decision Making*, Cambridge, MA: Harvard University Press, 1992, 416 pp.

Korhonen, P., and J. Wallenius, "A Pareto Race", <u>Naval Research Logistics</u>, 35, 615-623, (1988)

Lotfi, V., Stewart, T. J., and Zionts, S., "An Aspiration-Level Interactive Model for Multiple Criteria Decision Making", <u>Computers and Operations Research</u>, Vol. 19, No. 7, 671-681, 1992.

Roy, B., and D. Bouyssou, *Aide Multicritère à la Décision: Méthodes et Cas*, Paris: Economica, 1993, 695 pp. (In French).

Saaty, T. L., *The Analytic Hierarchy Process*, New York: McGraw-Hill, 1980, 287 pp.

Steuer, R. E., L. R. Gardiner, and J. Gray, "A Bibliographic Survey of the Activities and International Nature of Multiple Criteria Decision Making", <u>Journal of Multi-Criteria Decision Analysis</u>, September, 1996, Vol. 5, No. 3, pp. 195-217.

Steuer, R. E., "Implementing the Tchebycheff Method in A Spreadsheet", in Karwan, M. H., J. Spronk, and J. Wallenius (eds.), *Essays in Decision Making - A Volume in Honour of Stanley Zionts*, Berlin: Springer-Verlag, 1997, 93-103.

Part 2
Preference Modeling

Discrete Decision Problems, Multiple Criteria Optimization Classes and Lexicographic Max-Ordering

Matthias Ehrgott[1]

[1] Department of Mathematics, University of Kaiserslautern, PO Box 3049, 67653 Kaiserslautern, Germany, e-mail: ehrgott@mathematik.uni-kl.de; partially supported by the Deutsche Forschungsgemeinschaft (DFG) and grant ERBCHRXCT930087 of the European HC&M Programme

Abstract. The topic of this paper are discrete decision problems with multiple criteria. We first define discrete multiple criteria decision problems and introduce a classification scheme for multiple criteria optimization problems. To do so we use multiple criteria optimization classes. The main result is a characterization of the class of lexicographic max-ordering problems by two very useful properties, reduction and regularity . Subsequently we discuss the assumptions under which the application of this specific MCO class is justified. Finally we provide (simple) solution methods to find optimal decisions in the case of discrete multiple criteria optimization problems.

Keywords. Discrete decision problems, Classification, Lexicographic max-ordering

1 Discrete Decision Problems and Multiple Criteria Optimization Classes

A discrete decision problem consists of selecting from a finite set of alternatives $A = \{a_1, \ldots, a_m\}$ an 'optimal' one. In this paper we consider the context of multiple criteria decision making (MCDM). The solution of discrete decision problems with multiple criteria is often called multiple attribute decision making, see [HY81]. Every alternative is evaluated with respect to a certain number of criteria, or objective functions. We assume that $f_q : A \to I\!R$; $q = 1, \ldots, Q$ are Q real valued functions representing the different criteria. Therefore we can calculate the value of each alternative with respect to each criterion as given by $v_{qj} = f_q(a_j)$. Since the number of alternatives as well as the number of criteria is finite it is convenient to write all the values as a $Q \times m$ matrix:

$$V = (v_{qj}) = (f_q(a_j)).$$

Matrix V summarizes the performance of all alternatives with respect to all criteria. We understand an optimal alternative to be one with minimal performance with respect to the Q criteria. This implies that minima of a set of vectors, namely of $\{(f_1(a_j), \ldots, f_Q(a_j)) : j = 1, \ldots, m\}$, or, in other words, of the columns of V, have to be found. Due to the fact that, if $Q \geq 2$, there does not exist a canonical total order of $\mathbb{R}^Q$, there are lots of reasonable choices of a (partial) order of $\mathbb{R}^Q$ which will lead to as many different definitions of minima. This means that, besides specification of the criteria, the decision maker (DM) faces the task of selecting an order, which will define the optimal alternatives of the decision problem. Finally he also has to solve his decision problem. It is a straightforward assumption that the DM will have certain ideas about properties of this solution process.

In the remaining part of this section we will introduce a framework which allows to formalize the notion of the DM's ideas concerning desirable features of the solution process and properties of optimal solutions of decision problems. In Section 2 we will then introduce two such desirable properties and show that they uniquely define a specific pattern of solving decision problems. The corresponding (simple) solution method will be presented in Section 3.

Formally, let us consider a general multiple criteria optimization problem (MCOP). Let $\mathcal{F}$ be the set of feasible solutions and $f : \mathcal{F} \to \mathbb{R}^Q$ be the objective function. Function f is a vector valued function and can be written as $f = (f_1, \ldots, f_Q)$, i.e. $f_1, \ldots, f_Q$ are the Q criteria. The goal of the solution of MCOP is to find 'minimizers' of the objective function, considered to be optimal solutions of the MCOP at hand. It is quite common in MCDM that an optimal decision is found by optimizing not directly with respect to the criteria. Often a utility function U is applied to the vectors $f(x)$, $x \in \mathcal{F}$. Then any $x \in \mathcal{F}$ minimizing $U(f(x))$ is an optimal solution of the MCOP. For instance, U can be a scalarizing function defined by a weighting vector $\lambda = (\lambda_1, \ldots, \lambda_Q)$ where $\lambda_q \geq 0$ and $\sum_{q=1}^Q \lambda_Q = 1$. In this case U is defined by $U(f(x)) = \sum_{q=1}^Q \lambda_q f_q(x)$.

From these considerations we conclude that, in general, the solution of an MCOP cannot be determined from the feasible set $\mathcal{F}$ and the objective function f, mapping $\mathcal{F}$ to $\mathbb{R}^Q$, only. Information concerning the utility function U and the final comparison of the feasible solutions in the image space of U ($\mathbb{R}$ with the canonical order in in the case of a scalarizing function) is needed. However, this information is usually implicitly defined by the 'ideas' of the DM about the solution of 'his' MCOP. Making this information available the DM could possibly gain a more profound understanding of the structure of an MCOP. Furthermore, it would be easier for him to communicate with others or even help him to find existing solution procedures.

We therefore propose to include this 'hidden' information explicitly in the formulation of an MCOP. Our concept is more general and comprises utility functions, scalarizing functions and many other as special cases. This concept provides the possibility of a classification of multiple criteria optimization problems, see [Ehr97b], and also is a helpful tool for theoretical achievements in the area.

Let $\theta : \mathbb{R}^Q \to \mathbb{R}^P$ be a point-to-set map, i.e. $\theta(v) \subset \mathbb{R}^P$ is a subset of points in $\mathbb{R}^P$, where $P \geq 1$. This point-to-set map is applied to the objective values vectors $f(x)$, yielding a set of outcomes

$$O = \bigcup_{x \in \mathcal{F}} \theta(f(x)).$$

Let us assume that $\mathbb{R}^P$ is endowed with some kind of ordering $\preceq$, which we assume to be at least a strict partial order. Then we define the set of minima to be all $z \in O$ such that there does not exist another point $z' \in O$ satisfying $z' \preceq z$ and $z' \neq z$. An optimal solution of an MCOP is then any feasible solution x such that x is in the preimage of an element of the set of minima. The map θ will be called the **model** of the MCOP problem.

We illustrate the concept by showing that scalarizing functions mentioned above are special cases of models. If we define θ by $\theta(f(x)) = \sum_{q=1}^{Q} \lambda_q f_q(x)$ and let $P = 1$ we have $O = \{\sum_{q=1}^{Q} \lambda_q f_q(x) : x \in \mathcal{F}\}$ and the set of minima is simply the minimum of the scalarizing function and any minimizer of that same function is considered an optimal solution, as usual. Note that θ maps to $\mathbb{R}$ and solutions are compared by comparing $\theta(f(x))$ values in the ordered set $(\mathbb{R}, \leq)$. But, moreover, our concept of the model map allows the following definition: let

$$\Lambda := \left\{ \lambda \in \mathbb{R}^Q : \lambda_q \geq 0, \ q = 1, \ldots, Q; \ \sum_{q=1}^{Q} \lambda_q = 1 \right\}.$$

We define

$$\theta(f(x)) := \langle \Lambda, f(x) \rangle := \{\langle \lambda, f(x) \rangle : \lambda \in \Lambda\}.$$

Here $\langle \lambda, f(x) \rangle$ denotes the Euclidean scalar product of λ and $f(x)$ in $\mathbb{R}^Q$. By this definition the set of minima is the set of all minima with respect to any scalarizing function.

We introduce some notation and formalize the ideas of the above discussion. As mentioned earlier the basic information about the MCOP problem is given by the set of feasible solutions $\mathcal{F}$, the criteria, $f = (f_1, \ldots, f_Q)$, and the image space $\mathbb{R}^Q$ of f. We will call this triple $(\mathcal{F}, f, \mathbb{R}^Q)$ the **data** of the MCOP. The description of the MCOP is completed by the model map θ (which is possibly a point-to-set map) and the **ordered set** $(\mathbb{R}^P, \preceq)$, in which minima are actually sought. Therefore we have three basic elements defining an MCO problem, namely data, model, and ordered set. From now on MCOPs will be denoted by

$$(\mathcal{F}, f, \mathbb{R}^Q)/\theta/(\mathbb{R}^P, \preceq).$$

As an indication that $P > 1$ makes perfect sense we note that if we want to solve an MCOP problem in the commonly used sense of Pareto optimality (i.e.

a solution $x^* \in \mathcal{F}$ is Pareto optimal if there does not exist another solution $x \in \mathcal{F}$ such that $f_q(x) \le f_q(x^*)$ for all $q = 1, \ldots, Q$ and $f_j(x) < f_j(x^*)$ for at least one j). We choose $\theta = \mathrm{id}$, the identity map, $\mathbb{R}^P = \mathbb{R}^Q$ and $<$ for $\preceq$. Here we use $<$ in the following way: if $a, b \in \mathbb{R}^Q$ then $a < b$ if $a_q \le b_q$ for all $q = 1, \ldots, Q$ and $a_j < b_j$ for at least one j. Thus $(\mathcal{F}, f, \mathbb{R}^Q)/\mathrm{id}/(\mathbb{R}^Q, <)$ denotes an MCOP to be solved in the sense of Pareto optimality.

The solution of an MCOP $(\mathcal{F}, f, \mathbb{R}^Q)/\theta/(\mathbb{R}^P, \preceq)$ consists of two parts: the determination of the set of minima of $\theta(f(\mathcal{F}))$ and the determination of the set of optimal solutions, i.e. the preimage of the set of minima. The former set will be denoted by

$$\mathcal{V}_{\mathrm{opt}} \left((\mathcal{F}, f, \mathbb{R}^Q)/\theta/(\mathbb{R}^P, \preceq) \right)$$

and the latter by

$$\mathbf{Opt} \left((\mathcal{F}, f, \mathbb{R}^Q)/\theta/(\mathbb{R}^P, \preceq) \right).$$

The notation of MCOP we have introduced here can be used as a classification scheme for multiple criteria optimization problems. We refer to [Ehr97b] for more about this aspect. The classification makes it easily possible to compare various MCOP problems and to exhibit their interrelations. It may also provide a structure and guiding line through the vast amount of literature on multiple criteria optimization. A second main advantage of the previously introduced notation is the possibility of proving results which hold independent of the data of a specific problem. In [Ehr97b] various results of this type have been proven. They can be interpreted as structural properties of whole classes of MCOPs. We follow [Ehr97b] and define an **MCO class** to comprise the set of all MCOPs which have the same model map and ordered set. An MCO class is denoted by

$$(\bullet, \bullet, \bullet)/\theta/(\mathbb{R}^P, \preceq).$$

The dots indicate an arbitrary entry. We note that results which can be shown for an MCO class are valid for all problems in the class, i.e. for any specific problem data.

In Section 2 we introduce properties of MCO classes which are reasonable from a decision makers point of view. We have mentioned the 'ideas' of the DM about the solution of MCOPs earlier. Formally, these 'ideas' are properties of MCO classes. For two of these properties we show that they characterize a specific MCO class, the lexicographic max-ordering MCO class. Subsequently further properties of this class are established. Section 2 is concluded by a discussion of the circumstances under which the application of this specific MCO class seems to be justified. Section 3 is devoted to (simple) solution procedures for determining optimal decisions under the lex-MO MCO class when the MCOP is a discrete multiple criteria decision problem.

2   Reduction, Regularity and the Lexicographic Max-Ordering MCO Class

In this section we propose the lexicographic max-ordering MCO class. This class will be shown to be uniquely defined by two properties which we describe before proving the main result. Throughout this paper we only consider MCO classes which satisfy the following **normalization property**. For the special case that $Q = 1$ it is natural to assume that the usual single objective minimization with the canonical order of $\mathbb{R}$ should be applied. Therefore, if $(\bullet, \bullet, \bullet)/\text{sort}/(\mathbb{R}^P, \preceq)$ is any MCO class, we require that $Q = 1$ implies

$$(\bullet, \bullet, \mathbb{R})/\text{sort}/(\mathbb{R}^P, \preceq) = (\bullet, \bullet, \mathbb{R})/\text{id}/(\mathbb{R}, \leq),$$

which actually means ordinary minimization of one criterion. Note that the MCO classes defined by Pareto optimality $(\bullet, \bullet, \bullet)/\text{id}/(\mathbb{R}^Q, <)$ and by scalarizing the criteria $(\bullet, \bullet, \bullet)/\langle \lambda, . \rangle/(\mathbb{R}, \leq)$ both satisfy the normalization property. For other MCO classes we refer to [Ehr97b].

Let $(\bullet, \bullet, \bullet)/\theta/(\mathbb{R}^P, \preceq)$ be any MCO class. What could be desirable properties from a decision makers point of view? We propose two. The first one is related to a reduced optimization problem. Let us suppose that, for whatever reason, for some objective functions the values that are taken for some solutions in $\textbf{Opt}\left((\bullet, \bullet, \bullet)/\theta/(\mathbb{R}^P, \preceq)\right)$ are known. Then it should suffice to consider for the minimization only the remaining objectives, with the additional constraints that for the known objectives the known values are taken. If the optimal solutions of this reduced minimization problem are exactly those of the original problem which have the given values for the specified objectives, we say that the MCO class satisfies the **reduction property**. Formally we define the **reduced problem** as follows.

Definition 1 *Let* $(\bullet, \bullet, \bullet)/\theta/(\mathbb{R}^P, \preceq)$ *be an MCO class and let* $(\mathcal{F}, f, \mathbb{R}^Q)$ *be data to define any MCO problem of the given MCO class. Let* $y \in \mathbb{R}^Q$ *be such that there exists at least one* $x \in \textbf{Opt}\left((\mathcal{F}, f, \mathbb{R}^Q)/\theta/(\mathbb{R}^P, \preceq)\right)$ *with* $f(x) = y$. *Furthermore let* $\mathcal{K} := \{i_1, \ldots, i_k\} \subseteq \mathcal{Q}$ *be an index set. Then the reduced problem for* $\mathcal{K}$, *denoted by* $RP(\mathcal{K})$, *where* $f^{\mathcal{K}} := (f_{i_1}, \ldots, f_{i_k})$ *is defined by*

$$(\mathcal{F}^{\mathcal{K}}, f^{\mathcal{K}}, \mathbb{R}^k)/\theta/(\mathbb{R}^P, \preceq)$$

and $\mathcal{F}^{\mathcal{K}} := \{x \in \mathcal{F} : f_q(F) = y_q \; \forall q \in \mathcal{Q} \setminus \mathcal{K}\}$.

Thus, in the reduced problem the values of the functions with indices in $\mathcal{Q} \setminus \mathcal{K}$ are fixed, those with indices in $\mathcal{K}$ are still to be minimized. Using the definition of the reduced problem, we define the reduction property.

Definition 2 *We say that an MCO class* $(\bullet, \bullet, \bullet)/\theta/(\mathbb{R}^P, \preceq)$ *satisfies the* **reduction property** *if for all MCO problem data* $(\mathcal{F}, f, \mathbb{R}^Q)$ *defining problems in the class the following holds: for all index sets* $\mathcal{K} \subseteq \mathcal{Q}$ *and for all*

$(y_1, \ldots, y_Q) \in \mathbb{R}^Q$ as in Definition 1

$$\mathbf{Opt}\left((\mathcal{F}^\mathcal{K}, f^\mathcal{K}, \mathbb{R}^k)/\theta/(\mathbb{R}^P, \preceq)\right) =$$
$$\left\{x \in \mathbf{Opt}\left((\mathcal{F}, f, \mathbb{R}^Q)/\theta/(\mathbb{R}^P, \preceq)\right) : f_q(x) = y_q \ \forall q \notin \mathcal{K}\right\}.$$

The second property can be interpreted as a pessimistic point of view of the DM: he will at least want to have a solution such that the worst criterion is as good as possible. I.e. an optimal solution of an MCOP should be such that

$$\max_{q=1,\ldots,Q} f_q(x^*) \leq \max_{q=1,\ldots,Q} f_q(x) \ \forall x \in \mathcal{F}.$$

Evidently, minimizing the maximum of the Q criteria is again an MCO class. Let $\theta(f(x)) := \max_{q=1,\ldots,Q} f_q(x)$ and $(\mathbb{R}^P, \preceq) := (\mathbb{R}, \leq)$. The corresponding MCO class is

$$(\bullet, \bullet, \bullet)/\max/(\mathbb{R}, \leq)$$

and is called max-ordering MCO class. Formally, the pessimistic attitude of a DM is formulated as the regularity property.

Definition 3 *An MCO class* $(\bullet, \bullet, \bullet)/\theta/(\mathbb{R}^P, \preceq)$ *satisfies the* **regularity property** *if*

$$\mathbf{Opt}\left((\mathcal{F}, f, \mathbb{R}^Q)/\theta/(\mathbb{R}^P, \preceq)\right) \subseteq \mathbf{Opt}\left((\mathcal{F}, f, \mathbb{R}^Q)/\max/(\mathbb{R}, \leq)\right).$$

for all MCO problems $(\mathcal{F}, f, \mathbb{R}^Q)/\theta/(\mathbb{R}^P, \preceq)$ *in the class.*

The main purpose of this section is to show that reduction and regularity property uniquely define the lexicographic max-ordering MCO class. We will now provide the definition of this class. First, we have to define the model map.

Definition 4 *For any element* $x \in \mathbb{R}^Q$ *we define*

$$\mathrm{sort}(x) = (\mathrm{sort}_1(x), \ldots, \mathrm{sort}_Q(x))$$

to be the vector containing the components of x *in non-increasing order:* $\mathrm{sort}_1(x) \geq \ldots \geq \mathrm{sort}_Q(x)$, $\{x_1, \ldots, x_Q\} = \{\mathrm{sort}_1(x), \ldots, \mathrm{sort}_Q(x)\}$.

Since sort is a mapping sort $: \mathbb{R}^Q \to \mathbb{R}^Q$ it remains to specify the order. We shall use the lexicographic order. Therefore the ordered set is $(\mathbb{R}^Q, \leq_{lex})$ and the lexicographic max-ordering class is defined by

$$(\bullet, \bullet, \bullet)/\mathrm{sort}/(\mathbb{R}^Q, \leq_{lex}).$$

We will usually call that class the lex-MO MCO class for short. By this definition, a feasible solution $x^* \in \mathcal{F}$ of an MCOP is a **lex-MO solution** if its objective function vector is a lexicographically minimal element of $\mathrm{sort}(f(\mathcal{F}))$, i.e. if

$$\mathrm{sort}(f(x^*)) \leq_{lex} \mathrm{sort}(f(x)) \quad \forall x \in \mathcal{F}.$$

This definition of optimality has appeared in the literature earlier, although with different names, see e.g. [Ogr97] [MO92] and [Beh86, Beh81]. We also refer to related papers [CLT96] and [Beh77]. In none of these papers the relevance for general multiple criteria optimization has been considered. This was done for the first time in [Ehr95].

To prove our main result we first show that $(\bullet, \bullet, \bullet)/\text{sort}/(I\!\!R^Q, \leq_{lex})$ does indeed satisfy the reduction and regularity property.

Theorem 1 *The lex-MO MCO class* $(\bullet, \bullet, \bullet)/\text{sort}/(I\!\!R^Q, \leq_{lex})$ *satisfies reduction and regularity property.*

Proof:

We will write **Opt** and **Opt** $(RP(\mathcal{K}))$ for the optimal sets of original and reduced problem in the proof. First we show that $(\bullet, \bullet, \bullet)/\text{sort}/(I\!\!R^Q, \leq_{lex})$ satisfies the regularity property. Let $(\mathcal{F}, f, I\!\!R^Q)$ be data of an MCOP of the lex-MO class and let $x^* \in \mathbf{Opt}\left((\mathcal{F}, f, I\!\!R^Q)/\text{sort}/(I\!\!R^Q, \leq_{lex})\right)$ be an optimal solution. Assume there exists some feasible solution x such that

$$\max_{q=1,\ldots,Q} f_q(x) < \max_{q=1,\ldots,Q} f_q(x^*).$$

Then $\text{sort}_1(f(x)) < \text{sort}_1(f(x^*))$. This is a contradiction to the choice of x^* because it obviously implies $\text{sort}(f(x)) <_{lex} \text{sort}(f(x^*))$.

To prove that the lex-MO class satisfies the reduction property, too, consider any MCO problem $(\mathcal{F}, f, I\!\!R^Q)/\text{sort}/(I\!\!R^Q, \leq_{lex})$. We have to show that

$$\mathbf{Opt}\,(RP(\mathcal{K})) = \{x \in \mathbf{Opt} : f_q(x) = y_q \quad \forall q \notin \mathcal{K}\} \tag{1}$$

for every choice of $\mathcal{K}$ and $(y_1, \ldots, y_Q)$ as in Definition 1. We denote the latter set in (1) by $\mathbf{Opt}^*$ for brevity. We observe that for all solutions $x \in \mathbf{Opt}^*$ it holds that

$$f_q(\dot{x}) = f_q(\bar{x}) = y_q \quad \forall q \in \{1, \ldots, Q\} \setminus \mathcal{K} \quad \forall \bar{x} \in \mathbf{Opt}\,(RP(\mathcal{K})) \tag{2}$$

by the definition of the reduced problem.

To show (1) first let $\bar{x}$ be an element of $\mathbf{Opt}\,(RP(\mathcal{K}))$. By definition of $\mathbf{Opt}^*$ it follows that

$$\text{sort}(f(x)) \leq_{lex} \text{sort}(f(\bar{x})) \quad \forall x \in \mathbf{Opt}^*. \tag{3}$$

From (3) and (2) we conclude that

$$\text{sort}(f^{\mathcal{K}}(x)) \leq_{lex} \text{sort}(f^{\mathcal{K}}(\bar{x}))$$

which due to the choice of $\bar{x}$ as an optimal solution of the reduced problem must be satisfied with equality. Therefore we have shown that $\text{sort}(f(x)) = \text{sort}(f(\bar{x}))$ which clearly implies $\bar{x} \in \mathbf{Opt}^*$.

We proceed with the reverse inclusion. Let x be an element of $\mathbf{Opt}^*$. Note that x^* is feasible for the reduced problem by definition of $RP(\mathcal{K})$. Now assume the existence of a solution $\bar{x} \in \mathbf{Opt}\,(RP(\mathcal{K}))$ such that

$$\text{sort}(f^{\mathcal{K}}(\bar{x})) <_{lex} \text{sort}(f^{\mathcal{K}}(x)).$$

This implies that also $\text{sort}(f(\bar{x})) <_{lex} \text{sort}(f(x))$, contradicting the choice of x. Hence it must hold that

$$\text{sort}(f(\bar{x})) \leq_{lex} \text{sort}(f(x)). \tag{4}$$

(3) and (4) imply that $\text{sort}(f(x)) = \text{sort}(f(\bar{x}))$, therefore x belongs to the set $\mathbf{Opt}\,(RP(\mathcal{K}))$. We have proven both inclusions in (1) and the proof is complete. $\qquad\square$

We proceed to prove the converse of Proposition 1 and thus the main result of the paper, namely that reduction and regularity are characteristic properties of the lex-MO class. Theorem 2 has first been proven in [Ehr97a] and is also contained in [Ehr97b].

Theorem 2 *An MCO class $(\bullet, \bullet, \bullet)/\theta/(I\!\!R^P, \preceq)$ satisfies reduction and regularity property if and only if*

$$(\bullet, \bullet, \bullet)/\theta/(I\!\!R^P, \preceq) \;=\; (\bullet, \bullet, \bullet)/\text{sort}/(I\!\!R^Q, \leq_{lex}).$$

Proof:

Due to Theorem 1 we only have to show the 'only if' part. Let us assume that $(\bullet, \bullet, \bullet)/\theta/(I\!\!R^P, \preceq)$ is an MCO class satisfying both reduction and regularity property.

Let $(\mathcal{F}, f, I\!\!R^Q)$ be data of an MCOP of class $(\bullet, \bullet, \bullet)/\theta/(I\!\!R^P, \preceq)$. We prove the result by induction on the number Q of criteria. For $Q = 1$ the result follows immediately from the normalization property, since all MCO classes coincide in that case.

Let us now assume that the result holds for $P \leq Q - 1$ criteria and consider the case of Q criteria. We have to show that

$$\mathbf{Opt}\,((\mathcal{F}, f, I\!\!R^Q)/\theta/(I\!\!R^P, \preceq)) = \mathbf{Opt}\,((\mathcal{F}, f, I\!\!R^Q)/\text{sort}/(I\!\!R^Q, \leq_{lex})). \tag{5}$$

Let $\bar{x}$ be a solution in $\mathbf{Opt}\,((\mathcal{F}, f, I\!\!R^Q)/\theta/(I\!\!R^p, \preceq))$. Furthermore, denote by y the optimal value of a max-ordering solution: y is the unique point in $\mathcal{V}_{\text{opt}}\,((\mathcal{F}, f, I\!\!R^Q)/\max/(I\!\!R, \leq))$. Since $(\bullet, \bullet, \bullet)/\theta/(I\!\!R^P, \preceq)$ satisfies the regularity property there must exist an index q^* such that $f_{q^*}(\bar{x}) = y$. Therefore $\bar{x} \in \{x \in \mathbf{Opt}\,((\mathcal{F}, f, I\!\!R^Q)/\theta/(I\!\!R^p, \preceq)) : f_{q^*}(x) = y\}$.

Now we consider the reduced problem with $\mathcal{K} = \{1, \ldots, Q\} \setminus \{q^*\}$ and the value of f_{q^*} fixed at y. Then

$$\{x \in \mathbf{Opt}\,((\mathcal{F}, f, I\!\!R^Q)/\theta/(I\!\!R^P, \preceq)) : f_{q^*}(x) = y\}$$

$$\begin{aligned}
&= \ \mathbf{Opt}\left((\mathcal{F}^{\mathcal{K}}, f^{\mathcal{K}}, \mathbb{R}^{Q-1})/\theta/(\mathbb{R}^{P}, \preceq)\right) \\
&= \ \mathbf{Opt}\left((\mathcal{F}^{\mathcal{K}}, f^{\mathcal{K}}, \mathbb{R}^{Q-1})/\mathrm{sort}/(\mathbb{R}^{Q-1}, \leq_{lex})\right) \\
&= \ \left\{x \in \mathbf{Opt}\left((\mathcal{F}, f, \mathbb{R}^{Q})/\mathrm{sort}/(\mathbb{R}^{Q}, \leq_{lex})\right) : f_{q^{*}} = y\right\} \\
&\subseteq \ \mathbf{Opt}\left((\mathcal{F}, f, \mathbb{R}^{Q})/\mathrm{sort}/(\mathbb{R}^{Q}, \leq_{lex})\right).
\end{aligned}$$

The first equation follows from the reduction property for $(\bullet, \bullet, \bullet)/\theta/(\mathbb{R}^{P}, \preceq)$. The second follows from the induction hypothesis since $RP(\mathcal{K})$ is a multiple criteria optimization problem with $Q-1$ objective functions and feasible set $\mathcal{F}^{\mathcal{K}}$ as defined in Definition 1. Finally the third is implied by the reduction property for $(\bullet, \bullet, \bullet)/\mathrm{sort}/(\mathbb{R}^{Q}, \leq_{lex})$. The inclusion is trivial. Hence we have shown $\bar{x} \in \mathbf{Opt}\left((\mathcal{F}, f, \mathbb{R}^{Q})/\mathrm{sort}/(\mathbb{R}^{Q}, \leq_{lex})\right)$.

Proving the converse inclusion analogously completes the proof of (5). We let $\bar{x}$ be an element of $\mathbf{Opt}\left((F, f, \mathbb{R}^{Q})/\mathrm{sort}/(\mathbb{R}^{Q}, \leq_{lex})\right)$. By regularity we see that there exist q^{*} and y such that

$$\bar{x} \in \left\{x \in \mathbf{Opt}\left((F, f, \mathbb{R}^{Q})/\mathrm{sort}/(\mathbb{R}^{Q}, \leq_{lex})\right) : f_{q^{*}}(x) = y\right\}.$$

As above we conclude that $\bar{x} \in \mathbf{Opt}\left((\mathcal{F}, f, \mathbb{R}^{Q})/\theta/(\mathbb{R}^{P}, \preceq)\right)$. $\qquad\square$

The interpretation of Theorem 2 is that, if the DM agrees with the reduction property and has a pessimistic point of view (regularity property), he must choose sort as model map and the lexicographic order of $\mathbb{R}^{Q}$. In other words, he will have to solve problems of the lexicographic max-ordering MCO class. In Theorem 3 we summarize further properties of the lex-MO class.

Theorem 3 Let $(\mathcal{F}, f, \mathbb{R}^{Q})/\mathrm{sort}/(\mathbb{R}^{Q}, \leq_{lex})$ be a problem of the lex-MO class. Then the following hold.

(i) The set of minima $\mathcal{V}_{opt}\left((\mathcal{F}, f, \mathbb{R}^{Q})/\mathrm{sort}/(\mathbb{R}^{Q}, \leq_{lex})\right)$ is a singleton.

(ii) $\mathbf{Opt}\left((\mathcal{F}, f, \mathbb{R}^{Q})/\mathrm{sort}/(\mathbb{R}^{Q}, \leq_{lex})\right) \subseteq \mathbf{Opt}\left((\mathcal{F}, f, \mathbb{R}^{Q})/\mathrm{id}/(\mathbb{R}^{Q}, <)\right)$.

(iii) Let π be a permutation of $\{1, \ldots, Q\}$. Define the corresponding permutation of objective functions $f_{\pi}(x) := (f_{\pi(1)}(x), \ldots, f_{\pi(Q)}(x))$. Then

$$\mathbf{Opt}\left((\mathcal{F}, f, \mathbb{R}^{Q})/\mathrm{sort}/(\mathbb{R}^{Q}, \leq_{lex})\right) =$$
$$\mathbf{Opt}\left((\mathcal{F}, f_{\pi}, \mathbb{R}^{Q})/\mathrm{sort}/(\mathbb{R}^{Q}, \leq_{lex})\right).$$

(iv) Let $\tau : \mathbb{R} \to \mathbb{R}$ be a strictly monotone increasing mapping. Define the function $f_{\tau q} := \tau \circ f_{q}$ $q = 1, \ldots, Q$ and let $f_{\tau} := (f_{\tau 1}, \ldots, f_{\tau Q})$. Then

$$\mathbf{Opt}\left((\bullet, f, \bullet)/\mathrm{sort}/(\mathbb{R}^{Q}, \leq_{lex})\right) =$$
$$\mathbf{Opt}\left((\bullet, f_{\tau}, \bullet)/\mathrm{sort}/(\mathbb{R}^{Q}, \leq_{lex})\right).$$

Proof:
Let $(\mathcal{F}, f, \mathbb{R}^{Q})/\mathrm{sort}/(\mathbb{R}^{Q}, \leq_{lex})$ be an MCO problem of the lex-MO class.

(i) The first assertion is clear from the definition of the lex-MO class, because the lexicographic order is a total order.

(ii) Let $x^* \in \mathbf{Opt}\left((\mathcal{F}, f, \mathbb{R}^Q)/\mathrm{sort}/(\mathbb{R}^Q, \leq_{lex})\right)$ be a lex-MO solution. Assume there exists some feasible solution x such that $f(x) < f(x^*)$. It follows that $\mathrm{sort}(f(x)) \leq_{lex} \mathrm{sort}(f(x^*))$, with $\mathrm{sort}(f(x)) \neq \mathrm{sort}(f(x^*))$, contradicting the choice of x^*.

(iii) Clearly $\mathrm{sort}(f(x)) = \mathrm{sort}(f_\pi(x))$ holds for all feasible solutions $x \in \mathcal{F}$.

(iv) By the strict monotonicity of τ it holds that $f_q(x) < f_r(x)$ if and only if $\tau(f_q(x)) \leq \tau(f_r(x))$ for all $x \in \mathcal{F}$. $\square$

Besides the results of Theorem 3 it can also be shown that lex-MO solutions can be used to parametrize Pareto optimal solutions. We refer to [Ehr95] for details. We will now discuss the relevance of Theorem 3 for decision makers. First we note that (ii) means that lex-MO solutions are always Pareto optimal. Pareto optimality is quite natural a requirement of optimal solutions of MCOPs. If a solution which is not Pareto optimal represents an optimal decision it would be possible to find another one which is not worse with respect to all criteria an d strictly better with respect to at least one. Such a situation would raise the suspicion that the MCOP is not formulated correctly. Beyond this positive aspect, the fact that the set of minima is a singleton (i) provides an improvement to Pareto optimality: the value of a lex-MO solution is uniquely defined. The set of Pareto optimal solutions is often prohibitively large to define an optimal solution, it can even be shown that every feasible solution is Pareto optimal in some cases.

Results (iii) and (iv) will help us to identify situations in which the application of the lex-MO MCO class is appropriate. Note that due to (iii) the numbering of criteria is irrelevant for the optimal solution. Therefore we must assume the DM's indifference with respect to the criteria. Indifference implies that there are no rankings of criteria or relative importances present, not even implicitly. In these cases we would consider lexicographic or scalarized optimization the better choice. See also the definition of λ-extremal and (global) lexicographic MCO classes in [Ehr97b]. Finally, invariance under monotone transformations implies that performance with respect to the criteria can be measured on ordinal scales. As a conclusion we state that the lex-MO MCO class is the appropriate tool when the DM is totally indifferent with respect to the criteria and performance is measured (at least) on ordinal scales. Both are, as we conjecture, rather weak assumptions.

Moreover, the sorting of the criteria suggests that the scales should be the same for all criteria. If this is not the case a priori a normalization can be applied. For instance, we can define

$$v'_{qj} := \frac{\max_{j=1,\ldots,m} v_{qj} - v_{qj}}{\max_{j=1,\ldots,m} v_{qj} - \min_{j=1,\ldots,m} v_{qj}}.$$

Then all values v'_{qj} are in the interval $[0, 1]$. We refer to [HY81] for more on normalization of scales.

3 Optimality for Discrete Decision Problems

In this paper we will focus on discrete problems, as we have already indicated in Section 1. In this case note that the set of values of the criteria is finite. We let $R = \{v_{qj} : j = 1, \ldots, m; \ q = 1, \ldots, Q\}$. Then for all $a_j \in A$ we have that $v(a_j) \in R^Q$. Therefore we will replace $I\!R^Q$ by the finite set R^Q. Also, since all value vectors are explicitly given via the performance matrix V, there is no need to consider the criteria as objective functions (this is of course needed in general multiple objective optimization). It is therefore possible to combine f and R^Q and only note V instead, in the above notation for MCOPs. We will use the notation $v(a_j)$ or v_j for the columns of V. Thus, a multiple criteria discrete decision problem is denoted by

$$(A, V)/\theta/(I\!R^P, \preceq).$$

The following assumption guarantees that the problem is not trivial. Define

$$v_q^* := \min_{j=1,\ldots,Q} v_{qj}.$$

Then the vector $v^* = (v_1^*, \ldots, v_Q^*)$ is called the utopia point and, clearly, if for some alternative a_j it holds that $v(a_j) = v^*$ only this point, the ideal solution, can be considered as an optimal decision. We therefore assume that an ideal solution does not exist.

We now propose two approaches to solve a discrete multiple criteria decision problem of the lex-MO class. The first is to sort the vectors v_j and compare them lexicographically.

Algorithm 1 for $(A, V)/\text{sort}/(I\!R^Q, \leq_{lex})$

Input: performance matrix V
Output: set of optimal decisions Opt

for $j = 1, \ldots, m$ **do** $v'_j := \text{sort}(v_j)$
$min := 1$
$i := 1$
while $i \leq m$ **do**
       begin
              if $v'_i <_{lex} v'_{min}$
                  then $min := i$
              $i := i + 1$
       end
$Opt := \{a_j : v'_j = v'_{min}\}$

The time needed for this algorithm is bounded by $O(mQ \log Q)$ for the sorting of the m vectors v_j. Finding the lexicographic minimum of the sorted vectors v_j' can be done in $O(mQ)$ time. The second algorithm uses performance matrix V, finds the smallest maxima of the columns and constructs a smaller matrix. This step is applied iteratively.

Algorithm 2 for $(A, V)/\text{sort}/(\mathbb{R}^Q, \leq_{lex})$

Input: performance matrix V
Output: set of optimal decisions Opt

$count := 1$
for $i := 1, \ldots, m$ **do** $Q_i := \{1, \ldots, Q\}$
$Opt := \{a_1, \ldots, a_m\}$
1: **begin**
        for $i := 1, \ldots, m$ **do**
          begin
$$m_i := \max_{q=1,\ldots Q} v_{qi}$$
$$i^* := \text{argmin}\{m_i\}$$
$$Q_i := Q_i \setminus \{i^*\}$$
        end
$$m^* := \min_{i=1,\ldots m} m_i$$
        $Opt := Opt \setminus \{a_j : m_j > m^*\}$
        for $a_j \in Opt$ **do** $v_j' := (v_{qi} : q \in Q_i)$
        $count := count + 1$
  end
if $m > 1$ and $count \leq Q$
    then goto 1:
    else STOP

In each iteration m maxima of Q numbers are determined in $O(Qm)$ time. Due to the condition $count \leq Q$ and the fact that the cardinality of all Q_i, for which a_i remains in Opt, decreases at least by one in each iteration the algorithm stops after at most Q iterations. Therefore the overall time needed is $O(mQ^2)$.

Comparing both algorithms, we see that in the worst case the second one is slower, however, for the first the running time is always $O(mQ \log Q)$, even in the best case. The second one on the other hand may be much faster in practice, because Opt tends to become a singleton after only a few iterations.

We illustrate the methods with a small example. Let us assume somebody wants to buy a new car. There are five models a_1 to a_5 available and the relevant criteria for the decision are reputation of the brand (c1), petrol consumption (c2), price (c3), design (c4), and tenure of the guarantee (c5). To have a common ordinal scale for the criteria, she/he decides to evaluate the cars with respect to the criteria on a 10 point scale, the fewer points the better. Results are displayed in the matrix below

	a_1	a_2	a_3	a_4	a_5
(c1)	5	1	7	4	4
(c2)	4	4	3	5	2
(c3)	7	3	1	6	1
(c4)	3	5	2	1	1
(c5)	2	1	2	3	5

According to the first algorithm we would sort all the columns of this matrix V resulting in

$$\begin{pmatrix} 7 & 5 & 7 & 6 & 5 \\ 5 & 4 & 3 & 5 & 4 \\ 4 & 3 & 2 & 4 & 2 \\ 3 & 1 & 2 & 3 & 1 \\ 2 & 1 & 1 & 1 & 1 \end{pmatrix}.$$

Finding the lexicographic minimum of the columns would yield the optimal decision $Opt = \{a_5\}$.

Let us now apply the second method. In the first iteration we find $m_1 = 7, m_2 = 5, m_3 = 7, m_4 = 6$, and $m_5 = 5$. Therefore we would delete a_1, a_3, and a_4 from Opt. Furthermore $Q_2 = \{1, 2, 3, 5\}$ and $Q_5 = \{1, 2, 3, 4\}$. In the second iteration we therefore consider a much smaller problem, shown in the following matrix

$$\begin{pmatrix} 1 & 4 \\ 4 & 2 \\ 3 & 1 \\ 1 & 1 \end{pmatrix}.$$

We now find 4 as the maximal entry in both columns, hence cannot change Opt but make the problem still smaller since $Q_2 = \{1, 3, 5\}$ and $Q_5 = \{2, 3, 4\}$. In the third iteration we therefore consider

$$\begin{pmatrix} 1 & 2 \\ 3 & 1 \\ 1 & 1 \end{pmatrix}.$$

Now $m_2 = 3 > m_5 = 2$, a_2 is deleted from Opt, which contains only a single element and the algorithm stops.

References

[Beh77] F.A. Behringer. On optimal decisions under complete ignorance – A new criterion stronger than both Pareto and maxmin. *European Journal of Operational Research*, 1:295–306, 1977.

[Beh81] F.A. Behringer. A simplex based algorithm for the lexicographically extended linear maxmin problem. *European Journal of Operational Research*, 7:274–283, 1981.

[Beh86] F.A. Behringer. Linear multiobjective maxmin optimization and some Pareto and lexmaxmin extensions. *OR Spektrum*, 8:25–32, 1986.

[CLT96] F. Christensen, M. Lind, and J. Tind. On the nucleolus of NTU-games defined by multiple objective linear programs. *ZOR – Mathematical Methods of Operations Research*, 43(3):337–352, 1996.

[Ehr95] M. Ehrgott. Lexicographic max-ordering – a solution concept for multicriteria combinatorial optimization. In D. Schweigert, editor, *Methods of Multicriteria Decision Theory, Proceedings of the 5th Workshop of the DGOR-Working Group Multicriteria Optimization and Decision Theory*, pages 55–66, 1995.

[Ehr97a] M. Ehrgott. A characterization of lexicographic max-ordering solutions. In *Methods of Multicriteria Decision Theory, Proceedings of the 6th Workshop of the DGOR-Working Group Multicriteria Optimization and Decision Theory Alexisbad 1996*, volume 2389 of *Deutsche Hochschulschriften*, 1997.

[Ehr97b] M. Ehrgott. *Classification and Methodology in Multiple Criteria Optimization*. Shaker Verlag, Aachen, 1997. Ph.D. Thesis.

[HY81] C.L. Hwang and K. Yoon. *Multiple Attribute Decision Making – Methods and Applications*. Springer-Verlag, Berlin, Heidelberg, New York, 1981.

[MO92] E. Marchi and J.A. Oviedo. Lexicographic optimality in the multiple objective linear programming: The nucleolar solution. *European Journal of Operational Research*, 57:355–359, 1992.

[Ogr97] W. Ogryczak. On the lexicographic minimax approach to location problems. *European Journal of Operational Research*, 100:566–585, 1997.

EXPLOITATION OF A ROUGH APPROXIMATION OF THE OUTRANKING RELATION IN MULTICRITERIA CHOICE AND RANKING

Salvatore Greco[1], Benedetto Matarazzo[1], Roman Slowinski[2] and Alexis Tsoukiàs[3]

[1]Faculty of Economics, University of Catania, Corso Italia, 55,
  95129 Catania, Italy
[2]Institute of Computing Science, Poznan University of Technology,
  Piotrowo 3a, 60-965 Poznan, Poland
[3]LAMSADE, Université de Paris Dauphine,
  Place du Maréchal De Lattre de Tassigny, F-75775 Paris Cedex 16, France

Abstract. Given a finite set A of actions evaluated by a family of criteria, we consider a preferential information in the form of a pairwise comparison table (PCT) including pairs of actions from a subset $B \subseteq A \times A$ described by graded preference relations on particular criteria and a comprehensive outranking relation. Using the rough set approach to the analysis of the PCT, we obtain a rough approximation of the outranking relation by a graded dominance relation. Decision rules derived from this approximation are then applied to a set $M \subseteq A$ of potential actions. As a result, we obtain a four-valued outranking relation on set M. The construction of a suitable exploitation procedure in order to obtain a recommendation for multicriteria choice and ranking is an open problem within this context. We propose an exploitation procedure that it is the only one satisfying some desirable properties.

Keywords. Rough sets, multicriteria decision making, four-valued outranking, exploitation procedures.

1 Introduction

A rough set approach to multicriteria decision analysis has been proposed by Greco, Matarazzo and Slowinski (1996). This methodology operates on a pairwise comparison table (PCT) (Greco, Matarazzo and Slowinski, 1995), including pairs of actions described by graded preference relations on specific criteria and by a comprehensive preference relation. It builds up a rough approximation of the comprehensive preference relation using graded dominance relations. Furthermore, some decision rules in the "if ... then..." form are derived from the rough approximation of the preference relation. If the comprehensive preference relation is an outranking relation, the application of these decision rules to a set of actions gives a four-valued outranking relation (Tsoukias and Vincke, 1995, 1997), i.e. a binary relation which, with respect to any pair of actions (a,b), characterizes the proposition "a is at least as good as b" as true, contradictory, unknown or false. Finally, in order to obtain a recommendation (Roy, 1993) for the decision problem at hand, a suitable exploitation procedure of the four-valued outranking relation should be applied. This paper, which is a reduced version of Greco, Matarazzo, Slowinski and Tsoukias (1997), is focused on this exploitation procedure. More precisely, we consider multicriteria ranking and choice problems, and we propose an exploitation procedure, called scoring procedure, which we characterize by proving that it is the only one ensuring some desirable properties.

The paper is structured as follows. In section 2, we introduce the rough approximation of a preference relation and the generation of decision rules. In section 3, we describe the four-valued outranking relation. In section 4, we introduce the application of decision rules, showing how it defines a four-valued outranking relation. Furthermore, the scoring procedure is presented. Section 5 proposes a characterization of this scoring procedure. Section 6 groups conclusions.

2 Rough set analysis of a preferential information

2.1 Pairwise comparison table

In order to represent preferential information provided by the decision maker (DM) in form of a pairwise comparison of some actions, we shall use a pairwise comparison table, introduced in Greco, Matarazzo and Slowinski (1995).

Let A be a finite set of actions (feasible or not), considered by the DM as a basis for exemplary pairwise comparisons. Let also C be the set of criteria (condition attributes) describing the actions.

For any criterion $q \in C$, let T_q be a finite set of binary relations defined on A on the basis of the evaluations of actions of A with respect to the considered criterion q, such that $\forall (x,y) \in A \times A$ exactly one binary relation $t \in T_q$ is verified. More precisely, given the domain V_q of $q \in C$, if $v'_q, v''_q \in V_q$ are the respective evaluations of $x, y \in A$ by means of q and $(x,y) \in t$ with $t \in T_q$, then for each $w, z \in A$ having the same evaluations v'_q, v''_q by means of q, $(w,z) \in t$. For interesting applications it should be $card(T_q) \geq 2$, $\forall q \in C$.

Furthermore, let T_d be a set of binary relations defined on A (comprehensive pairwise comparisons) such that at most one binary relation $t \in T_d$ is verified, $\forall (x,y) \in A \times A$.

The *pairwise comparison table* (PCT) is defined as an information table $S_{PCT} = \langle B, C \cup \{d\}, T_C \cup T_d, g \rangle$, where $B \subseteq A \times A$ is a non-empty *sample of pairwise comparisons*, $T_C = \underset{q \in C}{Y} T_q$, d is a decision corresponding to the comprehensive pairwise comparison (comprehensive preference binary relation), and $g:B \times (C \cup \{d\}) \to T_C \cup T_d$ is a total function such that $g[(x,y),q] \in T_q \ \forall (x,y) \in A \times A$ and $\forall q \in C$, and $g[(x,y),d] \in T_d$, $\forall (x,y) \in B$. It follows that for any pair of actions $(x,y) \in B$ one and only one binary relation $t \in T_d$ is verified. Thus, T_d induces a partition of B. In fact, information table S_{PCT} can be seen as decision table, since the set of considered criteria C and decision d are distinguished.

In this paper, we consider S_{PCT} related to the choice and ranking problems (Roy, 1985) and assume that the exemplary pairwise comparisons provided by the DM can be presented in terms of *graded preference binary relations*:

$$T_q = \{ P_q^h, \ h \in H_q \},$$

where $H_q = \{h \in Z: h \in [-p_q, r_q]\}$ and $p_q, r_q \in N$, $\forall q \in C$ and $\forall (x,y) \in A \times A$;

- $x P_q^h y$, $h > 0$, means that action x is preferred to action y by degree h with respect to criterion q,

- $x P_q^h y$, $h < 0$, means that action x is not preferred to action y by degree h with respect to criterion q,

- $x P_q^0 y$ means that x is similar (asymmetrically indifferent) to y with respect to criterion q.

Let us remark that the similarity represented by the binary relation P_q^0 has been introduced by Slowinski and Vanderpooten (1995, 1996, 1998) in very general terms, i.e. without any specific reference to preference modeling. Let us remember that a *similarity* relation is only *reflexive* (i.e., with respect to P_q^0, we have $x P_q^0 x$ $\forall x \in A$ and $\forall q \in C$), relaxing therefore the properties of symmetry and transitivity. The abandon of the transitivity requirement is easily justifiable, remembering – for example – Luce's paradox of the cups of tea (1956). As for the symmetry, one should notice that yRx, which means "y is similar to x", is directional; there is a *subject* y and a *referent* x, and in general this is not equivalent to the proposition "x is similar to y", as maintained by Tversky (1977). This is quite immediate when the similarity relation is defined in terms of a percentage difference between evaluations of the actions compared on the attribute at hand, calculated with respect to the evaluation of the referent action. In terms of preference modeling, similarity relation, even if not symmetric, resembles indifference relation. Thus, in this context, we also call this similarity relation "asymmetric indifference".

Of course, $\forall x, y \in A$

$$[x P_q^h y, \ h \geq 0] \Leftrightarrow [y P_q^k x, \ k \leq 0].$$

Therefore, $\forall (x,y),(w,z) \in A \times A$ and $\forall q \in C$:

- if $xP_q^h y$ and $wP_q^k z$, $k \geq h \geq 0$, then w is preferred to z not less than x is preferred to y with respect to criterion q;

- if $xP_q^h y$ and $wP_q^k z$, $k \leq h \leq 0$, then w is not preferred to z not less than x is not preferred to y with respect to criterion q.

The set of binary relations T_d is defined analogously; however, $xP_d^h y$ means that x is comprehensively preferred to y by degree h.

2.2 Rough approximation of a preference relation

Let $H_P = \underset{q \in P}{I} H_q$ $\forall P \subseteq C$. Given $x,y \in A$, $P \subseteq C$ and $h \in H_P$, we say that x *positively dominates* y by degree h with respect to the set of criteria P iff $xP_q^f y$ with $f \geq h$, $\forall q \in P$. Analogously, $\forall x,y \in A$, $P \subseteq C$ and $h \in H_P$, x *negatively dominates* y by degree h with respect to the set of criteria P iff $xP_q^f y$ with $f \leq h$, $\forall q \in P$. Thus, $\forall h \in H_P$, every $P \subseteq C$ generates two binary relations (possibly empty) on A, which will be called *P-positive-dominance of degree h*, denoted by D_{+P}^h, and *P-negative-dominance of degree h*, denoted by D_{-P}^h, respectively. The relations D_{+P}^h and D_{-P}^h satisfy the following properties:

(P1)   if $(x,y) \in D_{+P}^h$, then $(x,y) \in D_{+R}^k$, for each $R \subseteq P$ and for every $k \leq h$;

(P2)   if $(x,y) \in D_{-P}^h$, then $(x,y) \in D_{-R}^k$, for each $R \subseteq P$ and for every $k \geq h$.

In the following, we consider a PCT where the decision d can have only two values on $B \subseteq A \times A$:

1) x outranks y, which will be denoted by xSy or $(x,y) \in S$,

2) x does not outrank y, which will be denoted by $xS^c y$ or $(x,y) \in S^c$,

where "x outranks y" means "x is at least as good as y" (Roy, 1985). Let us remember that the minimal property verified by the outranking relation S is reflexivity (see Roy, 1991; Bouyssou, 1996).

We propose to approximate the binary relation S by means of the D_{+P}^h binary dominance relations. Therefore, S is seen as a *rough binary relation* (see Greco, Matarazzo and Slowinski, 1995).

The P-lower approximation of S, denoted by $\underline{P}S$, and the P-upper approximation of S, denoted by $\overline{P}S$, are defined, respectively, as:

$$\underline{P} S = \underset{h \in H_P}{Y} \left\{ \left(D_{+P}^h \cap B \right) \subseteq S \right\},$$

$$\overline{P} S = \underset{h \in H_P}{I} \left\{ \left(D_{+P}^h \cap B \right) \supseteq S \right\}.$$

Taking into account property (P1) of the dominance relations D_{+P}^h, $\underline{P} S$ can be viewed as the dominance relation D_{+P}^h which has the largest intersection with B included in the outranking relation S, and $\overline{P} S$ as the dominance relation D_{+P}^h including S which has the smallest intersection with B.

Analogously, we can approximate S^c by means of the D_{-P}^h dominance relations:

$$\underline{P} S^c = \underset{h \in H_P}{Y} \left\{ \left(D_{-P}^h \cap B \right) \subseteq S^c \right\},$$

$$\overline{P} S^c = \underset{h \in H_P}{I} \left\{ \left(D_{-P}^h \cap B \right) \supseteq S^c \right\}.$$

Taking into account property (P2), the interpretation of $\underline{P} S^c$ and $\overline{P} S^c$ is similar to the interpretation of $\underline{P} S$ and $\overline{P} S$.

2.3 Decision rules

We can derive a generalized description of the preferential information contained in a given PCT in terms of decision rules.

We will consider the following kinds of decision rules:

1) D_{++}-decision rule, being a statement of the type: $x D_{+P}^h y \Rightarrow xSy$;

2) D_{+-}-decision rule, being a statement of the type: *not* $x D_{+P}^h y \Rightarrow xS^c y$;

3) D_{-+}-decision rule, being a statement of the type: *not* $x D_{-P}^h y \Rightarrow xSy$;

4) D_{--}-decision rule, being a statement of the type: $x D_{-P}^h y \Rightarrow xS^c y$.

Speaking about decision rules we will simply understand all the four kinds of decision rules together.

If there is at least one pair $(w,z) \in B$ such that $w D_{+P}^h z$ and wSz, and there is no $(v,u) \in B$ such that $v D_{+P}^h u$ and $vS^c u$, then $x D_{+P}^h y \Rightarrow xSy$ is accepted as a D_{++}-decision rule. A D_{++}-decision rule $x D_{+P}^h y \Rightarrow xSy$ will be called *minimal* if there is not any other rule $x D_{+R}^k y \Rightarrow xSy$ such that $R \subseteq P$ and $k \leq h$. Let us observe that, since each decision rule is an implication, a minimal decision rule represents an implication such that there is no other implication with an antecedent at least of the same weakness and a consequent of at least the same strength. The other rules can be characterized analogously.

Theorem 2.1. (Greco, Matarazzo, Slowinski, 1996). If

    1) $x D_{+P}^h y \Rightarrow xSy$ is a minimal D_{++}-decision rule, then $\underline{PS} = D_{+P}^h \cap B$,

    2) $x D_{-P}^h y \Rightarrow xS^c y$ is a minimal D_{--}-decision rule, then $\underline{PS}^c = D_{-P}^h \cap B$,

    3) *not* $x D_{+P}^h y \Rightarrow xS^c y$ is a minimal D_{+-}-decision rule, then $\overline{PS} = D_{+P}^h \cap B$,

    4) *not* $x D_{-P}^h y \Rightarrow xSy$ is a minimal D_{-+}-decision rule, then $\overline{PS}^c = D_{-P}^h \cap B$.

3 Four-valued outranking

The basic idea of the four-valued outranking model of preferences (Tsoukias and Vincke, 1995, 1997) is connected with the search of "positive reasons" and "negative reasons" (xSy and $xS^c y$) supporting a hypothesis of the truth of a comprehensive outranking relation for an ordered pair (x,y) of actions. The combination of presence and absence of the positive and the negative reasons creates four possible situations for the outranking:

1) *true outranking*, denoted by $xS^T y$, iff there exist sufficient positive reasons to establish xSy and there do not exist sufficient negative reasons to establish $xS^c y$;

2) *contradictory outranking*, denoted by $xS^K y$, iff there exist sufficient positive reasons to establish xSy and sufficient negative reasons to establish $xS^c y$;

3) *unknown outranking*, denoted by $xS^U y$, iff there do not exist sufficient positive reasons to establish xSy and there do not exist sufficient negative reasons to establish $xS^c y$;

4) *false outranking*, denoted by $xS^F y$, iff there do not exist sufficient positive reasons to establish xSy and there exist sufficient negative reasons to establish $xS^c y$.

By such definitions it is possible to apply the rough approximations of outranking relations S and S^c defined on B, in order to build a preference model on M×M, where M⊆A, which could further be exploited to get a recommendation (choice or ranking) with respect to a set of actions from M. In other words, we are able to move from a descriptive model of decision maker's preferences expressed on B to a prescriptive model on M⊆A.

4 Application of decision rules and definition of a final recommendation

Given a set D of decision rules, obtained in the way described in section 2, and two actions $v, u \in A$,

1) if $x D_{+P}^h y \Rightarrow xSy$ is a D_{++}-decision rule and $v D_{+P}^h u$, then we conclude that vSu,

2) if *not* $x D_{+P}^h y \Rightarrow xS^c y$ is a D_{+-}-decision rule and *not* $v D_{+P}^h u$, then we conclude that $vS^c u$,

3) if *not* $x D_{-P}^h y \Rightarrow xSy$ is a D_{-+}-decision rule and *not* $v D_{-P}^h u$, then we conclude that vSu,

4) if $x D_{-P}^h y \Rightarrow xS^c y$ is a D_{--}-decision rule and $v D_{-P}^h u$, then we conclude that $vS^c u$.

According to the four-valued logic, from the application of the decision rules to the pair of actions $(x,y) \in A \times A$ there may arise one of the four following states:

- *true outranking*, denoted by $xS^T y$: this is the case when there exists at least one D_{++}-decision rule and/or at least one D_{-+}-decision rule stating that xSy, and no D_{--}-decision rule or D_{+-}-decision rule stating that $xS^c y$;

- *false outranking*, denoted by $xS^F y$: this is the case when there exists at least one D_{--}-decision rule and/or at least one D_{+-}-decision rule stating that $xS^c y$, and no D_{++}-decision rule or D_{-+}-decision rule stating that xSy;

- *contradictory outranking*, denoted by $xS^K y$: this is the case when there exists at least one D_{++}-decision rule and/or at least one D_{-+}-decision rule stating that xSy, and at least one D_{--}-decision rule and/or at least one D_{+-}-decision rule stating that $xS^c y$;

- *unknown outranking*, denoted by $xS^U y$: this is the case when there is no D_{++}-decision rule or D_{-+}-decision rule stating that xSy, and no D_{--}-decision rule or D_{+-}-decision rule stating that $xS^c y$.

Theorem 4.1. (Greco, Matarazzo, Slowinski, 1996) The application of all the decision rules obtained for a given S_{PCT} on any pair of actions $(v,u) \in A \times A$ results in the same outranking relation as obtained by the application of the minimal decision rules only.

From Theorem 4.1, we conclude that the set of all decision rules is completely characterized by the set of the minimal rules. Therefore, only the latter ones are presented to the DM and applied in the decision problem at hand.

In order to define a recommendation with respect to the actions of $M \subseteq A$, we can calculate a particular score based on the outranking relations S and S^c obtained from the application of these rules to the actions of M.

$\forall M \subseteq A$ and $\forall x \in M$, let

- $M^{++}(x) = \{y \in M-\{x\}:$ there is at least one D_{++} -decision rule and/or at least one D_{-+}-decision rule stating that $xSy\}$,
- $M^{+-}(x) = \{y \in M-\{x\}:$ there is at least one D_{++}-decision rule and/or at least one D_{-+}-decision rule stating that $ySx\}$,
- $M^{-+}(x) = \{y \in M-\{x\}:$ there is at least one D_{+-} -decision rule and/or at least one D_{--}-decision rule stating that $yS^cx\}$,
- $M^{--}(x) = \{y \in M-\{x\}:$ there is at least one D_{+-} -decision rule and/or at least one D_{--}-decision rule stating that $xS^cy\}$.

To each $x \in M$ we assign a *score*

$$S(x,M) = S^{++}(x,M) - S^{+-}(x,M) + S^{-+}(x,M) - S^{--}(x,M)$$

where $S^{++}(x,M)=card[M^{++}(x)]$, $S^{+-}(x,M)=card[M^{+-}(x)]$, $S^{-+}(x,M)=card[M^{-+}(x)]$, $S^{--}(x,M)=card[M^{--}(x)]$.

We can use this score to work out a recommendation in the ranking and choice problems. For the ranking problem, $S(x,M)$ establishes a total preorder on M. For choice problems, the final recommendation is $x^* \in M$ such that $S(x^*,M)= \max_{x \in M} S(x,M)$. We call these exploitation procedures *scoring procedures*.

5   A characterization of the scoring procedure

The use of a score-based procedure in presence of a four-valued outranking relation is a problem which goes beyond the exploitation of rough approximations (see Tsoukias and Vincke, 1997). For this reason we start with some general remarks concerning the use of such procedures.

We want also to stress that such procedures are not the only possibility when four-valued outranking relations have to be exploited. Moreover, the reader may notice that the use of the score, as defined in this paper, conceals the difference between uncertainty due to contradictions (contradictory outranking S^K) and uncertainty due to lack of information (unknown outranking S^U) contributing in the same manner to the score $S(x,M)$. However, in our opinion, any exploitation procedure results in a loss of information since it reduces the rich form of knowledge contained in the outranking relations (S and S^c) to a poorer one which is the final choice or ranking. In favor of the scoring procedure play its intuitive nature (it is easy to understand by decision makers), its clear and straightforward characterization (as it will be demonstrated in the following) and its easiness in implementation. In other words, we sacrifice some richness of the information to the easiness of use.

5.1 Some previous results

The scoring procedure proposed in the previous section can be considered as an extension to the four-valued logic of the well-known Copeland ranking and choice method (see Goodman, 1954; Fishburn, 1973).

These procedures have been characterized by Rubinstein (1980) and Henriet (1985) and, with respect to valued binary relations, by Bouyssou (1992a and b).

In this subsection we remember synthetically the results of Bouyssou, while in the following subsection we extend them to the four-valued outranking relation.

A valued (binary) outranking relation on A is a function R associating an element of $[0,1]$ with each ordered pair of actions $(a,b) \in A \times A$, with $a \neq b$. Let $R(A)$ be the set of all valued binary relations on A and 2^A the set of all non-empty subsets on A. A *ranking method* (RM), denoted by $\geq$, is a function assigning a ranking $\geq(M,R)$ on $M \subseteq A$ to any valued relation $R \in R(A)$ and to any (non-empty) $M \subseteq A$. A *choice function* (CF) on A is a function

$$C: 2^A \times R(A) \rightarrow 2^A$$

such that $C(M,R) \subseteq M$, for each $M \in 2^A$ and $R \in R(A)$.

The following properties of ranking and choice exploitation procedure are considered (Bouyssou, 1992a and b):

1) *strong monotonicity*: an exploitation procedure is strongly monotonic iff it responds in the right direction to a modification of R. More formally,

1a) RM $\geq$ is strongly monotonic iff $\forall a,b \in M \subseteq A$ and $\forall R \in R(A)$

$$a \geq (M,R)b \Rightarrow a > (M,R')b,$$

where $>(M,R)$ is the asymmetric part of $\geq(M,R)$ and R' is identical to R except that $R(a,c) < R'(a,c)$ or $R(c,a) > R'(c,a)$ for some $c \in M-\{a\}$;

1b) a CF C is strongly monotonic iff $\forall R \in R(A)$ and all $M \in 2^A$

$$a \in C(M,R) \Rightarrow \{a\} = C(M,R')$$

where R' is defined as previously.

2) *neutrality*: an exploitation procedure is neutral iff it does not discriminate between actions just because of their labels. More formally,

2a) a RM $\geq$ is neutral iff for all permutations σ on A, $\forall R \in R(a)$ and $\forall a,b \in M \subseteq A$

$$a \geq (M,R)b \Leftrightarrow \sigma(a) \geq (\sigma(M),R^\sigma)\sigma(b)$$

where R^σ is defined by $R^\sigma(\sigma(a), \sigma(b))=R(a,b)$ $\forall a,b \in A$;

2b) a CF C is neutral iff for all permutations σ on A, $\forall R \in R(a)$ and $\forall M \in 2^A$

$$a \in C(M,R) \Leftrightarrow \sigma(a) \in C(\sigma(M),R^\sigma).$$

3) *independence of circuits*: a circuit of length q in a digraph is an ordered collection of arcs $(u_1 , u_2, ..., u_q)$ such that for i=1, 2, ...,q, the initial extremity of u_i is the final extremity of u_{i-1} and the final extremity of u_i is the initial extremity of u_{i+1} , where u_0 is interpreted as u_q and u_{q+1} as u_1. A circuit is elementary iff each node being the extremity of one arc in the circuit is the extremity of exactly two arcs in the circuit. A transformation on an elementary circuit consists of adding the same quantity to the value of all the arcs in the circuit. A transformation on an elementary circuit is admissible if all the transformed valuations are still between 0 and 1. An exploitation procedure is independent of circuits iff its results do not change after an admissible transformation of R. More formally,

3a) a RM $\geq$ is independent of circuits iff $R,R' \in R(A)$, R' is obtained from R through an admissible transformation on an elementary circuit of length 2 or 3 and $\forall a,b \in M \subseteq A$

$$a \geq (M,R)b \Rightarrow a \geq (M,R')b;$$

3b) a CF C is independent of circuits iff $\forall M \in 2^A$ and $\forall R,R' \in R(A)$, such that R' is obtained from R through an admissible transformation on an elementary circuit of length 2 or 3 on M,

$$C(M,R)=C(M,R').$$

The property of independence of circuits makes an explicit use of the cardinal properties of the valuations R(a,b). This is not the case of the neutrality and monotonicity (Bouyssou, 1992a and b).

Given $R \in R(A)$ and $M \subseteq A$, a net flow $S_{NF}(x,M,R)$ can be associated to each $x \in M$ as follows:

$$S_{NF} (x,M,R) = \sum_{b \in M-\{x\}} (R(x,b) -R(b,x)).$$

More specifically, the RM $\geq$ such that

$$a \geq (M,R)b \text{ iff } S_{NF}(a,M,R) \geq S_{NF}(b,M,R)$$

is called *net flow ranking method*, and the CF C such that

$$C(M,R) = \{a \in M: S_{NF}(a,M,R) \geq S_{NF}(b,M,R) \ \forall b \in M\}.$$

is called *net flow choice method*.

Theorem 5.1. (Bouyssou 1992a). The net flow method is the only RM that is neutral, strongly monotonic and independent of circuits.

Theorem 5.2. (Bouyssou 1992b). The net flow method is the only CF that is neutral, strongly monotonic and independent of circuits.

5.2 Properties of the exploitation procedures for the four-valued outranking

In order to characterize the scoring procedure we consider a four-valued outranking relation as a function R_{4v} associating an element of $\{S^T, S^U, S^K, S^F\}$ with each ordered pair of actions $(a,b) \in A \times A$. Now, RM $\geq$ and CF C are defined analogously for a four-valued outranking relation, i.e., for $R_{4v}(A)$ being the set of all possible four-valued relations on A, RM $\geq$ is a function assigning a ranking $\geq(M,R_{4v})$ on $M \subseteq A$ to any $R_{4v} \in R_{4v}(A)$ and to any $M \subseteq A$, and CF C on A is a function

$$C: 2^A \times R_{4v}(A) \rightarrow 2^A$$

such that $C(M,R_{4v}) \subseteq M$, for each $M \in 2^A$ and each $R_{4v} \in R_{4v}(A)$.

Moreover, the property of neutrality maintains the same formulation as in the exploitation procedure for the valued outranking relation, i.e.

- a RM $\geq$ is neutral iff for all permutations σ on A, $\forall M \subseteq A$, $\forall R_{4v} \in R_{4v}(A)$ and $\forall a,b \in M$

$$a \geq (M,R_{4v})b \Leftrightarrow \sigma(a) \geq (\sigma(M),R^\sigma_{4v})\sigma(b)$$

- a CF C is neutral iff for all permutations σ on A, $\forall M \in 2^A$ and $\forall R_{4v} \in R_{4v}(A)$

$$a \in C(M,R_{4v}) \Leftrightarrow \sigma(a) \in C(\sigma(M),R^\sigma_{4v})$$

where for any permutation σ and $\forall a,b \in A$, R^σ_{4v} is defined by

$$R^\sigma_{4v}(\sigma(a),\sigma(b)) = R_{4v}(a,b).$$

Instead, the strong monotonicity and the independence of circuits properties have a formal definition which is slightly different from the previous definition and requires some new concepts.

A *4v-transformation* on the pair $(a,b) \in A \times A$ consists of changing the outranking relation S^X into the outranking relation S^Y, where $S^X, S^Y \in \{S^T, S^U, S^K, S^F\}$, and it is denoted by

$$aS^Xb \rightarrow aS^Yb.$$

Let us denote by $S^X \rightarrow S^Y$ the class of all the transformations $aS^Xb \rightarrow aS^Yb$ with $(a,b) \in A \times A$ and $S^X, S^Y \in \{S^T, S^U, S^K, S^F\}$.

Let T be the set of all 4v-transformations on the pairs $(a,b) \in A \times A$. We introduce an equivalence binary relation E on T. More specifically,

$$[aS^Xb \rightarrow aS^Yb] \; E \; [aS^Wb \rightarrow aS^Zb]$$

means that the transformation $[aS^Xb \rightarrow aS^Yb]$ has the same "strength" as the transformation $[aS^Wb \rightarrow aS^Zb]$, where $S^X, S^Y, S^W, S^Z \in \{S^T, S^U, S^K, S^F\}$.

We define the following *equivalence classes* for E:

1. $E^0 = (S^T \rightarrow S^T) \cup (S^F \rightarrow S^F) \cup (S^U \rightarrow S^U) \cup (S^K \rightarrow S^K) \cup (S^K \rightarrow S^U) \cup (S^U \rightarrow S^K)$, i.e. the class of the transformations from an outranking S^X to an outranking S^Y of the same strength;

2. $E^1 = (S^U \rightarrow S^T) \cup (S^K \rightarrow S^T) \cup (S^F \rightarrow S^U) \cup (S^F \rightarrow S^K)$, i.e. the class of the transformations from an outranking S^X to an outranking S^Y having a greater strength;

3. $E^{-1} = (S^T \rightarrow S^U) \cup (S^T \rightarrow S^K) \cup (S^U \rightarrow S^F) \cup (S^K \rightarrow S^F)$, i.e. the class of the transformations from an outranking S^X to an outranking S^Y having a weaker strength;

4. $E^2 = (S^F \rightarrow S^T)$, i.e. the class of the transformation from an outranking S^X to an outranking S^Y having a far greater strength (from total absence of outranking to sure presence of outranking);

5. $E^{-2} = (S^T \rightarrow S^F)$, i.e. the class of the transformation from an outranking S^X to an outranking S^Y having a far weaker strength (from sure presence of outranking to total absence of outranking).

Within the context of a four-valued outranking relation,

1'a) a RM $\geq$ is strongly monotonic iff $\forall M \subseteq A$ and $\forall a, b \in M$

$$a \geq (M, R_{4v})b \Rightarrow a > (M, R'_{4v})b$$

where $>(M, R_{4v})$ is the asymmetric part of $\geq (M, R_{4v})$ and R'_{4v} is identical to R_{4v} except that R'_{4v} is obtained from R_{4v} by means of a 4v-transformation $aS^Xc \rightarrow aS^Yc$ with $(S^X \rightarrow S^Y) \subset E^1 \cup E^2$ or $cS^Xa \rightarrow cS^Ya$ with $(S^X \rightarrow S^Y) \subset E^{-1} \cup E^{-2}$ for some $c \in M - \{a\}$;

1'b) CF C is strongly monotonic iff $\forall M \in 2^A$ and $R_{4v} \in R_{4v}(A)$

$$a \in C(M, R_{4v}) \Rightarrow \{a\} = C(M, R'_{4v})$$

where R'_{4v} is defined as previously.

A 4v-transformation on an elementary circuit consists of performing a 4v-transformation of the same equivalence class in the arcs of the circuit. A 4v-transformation on an elementary circuit is admissible if all the transformed outranking relations belong to the set $\{S^T, S^U, S^K, S^F\}$; e.g., if we have aS^Tb, bS^Uc, cS^Ta, an admissible transformation on the elementary circuit $\{(a,b), (b,c), (c,a)\}$ is aS^Ub, bS^Fc, cS^Ka. Let us point out that the elementary transformation on the arcs are $aS^Tb \rightarrow aS^Ub$, $bS^Uc \rightarrow bS^Fc$, $cS^Ta \rightarrow cS^Ka$. Therefore a RM $\geq$ is independent of circuits if $R_{4v}, R'_{4v} \in R_{4v}(A)$, R'_{4v} being obtained from R_{4v} through an admissible transformation on an elementary circuit and

$$a \geq (M, R_{4v}) b \Rightarrow a \geq (M, R'_{4v}) b.$$

Analogously, a CF C is independent of elementary circuits iff, under the same hypotheses, $\forall M \in 2^A$ and $\forall R_{4v}, R'_{4v} \in R_{4v}(A)$

$$C(M, R_{4v}) = C(M, R'_{4v}).$$

Let us remark that the four-valued outranking R_{4v} expresses some possible preference situations without using any numerical evaluation. Therefore, the property of independence of circuits makes no use of cardinal properties of the relations, similarly to the property of neutrality and monotonicity.

5.3 An extension of the previous results to the four-valued outranking

To extend the results of Bouyssou (1992a and b), we associate an element of $\{0, 1/2, 1\}$ with each $(a,b) \in A \times A$ introducing the valued outranking binary relation $\vec{R}_{4v} : A \times A \to [0,1]$ by stating:

$$\vec{R}_{4v}(a,b) = \begin{cases} 0 & \text{if } aS^F b \\ 1/2 & \text{if } aS^U b \text{ or } aS^K b \\ 1 & \text{if } aS^T b. \end{cases}$$

This is a reduction to the $[0,1]$ interval of the lattice of the four truth values, where the values S^U and S^K are incomparable (no numerical value is used there). Such a reduction could be judged arbitrary, but the following result shows that $\vec{R}_{4v}$ satisfies some desirable properties, allowing us to say that $\vec{R}_{4v}$ is the only valued relation which faithfully represents R_{4v}. Let us consider F: $\{S^T, S^U, S^K, S^F\} \to [0,1]$. From each $R_{4v} \in R_{4v}(A)$ we can obtain one $R \in R(A)$ by stating $R(a,b) = F(R_{4v}(a,b))$ $\forall (a,b) \in A \times A$.

Let us consider the following properties $\forall (a,b), (c,d) \in A \times A$:

R1) $F(R_{4v}(a,b)) = 1$ iff $aS^T b$,

R2) $F(R_{4v}(a,b)) = 0$ iff $aS^F b$,

R3) $F(R_{4v}^1(a,b)) - F(R_{4v}^2(a,b)) = F(R_{4v}^3(c,d)) - F(R_{4v}^4(c,d))$ iff $aS^X b$ according to R_{4v}^1, $aS^Y b$ according to R_{4v}^2, $cS^W d$ according to R_{4v}^3, $cS^Z d$ according to R_{4v}^4 and

$$[aS^X b \to aS^Y b] \text{ E } [cS^W d \to cS^Z d].$$

Property R1) says that $\forall (a,b) \in A \times A$ the transformation of the four-valued outranking R_{4v} into the valued outranking R should give the maximum value, i.e., $R(a,b)=1$, iff $aS^T b$. Analogously, property R2) says that, $\forall (a,b) \in A \times A$, the same transformation should give the minimum value, i.e., $R(a,b)=0$, iff $aS^F b$. Finally, property R3) says that, if 4v-transformations $S^X \to S^Y$ and $S^W \to S^Z$ are of the same strength, then we should have $F(S^X)-F(S^Y)= F(S^W)-F(S^Z)$.

Theorem 5.3. (Greco, Matarazzo, Slowinski, Tsoukias, 1997) Properties R1), R2) and R3) are satisfied if and only if

$$F(R_{4v}(a,b)) = \vec{R}_{4v}(a,b).$$

Lemma 5.1. (Greco, Matarazzo, Slowinski, Tsoukias, 1997) The following relation between the overall score $S(x,M)$ and the net flow $S_{NF}(x,M,\vec{R}_{4v})$ holds:

$$S(x,M) = 2\, S_{NF}(x,M,\vec{R}_{4v}), \quad \forall M \subseteq A \text{ and } \forall x \in M.$$

Lemma 5.1 shows that the overall score $S(x,M)$ is a strictly positive monotonic transformation of the net flow $S_{NF}(x,M,\vec{R}_{4v})$. Therefore, we conclude that the ranking and the choice obtained from $S(x,M)$ are the same as those obtained from $S_{NF}(x,M,\vec{R}_{4v})$.

Lemma 5.2. (Greco, Matarazzo, Slowinski, Tsoukias, 1997) Given $R_{4v}, R'_{4v} \in R_{4v}(A)$, if R'_{4v} is obtained from R_{4v} by an admissible 4v-transformation on an elementary circuit, then $\vec{R}'_{4v}$ is obtained from $\vec{R}_{4v}$ by an admissible transformation on an elementary circuit.

Due to Lemmas 5.1 and Lemma 5.2, Theorems 5.1 and 5.2 imply, respectively, the following two theorems (Greco, Matarazzo, Slowinski, Tsoukias, 1997).

Theorem 5.4. With respect to a four-valued outranking relation established by a set of decision rules, the scoring procedure based on $S(x,M)$ is the only RM which is neutral, strongly monotonic and independent of circuits.

Theorem 5.5. With respect to a four-valued outranking relation established by a set of decision rules, the scoring procedure based on $S(x,M)$ is the only CF which is neutral, strongly monotonic and independent of circuits.

6 Conclusions

We have been using the rough set approach to the analysis of preferential information concerning multicriteria choice and ranking problems. This information is given by a decision maker as a set of pairwise comparisons among some reference actions using the outranking relation. The outranking relation is approximated by means of a special form of dominance relation and decision rules are derived from these approximations. They represent the preference model of the decision maker. In result of application of these rules to a new set of potential actions, we get a four-valued outranking relation.

In this paper, we dealt with the problem of obtaining a recommendation from the above four-valued outranking relation. With this aim we proposed an exploitation procedure for ranking and choice problems based on a specific net flow score. Furthermore, we proved that this procedure is the only one which is neutral, strongly monotonic and independent of circuits.

Acknowledgements. We wish to thank Denis Bouyssou for helpful discussions about exploitation procedures. The research of the first two authors has been supported by grant no. 96.01658.CT10 from Italian National Council for Scientific Research (CNR). The research of the third author has been supported by KBN grant No. 8T11F 010 08 p02 from State Committee for Scientific Research KBN (Komitet Badan Naukowych).

References

Bouyssou, D., "Ranking methods based on valued preference relations: a characterization of the net-flow method", *European Journal of Operational Research*, 60, 1992a, 61-68.

Bouyssou, D., "A note on the sum of differences choice function for fuzzy preference relations", *Fuzzy sets and Systems*, 47, 1992b, 197-202.

Bouyssou, D., "Outranking Relations: Do they have special properties?", *Journal of Multi-Criteria Decision Analysis*, 5 (2), 1996, 99-111.

Fishburn, P. C., *The Theory of Social Choice*, Princeton University Press, Princeton, NJ, 1973.

Goodman, L. A. , "On methods of amalgamation", in: R.M. Thrall, C.H. Coombs and R.L. Davis, Eds., *Decision Processes*, Wiley, New York, 1954, 39-48.

Greco S., Matarazzo, B., Slowinski, R., *Rough Set Approach to Multi-Attribute Choice and Ranking Problems*, ICS Research Report 38/95, Warsaw University of Technology, Warsaw, 1995. Also in: G. Fandel and T. Gal (Eds.), *Multiple Criteria Decision Making*, Springer-Verlag, Berlin, 1997, 318-329.

Greco S., Matarazzo, B., Slowinski, R., *Rough Approximation of Preference Relation by Dominance Relations*, ICS Research Report 16/96, Warsaw University of Technology, Warsaw, 1996. Also in: *European Journal of Operational Research*, to appear.

Greco S., Matarazzo, B., Slowinski, R., Tsoukias, A., *Exploitation of a rough approximation of the outranking relation*, Cahier du LAMSADE, no. 152, University of Paris-Dauphine, Paris, 1997.

Luce, R.D., "Semi-orders and a theory of utility discrimination", *Econometrica*, 24, 1956, 178-191.

Henriet, D., "The Copeland choice function - An axiomatic characterization", *Social Choice and Welfare*, 2, 1985, 49-64.

Roy, B, *Méthodolgie multicritère d'aide à la décision*, Economica, Paris, 1985.

Roy, B., "The outranking approach and the foundations of ELECTRE methods", *Theory and Decision*, 31, 1991, 49-73.

Roy, B., "Decision science or decision aid science?", *European Journal of Operational Research*, Special Issue on Model Validation in Operations Research, 66, 1993, 184-203.

Rubinstein, A., "Ranking the participants in a tournament", *SIAM Journal of Applied Mathematics*, 38, 1980, 108-111.

Slowinski, R., Vanderpooten, D., *Similarity relation as a basis for rough approximations*, ICS Research Report 53/95, Warsaw University of Technology, Warsaw, 1995.

Slowinski, R., Vanderpooten, D., *A generalized definition of rough approximation*, ICS Research Report 4/96, Warsaw University of Technology, Warsaw, 1996.

Slowinski, R., Vanderpooten, D., "A generalized definition of rough approximation based on similarity", to appear in *IEEE Transactions on Data and Knowledge Engineering*, 1998.

Tsoukias, A., Vincke, Ph., "A new axiomatic foundation of the partial comparability theory", *Theory and Decision*, 39, 1995, 79-114.

Tsoukias, A., Vincke, Ph., "A new axiomatic foundation of partial comparability", in: J. Climaco (Ed.): *Multicriteria Analysis*, Springer-Verlag, Berlin, 1997, 37-50.

Tversky, A., "Features of similarity", *Psychological Review*, 84 (4), 1977, 327-352

An Algorithm for Solving Intransitivities without Repeating Trials

Jacinto González-Pachón and Mª Isabel Rodríguez-Galiano[1]

[1] Department of Artificial Intelligence, Universidad Politécnica de Madrid. 28660-Boadilla del Monte. Madrid. Spain

Abstract. Based on a recent method for solving intransitivities in repeated pairwise choice, we present how this algorithm is applied when there are no repeated trials. The new method assumption is: The more repetitions of an arc there are in the MMQO, the more original transitivities are based on this element, i.e., the more stable this arc is.

Keywords. Paired comparisons, intransitivities, quasi order, preference modelling, transitive closure

1 Introduction

Transitivity is generally accepted as a principle of rationality in Decision Theory. The presence of intransitive preferences complicates matters, and these preferences are considered, in general, as a sign of irrationality. From this point of view, one of the primary tasks for the Decision Sciences is to solve intransitivities, although the transitivity assumption has been relaxed in a series of important papers published in recent years: Fishburn (1982, 1991), Kahneman and Tversky (1979), Loomes and Sugden (1982).

The context of this paper is preference modelling using the method of paired comparisons. This method is mostly used when the alternatives to be compared cannot be measured in a physically meaningful way. Furthermore, as with the ELECTRE methodologies, indifferences and incomparabilities will be allowed as possible Decision Maker (DM) attitudes in the pairwise comparison process, see Roy (1996). However, as opposed to these methods, we consider a general framework where DMs may be unable to evaluate the strength of preferences.

One widely known procedure for solving intransitivities in a general pairwise comparisons context consists of finding a linear order, based on original preferences, such that the number of preference reversals is as small as possible, Slater (1961). This procedure is known as Slater's procedure and that number is known as Slater's index.

Recently, Mass *et al.* (1995) have presented an operational method for deriving a linear ranking of alternatives using repeated paired comparisons. This method is based on the notion of stable preferences, i.e., preferences that remain the same throughout all repetitions. This stability is interpreted as an objective measure of the intensity of preferences.

In this paper, we present a method for solving intransitivities based on the method by Mass *et al.* As opposed to this method, however, repeated trials are not necessary here. The information obtained from repetitions is now obtained by analysing the Mixture of Maximal Quasi-Orders (MMQO) associated with a reflexive binary relation, defined in González-Pachón *et al.* (1996, 1997).

A basic definition is given in Section 2 and the use of the MMQO concept in Mass *et al.*'s algorithm is explained in Section 3. Section 4 sets out an example. Finally, concluding remarks are added.

2 Basic Definitions and Results

Let A be a finite set of alternatives in a decision making problem, $A = \{a_1, \ldots, a_n\}$. Let S be the characteristic binary relation of a preference structure in the sense of Roubens and Vincke (1985). This binary relation represents the DM attitude below

$$a_i S a_j \text{ if and only if } "a_i \text{ is at least as good as } a_j"$$

Definition 1 *A binary relation S on a finite set A is a quasi order (q.o.) iff S is reflexive and transitive.*

The following definitions appear in González-Pachón *et al.* (1996, 1997).

Definition 2 *Let A be a finite set, S a reflexive binary relation on A and $V = \{S_i\}_{i \in I}$ the partially ordered set of all q.o.s included in S, using inclusion, $\subset$, as the order. $S^* \in V$ is a maximal q.o. included in S, if there is no $S' \in V$ such that $S^* \subset S'$.*

Definition 3 *The family of all maximal q.o.s included in a reflexive binary relation S will be referred to as the Mixture of Maximal Quasi-Orders (MMQO) associated with a reflexive binary relation S and will be denoted by $\mathcal{F}$.*

The following result appears in González-Pachón *et al.* (1996).

Proposition 1 *If S is a reflexive binary relation on a finite set A, the associated MMQO, $\mathcal{F}$, is not empty and covers S, i.e.*

$$\mathcal{F} \neq \varnothing \quad and \quad S = \bigcup_{Q \in \mathcal{F}} Q$$

Example

Assume the binary relation S defined by the graph in Fig. 1, for representing DM preferences in pairwise comparison of the set of alternatives $A = \{a, b, c, d\}$.

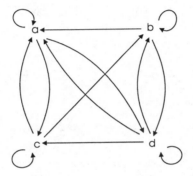

Fig. 1. Binary relation S

The MMQO associated with S is shown in Fig. 2 in the form of Hasse diagrams.

The binary relation S may be represented as a directed graph that contains objects, called nodes (alternatives), and where the relation between nodes is represented by arcs.

An arc will be referred to as $a_i a_j$, which means an arc from a_i to a_j, i.e., $a_i S a_j$. The indegree of a node a_i, $id(a_i)$ is the number of incoming arcs $a_j a_i$, where $a_i \neq a_j$. The outdegree of a node a_i, $od(a_i)$ is the number of outgoing arcs $a_i a_j$, where $a_i \neq a_j$.

We write $a_i P_k a_j$, if a_i is at least as preferred as a_j in k elements of $\mathcal{F}$, where $a_i \neq a_j$. Let us consider $k^* = max\, k$. For D_0, we take the directed graph corresponding to the transitive closure of P_{k^*}.

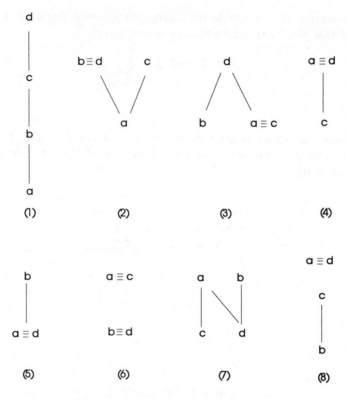

Fig. 2. MMQO associated with S

3 The Method

In this section, we show how Mass *et al.*'s algorithm is applied when there are no repeated trials. The only information is the MMQO associated with the characteristic binary relation of a preference structure.

Now, the more MMQO elements contain an arc, the more stable this arc will be. This new idea of stability is based on the following assumptions:

(a) Every DM would like to be rational (in our case, transitive).

(b) The more repetitions of an arc there are in the MMQO, the more original transitivities are based on this element, i.e., the more stable this arc is.

The first assumption is based on an observation published in González-Pachón *et al.* (1996). As a result, transitive closure will be applied after every step in the algorithm. The second assumption is due to the fact that an arc which participates in several revealed transitivities (transitivities included in

S), will appear in a sizeable subset of MMQO elements.

The following algorithm is a modified version of the algorithm published in Mass *et al.* (1995).

STEP 1 Create D_0, $i = 0$, $k = k^*$.

STEP 2 If not all aP_kb are included in D_i, go to step 5.

STEP 3 $k = k - 1$.

STEP 4 If $k > 0$, go to step 2. If $k = 0$, go to step 10.

STEP 5 Determine the score vector

$$\sigma(D_i) = (od(a_{(1)}), \, od(a_{(2)}), \ldots, \, od(a_{(n)}))$$

where nodes $a_1, \ldots, a_n$ have been reordered -$a_{(1)}, \ldots, a_{(n)}$- such that $od(a_{(1)}) \geq \ldots \geq od(a_{(n)})$.
The ranking of nodes with equal outdegrees is solved using their indegrees, as follows: if $od(a) = od(a')$ and $id(a) < id(a')$, the ranking of a is higher than the ranking of a'. If $id(a) = id(a')$, the ranking is arbitrary.

STEP 6 If $od(a_{(1)}) > od(a_{(2)}) > \ldots > od(a_{(n)})$, go to step 10.

STEP 7 Pairs of nodes are written with the highest outdegree first and are, subsequently, ranked in a reversed lexicographic way.

$(a_{(1)}, \, a_{(2)}), (a_{(1)}, \, a_{(3)}), \, \ldots, (a_{(1)}, \, a_{(n)}), (a_{(2)}, \, a_{(3)}), \, \ldots, (a_{(2)}, \, a_{(n)}), \, \ldots,$
$(a_{(n-1)}, \, a_{(n)})$.

STEP 8 Select the first pair of nodes (a, a') such that aP_ka' is not included in D_i.
D_{i+1} is the transitive closure of D_i with the arc aa'.

STEP 9 $i = i + 1$. Go to step 2.

STEP 10 STOP.

For an analysis of the computational complexity of this algorithm, see Mass *et al.* (1995).

4 Example

Let us consider the binary relation of the example set out in Section 2. In Fig. 1, we showed revealed preferences, which form cycles like $bSdScSb$, where b, c, d are not indifferent, because $no(bSc)$ and $no(cSd)$. The preference matrix of this example, in the sense of Mass *et al.*, is given in Table 1.

Table 1. Preference matrix

	a	b	c	d
a	-	0	2	1
b	3	-	0	2
c	2	2	-	0
d	3	2	3	-

The values of the cells represent the number of MMQO q.o.s, where the row alternative is at least as preferred as the column alternative.

In this case, $k^* = 3$ and $D_0 = P_3$, Fig. 3(a), because transitive closure does not imply any more arcs. This equality produces a new value for k, $k = 2$.

Now, $\sigma(D_0) = (2, 1, 0, 0)$, where the first 0 is the outdegree for node c, because the indegree of c is lower than the indegree of a, $id(c) = 1 < id(a) = 2$. The ordering of pairs to be selected is (d, b), (d, c), (d, a), (b, c), (b, a), (c, a). It is easily verified that an arc between d and b is not yet present in D_0, and dP_2b. So, (d, b) is selected . We get $D_1 = D_0 \cup \{(d, b)\}$, because transitive closure does not add any more arcs. The next arc in P_2 to be selected is (c, a). Similarly, $D_2 = D_1 \cup \{(c, a)\}$, as shown in Fig. 3(b). We determine $\sigma(D_2) = (3, 1, 1, 0)$, where the first 1 is the outdegree for c selected at random, because $id(c) = id(b)$. Now, the ordering of pairs to be selected is (d, c), (d, b), (d, a), (c, b), (c, a), (b, a). As an arc between c and b is not yet present in D_2 and cP_2b, then $D_3 = D_2 \cup \{(c, b)\}$ and $k = 1$. However, $\sigma(D_3) = (3, 2, 1, 0)$ and then stop. The final result, D_3, gives a solution, i.e., a transitive relation.

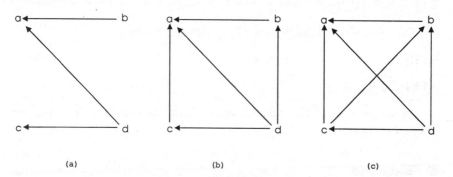

(a) (b) (c)

Fig. 3. (a) Preference relation $P_3 = D_0$, (b) Directed graph D_2, (c) Solution to our problem: directed graph D_3

5 Concluding Remarks

Using the family of MMQO associated with a reflexive binary, we have used the algorithm by Mass *et al.* (1995), when it is impossible to repeat trials. We propose a modified version of this algorithm.

The following assumptions are considered.

- Every DM would like to make transitive choices.

- The more times an element of the binary relation appears in the MMQO, the more original transitivities are based on this element.

On this basis, we observe that elements in the MMQO play the role of repeated trials in the above cited algorithm.

Acknowledgment

The authors would like thank two anonymous referees for their valuable suggestions and Mrs. Rachel Elliott for reviewing the English.

References

Fishburn, P.C. (1982) 'Nontransitive Measurable Utility', *J. Math Psycholg.*, 26, pp. 31-67.

Fishburn, P.C. (1991) 'Nontransitive Preferences in Decision Theory', *J. Risk and Uncert.*, 4, pp. 113-134.

González-Pachón, J. and Ríos-Insua, S. (1996) 'Mixture of Maximal Quasi Orders: A New Approach to Preference Modelling'. (submitted)

González-Pachón, J. and Ríos-Insua, S. (1997) 'A Method for Searching Rationality in a Pairwise Choices'. In G. Fandel and T. Gal (Eds.), *Multiple Criteria Decision Making*, LNEMS 448, Springer, pp. 374-382.

Kahneman, D. and Tversky (1979), A. 'Prospect Theory: An Analysis of Decision under Risk', *Econometrica*, 47, pp. 263-291.

Loomes, G. and Sugden, R. (1982) 'Regret Theory: An Alternative Theory of Rational Choice under Uncertainty', *Econ. J.*, 92, pp. 805-824.

Mass, A., Bezembinder, T. and Wakker, P. (1995) 'On Solving Intransitivities in Repeated Pairwise Choices', *Mathematical Social Sciences*, 29, pp. 83-101.

Roubens, M. and Vincke, Ph. (1985) *Preference Modelling*. Springer.

Roy, B. (1996) *Multicriteria Methodology for Decision Aiding*. Kluwer.

Slater, P. (1961) 'Inconsistencies in a Schedule of Paired Comparisons', *Biometrica*, 48, pp. 303-312.

The Approach to Multicriteria Decision Making Based on Pattern Recognition Algorithms

Anna Perekhod

Department of Mathematics, Simferopol State University
4, Yaltinskaya Street, Simferopol 333036, Crimea

1 Introduction

The paper introduces *Multiple Criteria Decision Making Problems with Incomplete Information (MCDMII)*. Usually such problems arise in *Intelligent Decision Support Systems (IDSS)* development or other cases of decision making practical realization. In most of cases, any practical problem is multicriteria, because a compromise between estimations by different criteria is necessary for the best solution choice [1]. On the other hand, it is a problem with incomplete information, because, as a rule, an exact setting of all data components in the decision making model is impossible [2].

An application of traditional multicriteria optimization methods is not enough in such situations. We propose to use *Artificial Intelligence (AI)* and *Pattern Recognition (PR)* methods or to combine these methods with known approaches to MCDM problem solving.

2 The Problem

Let us consider the following *multicriteria decision making problem*:

$$\begin{cases} \text{extr } f_j(\tilde{x}), \ j = \overline{1,r} \\ \quad \tilde{x} \in \Omega \subset B'' \end{cases} \tag{1}$$

where Ω is the admissible solutions set; $B'' = \{0,1\}''$ is a set of unit n-dimensional cube vertices; extr stands for „extremum". The criteria are *pseudo-Boolean functions* $\left\{ f_j : B'' \to R, \ j = \overline{1,r} \right\}$

We study the case, where the components of the model (1) exist, but there is no complete information about criteria $f_j, j = \overline{1,r}$ and admissible solutions set Ω. We call such problems *MCDMII problems with Boolean variables*.

We consider decision making as a choice of some element $\tilde{x} = \left(x_1, \ldots, x_n \right), x_i \in \{0,1\}, i = \overline{1,n}$ from the admissible solutions set Ω. The choice $\tilde{x}^* \in B''$ is obtained under initial information incompleteness and can be estimated. Thus, we have the information about the decisions, which are made, and their estimations. Experts estimate choices $\{\tilde{x}^*\}$ and these estimations are used to synthesize decision making model approximating (1).

3 Approach to the Problem Solving

Due to initial information incompleteness in MCDMII problems, application of methods which allow additional information acquisition is necessary. The most suitable for this purpose are inductive AI methods, for example machine learning and rule induction. The general aim of the methods is inducing logical rules from experience [12].

Mostly, such kind of algorithms have been developed for PR. PR problem may be formulated as decision making about membership of some objects to some classes (patterns). Objects are distinct examples (events, cases, states, etc.). Any object in PR theory is given by feature values collection (a vector containing values of the features). Features are attributes describing object properties. Classes (patterns) are disjoint sets containing of objects with common properties.

The main ideas used in this approach are as follows.

1. Due to initial information incompleteness, there is an indeterminacy in the problem. This indeterminacy should be decreased as much as possible according to the following principle. Let Z be a given PR problem. Let $I_0 = I_0(X)$ be initial information for any object $x \in X$. The information is incomplete and used to generate a set $\mathfrak{I}_0 = \mathfrak{I}(I_0)$ of possible logical inferences. The set $\mathfrak{I}_0$ is called an *indeterminacy region*. An indeterminacy region always includes a correct logical inference $g^*(I_0, x)$ for any $x \in X$ (iff such an inference exists in general). This principle is called the *principle of maximal indeterminacy reduction*.

2. A suitable way for indeterminacy decreasing is extraction of additional information. Additional information may be obtained by initial data model analysis or studying of properties of the functions which should be reconstructed or adding some constraints in optimization problems for admissible solutions set approximation. Any additional information which satisfies the condition

$$I_1 : \left(\mathfrak{I}_1 = \mathfrak{I}\left(\{I_0 \cup I_1\} \right) \subset \mathfrak{I}_0 \right) \& \left(g^*(I_0, x) \in \mathfrak{I}_1 \right), x \in X$$

is called a *reductional information*.

Process of the indeterminacy reduction may be generalized as follows

$$I_2, \ldots, I_k : \mathfrak{I}_2 = \mathfrak{I}_2(I_0 \cup I_1 \cup I_2), \ldots, \mathfrak{I}_k = \mathfrak{I}_k(I_0 \cup I_1 \cup \ldots \cup I_k); \mathfrak{I}_0 \supset \mathfrak{I}_1 \supset \ldots \supset \mathfrak{I}_k$$

This generalization illustrates *further extraction of additional information*. According to the above principle, we consider only those PR algorithms $\{A\}$:

$$(\forall x)\ A\!\left(\left\{\bigcup_{j=0}^{k} I_j\right\}, x\right) \in \mathfrak{I}_k, k = 0,1,\dots\ .$$

Such algorithms are called *initial information concordant* [2, 3].

Using of PR algorithms for function reconstruction and set approximation makes sense if the following conditions hold.

1. Any logical inference from the set $\mathfrak{I}_0$ of possible inferences can be computed.

   Any computed logical inference does not contradict the initial information. This condition is called *completeness*.

2. Computation of the analytical description is possible for any pattern (class of objects) [2].

   The MCDMII problem solving includes:

- approximation (recognition) of the admissible solutions set Ω;

- approximation of the Pareto set $P \subset \Omega$.

   We apply PR algorithms for these goals.

4 Subset Recognition Principles

We consider subset recognition of some finite set as dichotomy problem of PR (two patterns are given) and apply learning models for this purpose [4].

Let the inclusion $A \subset C$ be true for two sets. The finite set C is covered by two subsets: $C = A \cup \overline{A}$, where $\overline{A} = C \setminus A$. We can determine uniquely elements of one subset by known elements of other one if the set C is exactly defined [5].

Let a given partition be represented by incomplete information $I_0(A, \overline{A})$ as a Boolean learning table T:

$$T = A_1 \cup A_2:$$
$$(\tilde{a} \in A_1) \Rightarrow (\tilde{a} \in A);$$
$$(\tilde{a} \in A_2) \Rightarrow (\tilde{a} \in \overline{A})$$

A learning table T is non-contradictory, i.e. $A_1 \cap A_2 = \varnothing$.

Any object $\tilde{x} \in C$ is defined by values of the characteristics. These values give its description $I(\tilde{x})$. Subset recognition problem is a membership determination for any element of the set C to the subset A using the information $I_0(A, \overline{A})$ and descriptions for any object $\tilde{x} \in C$, i.e. computation of the predicate "$\tilde{x} \in A$" (or "$\tilde{x} \in \overline{A}$") [6,7].

For such problems, we propose to use *binary decision tree (BDT)* based learning algorithms. These algorithms make it possible to get generalized descriptions of classes (in this case classes are subsets) as logical formulas.

Definition 1. A *binary tree* is called a *decision tree* if the following properties are fulfilled

- any inner vertex is marked by feature predicate
- edges, which go from the vertex out, are marked by predicate values in the vertex
- end vertices (leaves) are marked by class labels
- there is no tree branch having two equal inner vertices

BDT for the problem solving realizes a function $BDT_{set}: C \rightarrow \{A, \overline{A}\}$. The mapping represents a system of two functions of two-valued logic. Thus, we can synthesize any Boolean function using BDTs. Also, any BDT may be repesented as *disjunctive normal form (DNF)*. This property is called *completeness* [2].

4.1 Recognition of Admissible Solutions Set

We consider elements $\tilde{x} = (x_1, \ldots, x_n)$ of the set B'' as feature values collections [8,9]. In other words, let us consider the feature predicates set $\{x_i : Y \rightarrow \{0,1\}\}$, $i = \overline{1,n}$; where Y is the set of states discribing the problem region.

Admissible solutions set Ω is a subset of the set of unit n-dimensional cube vertices. A learning table $T_{m_0 n} \subset B''$ contains elements of a set $\mathfrak{R} = \{\tilde{x}^*\}$ of made decisions. The set $\mathfrak{R}$ answers the requirements for learning tables:

$$\mathfrak{R} = \mathfrak{R}_0 \bigcup \mathfrak{R}_1$$
$$\mathfrak{R}_0 \bigcap \mathfrak{R}_1 = \varnothing$$
$$(\tilde{x} \in \mathfrak{R}_0) \Rightarrow (\tilde{x} \in \Omega)$$
$$(\tilde{x} \in \mathfrak{R}_1) \Rightarrow (\tilde{x} \notin \Omega)$$

BDT, which represents a logical description of the admissible solutions set Ω, must be admissible.

Definition 2. *BDT*, corresponding to admissible partition, is called *admissible*.

Definition 3. A set of disjoint intervals $N_1, \ldots, N_\mu \subset B''$ is called an *admissible partition*, if any interval contains learning table elements, corresponding to one class only and $\bigcup\limits_{q=1}^{\mu} N_q = B''$

Any BDT represents some partition, because any BDT branch corresponds to some elementary conjunction. Elementary conjunctions $K_1, \ldots, K_\mu : \bigvee\limits_{q=1}^{\mu} K_q = 1$ correspond to the intervals $N_1, \ldots, N_\mu \subset B''$. This fact makes it possible to synthesize any BDT using the *algorithm for admissible partition construction*.

1. Divide B'' into two disjoint intervals N_1^1 and N_2^1 (range of both intervals is equal to 1); $N_1^1 \bigcup N_2^1 = B''$; such that at least one pair $\tilde{\alpha}, \tilde{\beta} \in T_{m_0 n}$ exists which contains elements from different classes (patterns) and these elements belong

to different intervals of obtained in this step partition (this property is called the *main partition rule*).

2. If an interval N' (range is equal to r_i) containing elements from different classes exists in obtained partition, then repeat previous step for this interval such that the main partition rule holds, else end.

The main idea of the algorithm is partition the training data recursively into disjoint sets.

BDT, which represents subset logical description, is not unique. We can obtain several BDTs with various number m of inner vertices. BDT type depends on the using features and their order. It was proved that more preferable choice of a decision rule is among BDTs with the least number μ of leaves. A feature choice for a current tree vertex construction is a variable for the interval partition. The feature choice determines BDT properties and BDT complexity. BDT synthesis with minimal number μ of leaves for subset recognition is equivalent to the shortest orthogonal DNF search for a Boolean function. Computational complexity of the problem may be estimated as $O(3^n)$ by number of all possible elementary conjunctions [2,6].

Feature for branching must be chosen so that the main partition rule holds for more pairs $\tilde{\alpha}, \tilde{\beta} \in T_{m_0,n}$ of elements; moreover, any BDT branch must have no feature, which occurs twice. Such branching criteria may be used in the algorithm for admissible partition construction. Complexity of the algorithm is linear by number of features. If the number of features is not large ($n<20$), exact methods may be used, for example *branch-and-bound method*. [2,6].

BDT makes it possible to construct a logical description $\hat{\psi}(\tilde{x})$ of the admissible solutions set Ω as a DNF and compute the predicate "$\tilde{x} \in \Omega$" for any element of the set B'' (check an admissibility for the corresponding solution).

4.2 Pareto Set Recognition

A final solution must be chosen from the Pareto set $P \subset \Omega$. Most of all multicriteria problems may be formulated as Pareto set approximation [1]. Under conditions of initial information incompleteness the Pareto set approximation may be realized as a subset recognition.

Let the information about criteria $f_j, j = \overline{1,r}$ in MCDMII problem be partially specified by a set $\tilde{\rho}$ of *binary relations (BR)*, $\tilde{\rho} = \{\rho_1,...,\rho_r\}$. In other words, experts compare by preferences all possible solution pairs $\tilde{x}, \tilde{y} \in \Omega$.

Definition 4. If the solution pair $\tilde{x}, \tilde{y} \in \Omega$ belongs to the BR $\rho_j : \tilde{x} \overset{j}{>} \tilde{y}$ ($\tilde{x} \overset{j}{\geq} \tilde{y}$) then $\tilde{x}$ "better" ("no worse") $\tilde{y}$ by criterion j [10].

Using given BR of preferences by all criteria and a learning table

$$T_{m,n} = W_0 \cup W_1; W_0 \cap W_1 = \varnothing:$$
$$\left(\tilde{x} \in W_0\right) \Rightarrow \left(\tilde{x} \in P\right)$$
$$\left(\tilde{x} \in W_1\right) \Rightarrow \left(\tilde{x} \notin P\right),$$

we should recognize the Pareto set $P \subset \Omega$.

It is convenient to represent the information about preferences as a full oriented *multigraph (MG)*. The MG vertices denote the solutions $\tilde{x} \in \Omega$, the MG edges specify BR between all solution pairs (the direction from less preferable to more preferable solution) [10].

We propose to consider bicriteria problems as follows

$$\begin{cases} \text{extr } f_i(\tilde{x}) \\ \text{extr } f_j(\tilde{x}) \\ 1 \le i < j \le r \\ \tilde{x} \in \Omega \subset B'' \end{cases} \qquad (2)$$

From the multicriteria problem (1) may be extracted $C_r^2 = \dfrac{r(r-1)}{2}$ above problems [11].

We represent a MG, which is in accordance with the problem (2), as a square binary matrix:

$$a_{pq} = \begin{cases} 1, \text{ if } q \text{ is more preferable} \\ 0, \text{ if } p \text{ is more preferable} \end{cases}$$

$$\forall \; p = q; \; a_{pq} = 0$$

The upper matrix triangle contains the information about a criterion i, the lower matrix triangle contains the information about a criterion j.

Approximation of the Pareto sets P_{ij} for all possible problems (2) (the algorithm is based on the MG analysis and matrix representing MG) makes it possible to define the upper and the lower bounds for the Pareto set as follows

$$P_{inf} \subseteq P \subseteq P_{sup}$$
$$P_{inf} = \bigcap_{\{(i,j)\}} P_{ij}$$
$$P_{sup} = \bigcup_{\{(i,j)\}} P_{ij}$$

[10].

The set P_{sup} can contain non-Pareto points $\tilde{x} \in P_{sup} : \tilde{x} \notin P$. Using the learning algorithms based on BDT construction we can get a logical description for the Pareto set, check unimprovability for any $\tilde{x} \in P_{sup}$ and delete non-Pareto points (if it is necessary).

We suggest the following *procedure for the Pareto set* $P \subset \Omega$ *approximation.*

1. *Form* the learning set $W = \{W_0 \cup W_1\}$, where W_0 contains only Pareto, W_1 only non-Pareto points (in expert opinion).

2. *Synthesize* the BDT_p: $\Omega \to \{P, \overline{P}\}$; *build* the decision rule $\hat{\varphi}(\tilde{x}) = $ " $\tilde{x} \in P$ " as a DNF.

3. *Compute* the predicate " $\tilde{x} \in P$ " for any $\tilde{x} \in P_{sup}$.

The final rule for the Pareto solutions choice is presented in the form $F(\tilde{x}) = \hat{\varphi}(\tilde{x}) \& \hat{\psi}(\tilde{x})$, where $\hat{\varphi}(\tilde{x}) = $ " $\tilde{x} \in P$"; $\hat{\psi}(\tilde{x}) = $ " $\tilde{x} \in \Omega$".

5 Conclusions

We present one of possible approaches to *MCDMII with boolean variables*. A merit of such approaches is as follows. Only a set of precedents and expert estimations are necessary for decision rules synthesis. Usually it is not difficult to get such information.

References

[1] Larichev, O.: *Objective models and subjective solutions*. Nauka, Moscow. 1987.

[2] Donskoy, V., Bashta, A.: *Discrete models for decision making with incomplete information*. Tavria, Simferopol. 1992

[3] Donskoy, V., Perekhod, A.: On the multicriteria optimization problem with incomplete information about criteria. Abstracts of the International Conference in Multi-Objective Programming, Malaga, Spain, May 16-18, 1996, pp. 52-53

[4] Perekhod, A.: Multicriteria decision making models with incomplete information for intelligent systems. Abstracts of the International Conference on Intelligent Data Processing, Alushta, Ukraine, June 3-7, 1996, pp. 27-28

[5] Zakrevski, A.: *Logic of the recognition*. Nauka, Minsk. 1988

[6] Donskoy, V.I.: Decision trees learning algorithms. Zh. Vychisl. Math. Math. Fiz. 22, 4 (1982) 963-974

[7] Donskoy, V.I.: Weakly defined problems of Boolean linear programming with a partially specified set of admissible solutions. Zh. Vychisl. Math. Math. Fiz. 28, 9 (1988) 1379-1385

[8] Tou, J., Gonzalez, R.: *Pattern recognition principles*. Addison-Wesley Publishing Company Inc., Advanced Book Program Reading, Massachusetts, London, etc. 1974

[9] Orlov, V.: *Graph-schemes of recognition algorithms*. Nauka, Moscow. 1982

[10] Perekhod, A.: The algorithm for Pareto sets approximation on multigraphs. Programs, Systems, Models. (1996) 18-26

[11] Gamkrelidze, L.Ch., Ostroukh, E.N.: Methods for a solving of the discrete multicriteria optimization problems. Izv. Ross. Acad. Nauk, Tech. Cybern. 3 (1989) 150-155

[12] Langley, P., Simon, H.: Application of Machine Learning and Rule Induction. Communications of the ACM. 38,11 (1995) 55-64

Decision Analysis
under Partial Information

Vladislav V. Podinovski[1] and Victor V. Podinovski[2]

[1] Academy of Labour and Social Relations, Moscow, 117454, Russia
[2] Warwick Business School, University of Warwick, Coventry, CV4 7AL, UK

Abstract. The use of partial (incomplete) qualitative data in multiple criteria decision making problems under uncertainty is considered. It is assumed that the preferences of a decision maker (DM) between certain consequences and the probabilities of the uncertain factor values are given in the form of binary relations. The following three main issues are addressed. First, the DM's preference relation for uncertain consequences is deduced from the available information. Secondly, we discuss how this, normally partial, preference relation can be further extended given the information about the DM's attitude towards risk. Thirdly, we show how the available qualitative information can be used in conjunction with the maxmin optimality principle.

Keywords: Multiple criteria, uncertainty, qualitative probability, preference relations, lexicographic maxmin

1 Problem Description

Consider a decision making problem under uncertainty

$$\langle S, \Lambda, X, f, \rhd_- , \succ_\approx \rangle, \tag{1}$$

where S is a set of decision alternatives, or strategies; $\Lambda = \{\lambda^1,...,\lambda^n\}$ is a finite set of values of an uncertain factor (its elements are called "states of nature" or "elementary events"; its subsets are called "events"); X is a set of certain consequences which can be of any nature, in a multiple criteria case its elements are vectors $x = (x_1 , ..., x_m)$; f is a mapping $S \times \Lambda \to X$, where $f(s, \lambda^j)$ is the consequence of alternative $s \in S$ occurring for $\lambda^j \in \Lambda$, $j = 1, ..., n$.

The other two elements of (1) are binary relations describing the DM's preferences on X and the probabilities of events. The quasi-order $\rhd_-$ is a (partial) non-strict preference relation on X: for any x^1, x^2 from X relation $x^1 \rhd_- x^2$ means that x^1 is at least as preferable as x^2. The quasi-order $\rhd_-$ induces a strict preference relation $\rhd$ (asymmetrical part of $\rhd_-$) and an indifference relation $\sim$ (symmetrical part of $\rhd_-$) on X.

Relation $\succ_\approx$ is a partial qualitative probability (PQP) on the set of events 2^Λ. A PQP was formally introduced in [1] as a generalisation of a (complete) qualitative probability. It is not required that any two events are comparable by $\succ_\approx$. Relation $A \succ_\approx B$ means that event A is at least as probable as B. Since a PQP $\succ_\approx$ is a quasi-order, we can define its asymmetrical part $\succ$ and its symmetrical part $\approx$ with an obvious meaning of being strictly more or equally probable.

Note that, if numerical probabilities p_j of elementary events λ^j are known, a complete qualitative probability is defined by the following rule:

$$A \succ_\approx B \quad \text{if and only if} \quad \sum_{j:\lambda^j \in A} p_j \geq \sum_{j:\lambda^j \in B} p_j. \tag{2}$$

Not any qualitative probability can be represented by numerical probabilities [2]. Hence a model with known numerical probabilities is a special case of model (1).

Multiple criteria decision problems under uncertainty are an important class of problems described by model (1). In this case, the elements of X are n-dimensional vectors and preference relation $\triangleright_-$ on X is often partial, e.g. the Pareto relation.

Based on the information contained in model (1), we aim to define the DM's preference relation (a quasi-order) R^S on the set of alternatives S. Any decision alternative $s \in S$ can be fully characterised by a vector $y(s) = (f(s, \lambda^1), ..., f(s, \lambda^n))$. Such a vector $y(s)$ is a lottery with certain consequences from X whose probabilities are given by PQP $\succ_\approx$. We call such lotteries uncertain consequences.

Let $Y = X^n$ be the set of all feasible uncertain consequences $y = (y_1, ..., y_n)$. Note that the problem of defining relation R^S on S will be solved if a non-strict preference relation (quasi-order) R on the set Y is defined, because

$$s' \, R^S \, s'' \quad \text{if and only if} \quad y(s') \, R \, y(s'').$$

If relation R is defined, then the set of decision alternatives S can be narrowed to the set of non-dominated alternatives with respect to P - the asymmetrical part of R. Below we consider the problem of defining relation R on the set Y.

2 Utility Function Model as a Special Case

Model (1) with the preference relation $\triangleright_-$ and the PQP $\succ_\approx$ is more general than the utility function model. If a utility function u on X and a probability distribution $p = (p_1, ..., p_n)$ on Λ are known, then the utility of any $y = (y_1, ..., y_n) \in Y$ is

$$u_{(p)}(y) = \sum_{j=1}^n p_j u(y_j),$$

and the preference relation $R = R_{u,p}$ on Y is a complete quasi-order defined as

$$y' \, R_{u,p} \, y'' \quad \text{if and only if} \quad u_{(p)}(y') \geq u_{(p)}(y'').$$

The case of incomplete information about utilities and probabilities can be addressed in the following way. Let U be a feasible set of utility functions u. Let D

be a feasible set of probability distributions $p = (p_1, ..., p_n)$ on Λ. Then the preference relation $R = R_{U,D}$ on Y is a (partial) quasi-order defined as

$$y' R_{U,D} y'' \quad \text{if and only if} \quad u_{(p)}(y') \geq u_{(p)}(y'') \text{ for any } u \in U, p \in D. \qquad (3)$$

Definition (3) is not constructive. However, for a number of special cases efficient methods of determining relation $R_{U,D}$ have been developed. The case of a known utility function and incomplete knowledge of probabilities was first addressed by Fishburn [3] and later discussed by other authors (see a survey in [4]). The case of an unknown utility function but known probabilities is addressed by the stochastic dominance theory [5].

As noted in [6, 7], the general case of model (1) with an unknown utility function and unknown probabilities can be addressed by the methods of the criteria importance theory [8, 9]. It is worth noting that these methods are applicable to the general case of model (1) without the assumption about the existence of a utility function and probabilities. Table 1 presents a double dichotomy of decision models with incomplete information and relevant methods of their solution.

	known utility	unknown utility
known probability	utility theory methods	stochastic dominance methods
unknown probability	methods of Fishburn and other special methods	criteria importance theory methods

Table 1. Methods for models with incomplete information

3 Construction of the Preference Relation

In this section we show how the criteria importance theory methods can be applied to constructing a DM's preference relation R on Y using the known relations $\triangleright_{\_}$ and $\succ_{\approx}$.

3.1 Preference Relation in the General Case

The preference relation $\triangleright_{\_}$ defined on X induces the Pareto relation R^0 on the set of uncertain consequences Y:

$$y' R^0 y'' \quad \text{if and only if} \quad y'_j \triangleright_{\_} y''_j, \ j = 1,..., n. \qquad (4)$$

Let two events A and B be comparable by PQP $\succ_{\approx}$, so that either $A \succ B$ or $B \succ A$, or $A \approx B$ holds. Without a loss of generality we assume that $A \cap B = \varnothing$, otherwise we shall consider events $A\backslash C$ and $B\backslash C$, where $C = A \cap B$. We also assume that A and B are not empty and that

$$A = \{ \lambda^{j_1^A}, ..., \lambda^{j_r^A} \}, \quad B = \{ \lambda^{j_1^B}, ..., \lambda^{j_q^B} \}. \tag{5}$$

For simplicity, we shall identify λ^j with its ordinal number j. Thus, $j \in A$ is the same as $\lambda^j \in A$. Consider any two uncertain consequences

$$y = (y_1, ..., y_n) \quad \text{and} \quad y^{AB} = (y_1^{AB}, ..., y_n^{AB}) \tag{6}$$

satisfying the following conditions: (i) the DM is indifferent between the consequences y_j with $j \in A$, as well as between y_j with $j \in B$; (ii) y^{AB} is such that: $y_j^{AB} \sim y_k$ if either $j \in A$ and $k \in B$, or $j \in B$ and $k \in A$; (iii) $y_j^{AB} = y_j$ for all $j \in \Lambda \backslash (A \cup B)$.

For example, let $a_1, a_2, a_3, b_1, b_2, b_3, c \in X$. Assume that $a_1 \sim a_2, a_2 \sim a_3, b_1 \sim b_2, b_2 \sim b_3$ hold, and $y = (a_1, b_1, b_2, c, a_2, a_3)$. If $A = \{1, 5, 6\}$ and $B = \{2, 3\}$, then y and $y^{AB} = (b_1, a_1, a_2, c, b_2, b_3)$ satisfy the above conditions.

Let A and B be equally probable events: $A \approx B$. Then the DM should be indifferent between y and y^{AB} regardless of the preference $\rhd_{-}$ between certain consequences $y_{j_1^A}$ and $y_{j_1^B}$, which we denote $y\, I^{A \approx B}\, y^{AB}$. Therefore, a probability relation $A \approx B$ induces an indifference relation $I^{A \approx B}$ on Y, which includes all pairs of uncertain consequences (6).

Now let A be more probable than B: $A \succ B$. Then if $y_{j_1^A} \rhd y_{j_1^B}$ holds, the DM should prefer y to y^{AB}, which we denote $y\, P^{A \succ B}\, y^{AB}$. Therefore, a probability relation $A \succ B$ induces a strict preference relation $P^{A \succ B}$ on Y, which includes all pairs of uncertain consequences (6).

Denote by $R^{\succ \approx}$ the union of the Pareto relation R^0 and the above defined relations $I^{A \approx B}$ and $P^{A \succ B}$ for all non-empty pairs of event A, B comparable by PQP $\succ_{\approx}$. The DM's preference relation R can be found as a transitive closure of $R^{\succ \approx}$. Namely, $y' R y''$ holds if and only if there exists a chain

$$y' R^{\pi_1} y^1, \ y^1 R^{\pi_2} y^2, \ ... \ , y^T R^{\pi_{T+1}} y'', \tag{7}$$

where $y^1, y^2, ..., y^T \in Y$, and each R^{π_t}, $t = 1, ..., T+1$, is one of the following: $I^{A \approx B}$, $P^{A \succ B}$ or R^0.

Definition (7) of the preference relation R is not constructive, and it is worth discussing its determination in practical situations. In the general case a method similar to that introduced in [10] can be used. In a single criterion case ($m = 1$), this can be used without any modifications.

3.2 Special Case I

Assume that preference relation $\triangleright_{\_}$ on X is complete and that only judgements about the comparative probabilities of some elementary events $\{\lambda^j\}$ are available. A PQP $\succ_{\approx}$ can be deduced from these judgements using its transitivity and additivity properties. In this case only single element sets A and B in (5) may be taken into account. For any $y'' \in Y$ define

$$Y^{\succ_{\approx}}(y'') = \{\, y' \in Y \mid \text{there exists a chain (7) in which no } R^{\pi_t} \text{ is } R^0 \}. \quad (8)$$

Obviously, this set includes y'' and consists of no more than $n!$ elements. The following lemma and theorem generalise results from [11]:

Lemma. If $y' R^{\pi} y''$, $y'' R^0 y'''$, where R^{π} is an $I^{A=B}$ or $P^{A\succ B}$ type relation, then either $y' R^0 y'''$ or there exists $y* \in Y$ such that the following holds: $y' R^0 y*$, $y* R^{\pi} y'''$.

Theorem 1. Relation $y' R y''$ holds if and only if there exists $y''' \in Y^{\succ_{\approx}}(y'')$ such that $y' R^0 y'''$.

According to this theorem, checking relation $y' R y''$ is reduced to the construction of set (8) and checking the correctness of relation $y' R^0 y'''$ for each y''' from (8). In practical implementations, set (8) is being constructed gradually and may not be entirely constructed in a particular case.

3.3 Special Case II

Consider an even more special case than case I, namely assume that all elementary events $\{\lambda^j\}$ are ordered by their probabilities. In this case PQP $\succ_{\approx}$ can be represented by ordinal probabilities $p_1 > 0 ,..., p_n > 0$ such that inequality $p_i \geq p_j$ holds if and only if $\{\lambda^i\} \succ_{\approx} \{\lambda^j\}$. Let $y', y'' \in Y$. Introduce the set

$$W = \{y'_1\} \cup ... \cup \{y'_n\} \cup \{y''_1\} \cup ... \cup \{y''_n\}.$$

Without the loss of generality we shall assume that no two elements of W are indifferent by $\sim$, otherwise we delete indifferent elements from W leaving only one representative from each indifference class. Let $W = \{w_1, ..., w_r\}$ and $w_1 \succ ... \succ w_r$. Obviously, $r \leq 2n$.

For any $y \in Y$ define $p^j(y) = (p^j_1(y), ..., p^j_n(y))$, where $p^j_i(y) = p_i$ if $y_i \triangleright_{\_} w_j$ and $p^j_i(y) = 0$ otherwise. Denote $p^j_{\downarrow}(y)$ the vector obtained from $p^j(y)$ by permuting its components in the non-increasing order: $p^j_{\downarrow}(y)_1 \geq ... \geq p^j_{\downarrow}(y)_n$.

The following theorem is analogous to the results presented in [9, 12].

Theorem 2. Relation $y' R y''$ holds if and only if $p_{\downarrow}^{j}(y') \geq p_{\downarrow}^{j}(y'')$ for all $j = 1,..., r$.

A simpler condition for relation $y' R y''$, in the case where all elementary events are equally probable, is formulated in different terms in [9, 13].

4 Preference Relations Incorporating Risk Attitude

In the previous section we discussed the extension of relation $R^{\succ=}$ to a preference relation R on the set of uncertain consequences Y. In most cases, R is only a partial relation. Below we show how R can be further extended to a preference relation R^* taking into account the information about the DM's attitude towards risk.

Assume that set X of certain consequences is a convex set. A DM is risk averse if the expected value of any nondegenerate lottery (with no single consequence $x \in X$ having a probability of one) is preferred to the lottery. If any lottery is preferred to its expected value, then the DM is risk prone, and if any lottery and its expected value are indifferent, then the DM is risk neutral [14].

We start with analysing how the known DM's attitude towards risk reflects in the properties of relation R^* on Y. For any $y^0 \in Y$ define the upper cut of R^* as the set $\overline{R}^*(y^0) = \{y \in Y \mid y R^* y^0\}$ and the lower cut of R^* as $\underline{R}^*(y^0) = \{y \in Y \mid y^0 R^* y\}$. Quasi-order R^* is called concave if for any $y^0 \in Y$ the upper cut $\overline{R}^*(y^0)$ is a convex set. Similarly, R^* is convex if for any $y^0 \in Y$ the lower cut $\underline{R}^*(y^0)$ is a convex set.

Let $y'=(y'_1, ..., y'_n)$ and $y''=(y''_1, ..., y''_n)$ from Y be preferred to $y^0=(y^0_1, ..., y^0_n)$ from Y, namely $y' R^* y^0$ and $y'' R^* y^0$ hold. Let β and γ be two positive numbers such that $\beta + \gamma = 1$. We are going to show that, if a DM is risk averse, then the convex combination $\beta y' + \gamma y''$ is preferred to y^0.

As noted in Section 1, y', y'' and y^0 can be considered as lotteries with probabilities of consequences given by $\succ_{\approx}$. Assume that probabilities $p_1, ..., p_n$ represent $\succ_{\approx}$ as defined by (2). Consider two lotteries L_1 and L_2 (Fig. 1 and Fig. 2).

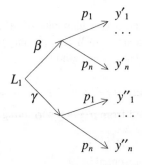

Fig. 1. Lottery L_1

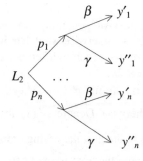

Fig. 2. Lottery L_2

The certain consequences of these lotteries y'_j and y''_j have the same probabilities of occurrence $\beta\, p_j$ and $\gamma\, p_j$ ($j = 1, ..., n$). Therefore L_1 and L_2 are indifferent. Define lottery L_3, substituting n sublotteries in lottery L_2 by their expected values as shown in Fig. 3. If the DM is risk averse, then lottery L_3 is preferred to L_2. Therefore, by transitivity, L_3 is preferred to L_1.

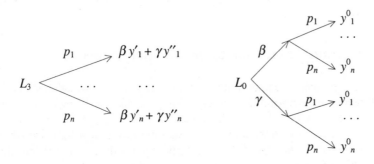

Fig. 3. Lottery L_3 **Fig. 4.** Lottery L_0

Now consider lottery L_0 (Fig. 4). It is indifferent to y^0 (which is a lottery itself). Also L_1 is preferred to L_0. Therefore, L_3 is preferred to L_0. Hence, $\beta\, y' + \gamma\, y''$ is preferred to y^0.

This observation demonstrates that the uncertain consequence $\beta\, y' + \gamma\, y''$ belongs to the upper cut of relation R^* for y^0, and thus, R^* is a concave quasi-order. We shall define R^* as the minimal concave quasi-order that extends $R^{\succ=}$ on Y. Relation R^* is called the closure of $R^{\succ=}$ in the class of concave quasi-orders and can be defined as the intersection of all concave quasi-orders extending $R^{\succ=}$ on the set Y.

If the DM is risk prone, then similar considerations lead to the conclusion that the preference relation R^* is the closure of $R^{\succ=}$ in the class of convex quasi-orders. Finally, if the DM is risk neutral, then R^* is the closure of $R^{\succ=}$ in the class of quasi-orders which are both concave and convex.

Note that, in order to be consistent with the information about the DM's attitude towards risk, quasi-order $\triangleright_{\_}$ on X must be concave (convex, both concave and convex) if the DM is risk averse (risk prone, risk neutral).

Without the assumption of the existence of probabilities p_j representing $\succ_{\approx}$, but requiring that relation $\triangleright_{\_}$ complies with the DM's attitude towards risk as discussed above, we can give the following definition of the DM's preference relation R^* on Y:

Definition 1. For a risk averse (risk prone, risk neutral) DM, the preference relation R^* is the closure of $R^{\succ_\approx}$ in the class of concave (convex, both concave and convex) quasi-orders.

Note that preference relation R^*, so defined, extends relation R from the previous section using the information about the DM's attitude towards risk. The problem of practical construction of relation R^* is not a simple task [8] and is not discussed here.

5 A Lexicographic Extension to the Maxmin Principle

In the previous sections we demonstrated how the qualitative information about the DM's preferences and probabilities can be used in the analysis of model (1). Our approach is based on the ideas of the criteria importance theory, and can be considered as an extension to the utility function approach. Below we discuss how qualitative information in (1) can be used in conjunction with the maxmin optimality principle.

5.1 The Case of Complete Information

Assume that both the preference relation $\triangleright_\sim$ and the qualitative probability $\succ_\approx$ are complete relations. For an uncertain consequence $y \in Y$ define $\min_1(y)$, a component minimal by $\triangleright$ among all components y_j of y. There may be other components of y indifferent by $\sim$ to $\min_1(y)$. Denote $A_1(y)$, the set of all such components which is an event - subset of Λ. Define $\min_2(y)$, a component minimal by $\triangleright$ among all components of y which are not in $A_1(y)$. Denote $A_2(y)$ the corresponding subset of Λ. Similarly we can define $\min_r(y)$ and $A_r(y)$ using components of y which are not in $A_1(y) \cup ... \cup A_{r-1}(y)$. Let $A_1(y)$, ..., $A_N(y)$ be all sets, so defined, where N may be different for different $y \in Y$: $N = N(y)$.

Define a non-strict preference relation $R^{lex} = I^{lex} \cup P^{lex}$ on Y where the indifference relation I^{lex} and the strict preference relation P^{lex} are defined as follows:

Definition 2. Relation $y' \; I^{lex} \; y''$ holds if and only if $N(y') = N(y'')$, and for all $j = 1, ..., N(y')$ the following holds: $\min_j(y') \sim \min_j(y'')$, $A_j(y') \approx A_j(y'')$.

Definition 3. Relation $y' \; P^{lex} \; y''$ holds if and only if there exists $1 \le k \le \min(N(y'), N(y''))$, such that

(i) $\min_j(y') \sim \min_j(y'')$, $A_j(y') \approx A_j(y'')$ for all $j = 1, ..., k - 1$;

and

(ii) $\min_k(y') \triangleright \min_k(y'')$, or $\min_k(y') \sim \min_k(y'')$ and $A_k(y'') \succ A_k(y')$.

Clearly, relation R^{lex} on Y, so defined, can be considered as the lexicographic extension to the maxmin optimality principle [15].

5.2 The Case of Partial Information

Now we do not assume that relations $\rhd_\sim$ and $\succ_\approx$ are complete relations. For any complete preference relation $\rhd_\sim^\alpha$ extending $\rhd_\sim$ on X, and for any complete qualitative probability $\succ_\approx^\beta$ extending PQP $\succ_\approx$ on 2^Λ, let $R_{\alpha,\beta}^{lex}$ be the lexicographic maxmin relation on Y as defined above. (The possibility of extending a PQP to a complete qualitative probability is not a trivial issue and is discussed in [1, 16].)

In a similar way to (3), we can define relation R^{lex} on Y as follows: $y' \ R^{lex} \ y''$ holds if and only if $y' \ R_{\alpha,\beta}^{lex} \ y''$ holds for any complete preference relation $\rhd_\sim^\alpha$ extending $\rhd_\sim$ on X, and for any complete qualitative probability $\succ_\approx^\beta$ extending PQP $\succ_\approx$ on 2^Λ.

The above definition is not constructive. A simple condition for checking relation $y' \ R^{lex} \ y''$ can be proved using the approach from [17]. For an uncertain consequence y, define $\overline{y}$ the set of classes of indifferent by $\sim$ components y_j of y. For any indifference class z from $\overline{y}$, define $\pi(z) = \{ \lambda^j \mid y_j \in z \}$ - the set of elementary events corresponding to consequences from z.

Now consider $y', y'' \in Y$ and the corresponding sets $\overline{y}'$ and $\overline{y}''$. Simultaneously delete any indifference class z' from $\overline{y}'$ and z'' from $\overline{y}''$ such that elements of z' and z'' are indifferent by $\sim$ and $\pi(z') \approx \pi(z'')$ holds. Denote the remaining sets $\overline{\overline{y}}'$ and $\overline{\overline{y}}''$.

Theorem 3. Relation $y' \ I^{lex} \ y''$ holds if and only if $\overline{\overline{y}}'$ and $\overline{\overline{y}}''$ are empty sets.

Theorem 4. Relation $y' \ P^{lex} \ y''$ holds if and only if the sets $\overline{\overline{y}}'$ and $\overline{\overline{y}}''$ are not empty and for any $z' \in \overline{\overline{y}}'$ there exists $z'' \in \overline{\overline{y}}''$ such that either (i) elements from z' are preferred by $\rhd$ to elements from z'', or (ii) elements from z' and from z'' are indifferent by $\sim$ and $\pi(z'') \succ \pi(z')$.

Let relations $I^{A=B}, P^{A\succ B}, P^0, R, P, I$ be as defined in section 3.1. Evidently, $x \ I^{A=B} \ y$ implies $x \ I^{lex} \ y$, and $x \ P^{A\succ B} \ y$ or $x \ P^0 \ y$ implies $x \ P^{lex} \ y$. Therefore, the following theorem holds:

Theorem 5. If $x \ R \ y$ holds, then $x \ R^{lex} \ y$ holds. Moreover, $x \ I \ y$ implies $x \ I^{lex} \ y$, and $x \ P \ y$ implies $x \ P^{lex} \ y$.

References

1. Naumov, G.Ye., Podinovski, V.V., and Podinovski, Vic.V. (1993) Subjective Probability: Representation and Derivation Methods. *Journal of Computer and System Sciences International*, **31**, 12 - 23.
2. Kraft, C.H., Pratt, J.W., and Seidenberg, A. (1959) Intuitive probability on finite sets. *Annals of Mathematical Statistics*, **30**, 408 - 419.
3. Fishburn, P.C. (1964) *Decision and Value Theory*, New York, John Wiley and Sons.
4. Athanassopoulos, A.D., and Podinovski, Vic.V. Dominance and Potential Optimality in Multiple Criteria Decision Analysis with Imprecise Information. (forthcoming in *Journal of Operational Research Society*).
5. Whitmore, G.A., and Findlay, M.C., eds. (1978) *Stochastic Dominance*, Lexington, Mass., D.C. Heath.
6. Podinovski, V.V. (1978) On Relative Importance of Criteria in Multiobjective Decision Making Problems. In: *Multiobjective Decision Problems*, Moscow, Mashinostroenie Publishing House, 48 - 92. (in Russian)
7. Podinovski, V.V. (1979) An Axiomatic Solution of the Problem of Criteria Importance in Multicriterial Problems. In: *Operations Research Theory: State of the Art*, Moscow, Nauka Publishing House, 117 - 149. (in Russian)
8. Podinovski, V.V. (1994) Criteria Importance Theory. *Mathematical Social Sciences*, **27**, 237 - 252.
9. Podinovski, V.V. (1993) Problems with Importance-Ordered Criteria. In: J. Wessels and A.P. Wierzbicki, eds., *User-Oriented Methodology and Techniques of Decision Analysis and Support*, Berlin, Springer Verlag, 150 - 155.
10. Osipova, V.A., Podinovski, V.V., and Yashina, N.P. (1984) On Non-contradictory Extension of Preference Relations in Decision Making Problems. *USSR Computational Mathematics and Mathematical Physics*, **24**, 128 - 134.
11. Podinovski, V.V. (1976) Multicriterial Problems with Importance-Ordered Criteria. *Automation and Remote Control*, **37**, 1728 - 1736.
12. Podinovski, V.V. (1978) Importance Coefficients of Criteria in Decision-Making Problems. *Automation and Remote Control*, **39**, 1514 - 1524.
13. Podinovski, V.V. (1975) Multicriterial Problems with Uniform Equivalent Criteria, *USSR Computational Mathematics and Mathematical Physics*, **15**, 47 - 60.
14. Keeney, R.L., and Raiffa, H. (1976) *Decisions with Multiple Objectives*, New York, John Wiley and Sons.
15. Podinovski, Vic.V. (1983) Criterion of Probabilistic-Lexicographic Maximin. *Moscow University Computational Mathematics and Cybernetics*, 39 - 45.
16. Podinovski, Vic.V. (1993) Extension of Partial Binary Relations in Decision-Making Models. *Computational Mathematics and Modeling*. **4**, 263 - 264.
17. Podinovski, Vic.V. (1985) A Lexicographic Approach to Decision Making under Uncertainty. In: *The Software of Computing Systems*. Moscow, Moscow University Press, 104 - 119. (in Russian)

Part 3
Methodologies of MCDA

Interval Judgements in the Analytic Hierarchy Process: A Statistical Perspective

Linda M. Haines

Department of Statistics and Biometry, University of Natal Pietermaritzburg, Private Bag X01, Scottsville 3209, South Africa

Abstract. Two broad approaches are commonly used in the Analytic Hierarchy Process for deriving a suitable ranking of alternatives from an interval judgement matrix. In the present study these approaches are examined from a statistical perspective and are extended and developed accordingly. The ideas are introduced and illustrated by means of a simple example.

Keywords. Analytic Hierarchy Process; interval judgements; random convex combinations; uniform distribution; log uniform distribution.

1 Introduction

This article is concerned with the step in the Analytic Hierarchy Process (AHP) termed preference elicitation in which paired comparisons between alternatives made on the basis of a particular criterion are elicited from the decision maker and a set of weights, or local priorities, for the alternatives are extracted from these comparisons. Specifically attention is focussed on a single comparison matrix within the AHP structure and not on a full hierarchy of criteria and attendant comparison matrices. In the AHP pairwise comparisons between alternatives are usually elicited as point judgements on a ratio scale comprising the integers between 1 and 9 together with their reciprocals, and this form of preference elicitation is assumed throughout the present study. Methods for extracting weights from such point judgements are well documented (Saaty, 1980; Saaty and Vargas, 1984; Crawford and Williams, 1985). It can happen however that the decision maker is uncertain about his or her preferences and is therefore reluctant to assign single point scores to the pairwise comparisons. In such cases it is natural to follow the procedure suggested by Saaty and Vargas (1987) and to elicit upper and lower bounds on the decision maker's preferences. Methods of extracting weights for the alternatives from such interval judgements have been proposed by a number of researchers and are described in the seminal works of Saaty and Vargas (1987), Arbel (1989) and Salo and Hämäläinen (1995).

The aim of the present study is to compare and contrast, within a statistical framework, the methods of extracting preference information from pairwise interval judgements developed by Saaty and Vargas (1987) and by Arbel (1989). In particular a simple example involving interval judgements is introduced in Section 2 and in Sections 3 and 4 this example is used to illustrate and to appraise the two approaches. Some broad conclusions and pointers for future research are given in Section 5.

2 An Example

Suppose that a decision maker is invited to compare three wines and to choose the best of those wines solely on the basis of taste. This problem can be formulated quite simply within the AHP framework as follows:-

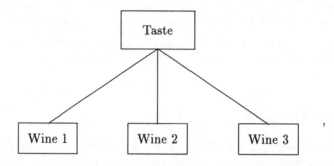

and the decision maker is then required to compare the wines pairwise using the appropriate ratio scale. Suppose however that the decision maker is uncertain about his opinions and that he therefore expresses his pairwise judgements as intervals rather than as point scores. In particular suppose that he feels that wine 1 is not more than twice as good as wine 2, and vice versa, that wine 1 is between 2 and 4 times better than wine 3, and that wine 2 is between 3 and 5 times better than wine 3. Then his preferences can be summarized succinctly in the matrix,

$$
A = \begin{bmatrix}
1 & [\frac{1}{2}, 2] & [2, 4] \\[2mm]
[\frac{1}{2}, 2] & 1 & [3, 5] \\[2mm]
[\frac{1}{4}, \frac{1}{2}] & [\frac{1}{5}, \frac{1}{3}] & 1
\end{bmatrix}, \tag{2.1}
$$

which is an example of an interval judgement matrix.

The problem of extracting information about the decision maker's preferences from the matrix (2.1) now arises and the following two questions in particular are clearly of immediate interest.

1. Can a sensible ranking of the wines be obtained and, more specifically, can meaningful estimates of the preference weights associated with the wines be found?

2. Can the fact that the decision maker is unclear as to whether or not he prefers wine 1 to wine 2 be quantified, for example by means of a probability?

These questions will be addressed specifically in the ensuing sections.

3 Feasible Regions

Suppose that the weights, w_1, w_2, and w_3, are associated with the wines 1, 2 and 3 respectively. Then, following the approach of Arbel (1989), the interval judgement matrix (2.1) can be interpreted in a simple but elegant manner as specifying a set, S, of weight vectors, $\boldsymbol{w} = (w_1, w_2, w_3)$, satisfying the constraints $\frac{1}{2} \leq w_1/w_2 \leq 2$, $2 \leq w_1/w_3 \leq 4$, and $3 \leq w_2/w_3 \leq 5$ in addition to the usual normalization constraints, $w_i \geq 0, i = 1, 2, 3$ and $\sum_{i=1}^{3} w_i = 1$. The set, S, is referred to as the feasible region within weight space and is illustrated in a ternary diagram in Figure 3.1.

Fig 3.1. Ternary diagram of the constraints on the weights

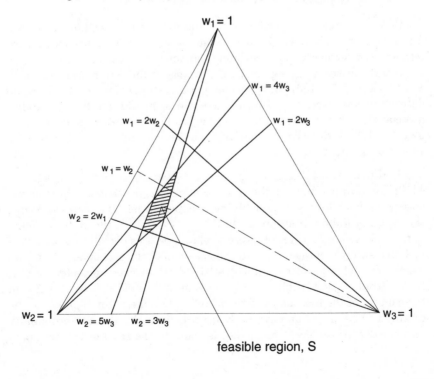

feasible region, S

It should be noted that each vector of weights in this region corresponds to a consistent point judgement matrix, A, with elements, $a_{ij} = \frac{w_i}{w_j}, i, j = 1, 2, 3$, and hence that Arbel's approach is compatible with such matrices.

A rich and rewarding approach to obtaining summary measures for the priority weights of the alternatives in the feasible region is to place a distribution on that region. Then, for example, the mean of the distribution can be taken as a point assessment of the ranking of the wines and probabilities associated with events of interest such as $w_1 > w_2$ can be computed. Two such distributions on S, which are deemed to be particularly interesting and appropriate, are now considered.

3.1 Random Convex Combinations

The feasible region, S, is a polytope and any weight vector, w, in that region can therefore be expressed as a convex combination of its five extreme vertices. Thus, following the algorithm of Arbel and Vargas (1992, 1993), it is possible to generate a weight vector randomly on S according to the scheme:-

1. Simulate four points uniformly on the interval $[0, 1]$, i.e. simulate a random partition of $[0, 1]$ comprising five subintervals with lengths $d_1, \ldots, d_5$ where $d_i \geq 0$, $i = 1, \ldots, 5$ and $\sum_{i=1}^{5} d_i = 1$.

2. Form the random weight vector, $w = \sum_{i=1}^{5} d_i v_i$, where the vectors, $v_1, \ldots, v_5$, denote the extreme vertices of S.

This procedure clearly specifies a distribution for the weights, w, in the feasible region, S, and weight vectors emanating from this distribution are termed here, somewhat appropriately, random convex combinations. The mean of the distribution is given by the average of the extreme vertices of S, viz. $(0.363, 0.507, 0.130)$, and provides a point estimate for the priorities the decision maker associates with the three wines. In addition the probability associated with the event, $w_1 > w_2$, is readily evaluated as $\text{Prob}(w_1 > w_2) = 2.16\%$ and serves to quantify the decision maker's preference for wine 1 over wine 2.

3.2 Uniform Distribution

The most obvious and, in the sense that consistent matrices within the feasible region, S, may reasonably be assumed to be generated randomly, the most natural distribution on that region to adopt is the uniform distribution. Certain quantities of interest relating to this distribution can then be immediately identified from the geometry of S. In particular, the mean, which provides a point assessment of the decision maker's preferences, is the centroid of the feasible region, and is given by $(0.376, 0.494, 0.131)$. In the context of the mechanics of rigid bodies, this mean corresponds to the centre of mass of the uniform body described by S. Also the probability associated with the event, $w_1 > w_2$, is equal to the ratio of the area of S

corresponding to $w_1 > w_2$ to the area of S itself, as shown in the ternary diagram of Figure 3.1, and is given by $\text{Prob}(w_1 > w_2) = 12.71\%$.

As a point of interest it can be readily shown that if the feasible region, S, is a simplex in weight space, then the distribution of random convex combinations on S is uniform. The converse of this statement is not necessarily true however. For example in the present case, the feasible region, S, is not a simplex and, since the means of the distribution of random convex combinations and of the uniform distribution do not coincide, the two distributions themselves are not the same. It is also interesting to note that the method for assessing weights associated with alternatives in a non-hierarchical decision making process introduced by Solymosi and Dombi (1986) and developed by Olson and Dorai (1992), and termed the centroid method, and also the geometric method presented by Islei and Lockett (1988), are closely related to the approach described here. However the decision making philosophies underpinning these methods are fundamentally very different to that upon which the AHP is based.

4 Point Judgement Matrices

Consider a point judgement matrix, A, with elements which fall within the intervals specified by the interval judgement matrix (2.1). Then a vector of preference weights associated with the three wines can be estimated from A by invoking the method of log least squares, i.e. by minimizing the criterion, $\sum_{i=1}^{3} \sum_{j=1}^{3} (\ln a_{ij} - \ln \frac{w_i}{w_j})^2$, with respect to w_1, w_2, and w_3 subject to the constraint, $\sum_{i=1}^{3} w_i = 1$, to give optimum weights satisfying

$$w_i \propto (\prod_{j=1}^{3} a_{ij})^{\frac{1}{3}}, i = 1, 2, 3. \tag{4.1}$$

Thus, following Saaty and Vargas (1987), the interval judgement matrix (2.1) can be interpreted as representing all point judgement matrices, both consistent and inconsistent, with elements which fall within the intervals specified by (2.1) and hence, in turn, as specifying a set, W, of priority weights obtained from these point judgement matrices by the method of log least squares and given by (4.1). The set, W, corresponding to the interval judgement matrix, (2.1), is shown in the weight space for w_1 and w_2, with $w_3 = 1 - w_1 - w_2$, in Figure 4.1.

In order to obtain information on the decision maker's preferences for the three wines from the set, W, it is sensible, again following Saaty and Vargas (1987), to adopt a suitable distribution for the elements of point judgement matrices which are compatible with the interval judgement matrix, (2.1). This distribution then in turn implies a distribution for the set of weight vectors, W. Two such scenarios are now considered.

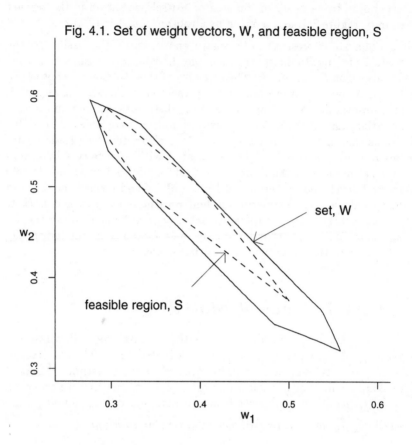

Fig. 4.1. Set of weight vectors, W, and feasible region, S

4.1 Uniform Distribution

Suppose, following Saaty and Vargas (1987), that a point judgement matrix, A, compatible with the interval judgement matrix, (2.1), is generated by taking its elements, a_{12}, a_{13} and a_{23}, to be independently and uniformly distributed on the corresponding intervals, $[\frac{1}{2}, 2]$, $[2, 4]$ and $[3, 5]$ respectively and by setting $a_{21} = 1/a_{12}$, $a_{31} = 1/a_{13}$, and $a_{32} = 1/a_{23}$. Such an approach is not completely satisfactory. Thus, for example, the element, a_{13}, is assumed to be uniformly distributed on $[2, 4]$ but the associated distribution of $a_{31} = 1/a_{13}$ is *not* uniform on $[\frac{1}{4}, \frac{1}{2}]$ and in fact a_{31} has a probability density function (p.d.f.) given by

$$
\begin{aligned}
f(a_{31}) &= \frac{1}{2a_{31}^2}, & \tfrac{1}{4} \le a_{31} \le \tfrac{1}{2}, \\
&= 0 & \text{elsewhere.}
\end{aligned}
\tag{4.2}
$$

Essentially, therefore, generating a_{13} uniformly on $[2, 4]$ is *not* compatible with generating $1/a_{13}$ uniformly on $[\frac{1}{4}, \frac{1}{2}]$, and vice versa. It is of course possible, following Moreno-Jimenez and Vargas (1993), to assume that a_{13} is uniform and that a_{31} has the p.d.f. (4.2), but this is again unsatisfactory in that the choice of whether to take a_{13} or a_{31} to be uniformly distributed would seem to be arbitrary.

For the purposes of comparison at a later stage, 10 000 point judgement matrices were generated from the interval judgement matrix, (2.1), in accordance with the distribution described above, and the associated weight vectors were calculated using the result (4.1). The mean weight vector for this simulated sample was found to be $(0.438, 0.435, 0.127)$ and gives an estimate of the decision maker's preferences for the wines based on the underlying distribution. In addition the proportion of simulated weight vectors for which $w_1 > w_2$ was found to be 55.2 % and provides an estimate of the probability that the decision maker prefers wine 1 to wine 2.

4.2 Log Uniform Distribution

The problems associated with the distributional assumptions introduced in Section 4.1 emanate from the fact that the intervals of the matrix (2.1) specify *ratios* of weights. One possible solution to this is therefore to invoke a log scale and to assume that, for example, the element, a_{13}, of a point judgement matrix compatible with (2.1) is generated as $\exp(u)$, where u is distributed uniformly distributed on the interval, $[\ln 2, \ln 4]$. In this way the p.d.f.'s of a_{13} and $a_{31} = \frac{1}{a_{13}}$ have the same kernel,

$$f(x) = \frac{1}{[\ln 4 - \ln 2]} \frac{1}{x},$$

and generating a_{13} is compatible with generating a_{31}. A sample of 10 000 point judgement matrices with elements independently and log uniformly distributed on the corresponding intervals of the matrix (2.1), was obtained by simulation and the associated weight vectors were calculated using the result (4.1). The mean vector for this simulated sample of weights was found to be $(0.411, 0.458, 0.131)$, and the probability that $w_1 > w_2$ was estimated as 38.3%.

5 Conclusions

The main aim of this paper has been to examine, from a statistical perspective, two approaches to the analysis of interval judgements in the AHP. The one approach, due to Arbel (1989), is based on a feasible region of preference weights specified by the interval judgements and the second, developed by Saaty and Vargas (1987), involves the generation of point judgement matrices, and hence weight vectors, compatible with the interval judgement matrix. The methods were implemented for a simple example involving the pairwise comparison of three wines and the results are conveniently drawn together in Table 5.1. It is clear from this table that the point es-

Table 5.1: Results for the wine example of Section 2

Method and distribution	w_1	w_2	w_3	$\text{Prob}(w_1 > w_2)$
Feasible region				
Random convex combinations	.363	.507	.130	2.16 %
Uniform	.376	.494	.131	12.71 %
Point judgement matrices				
Uniform	.438	.435	.127	55.2 %
Log uniform	.411	.458	.131	38.3 %

timates of the preference weights, and more particularly the estimates of the probability that wine 1 is preferred to wine 2, depend not only on the approach adopted but also on the distributional assumptions made within that approach.

Overall two important points have emerged from the present study and both merit further attention. Firstly great care must be exercised in formulating the assumptions to be made in the analysis of an interval judgement matrix in the AHP. Secondly, statistics provides a powerful, rigorous and rewarding framework within which to analyze decision making problems.

Acknowledgements : The author wishes to thank the University of Natal and the Foundation for Research Development, South Africa, for funding this work and Paul Fatti, Raimo Hämäläinen, Iain MacDonald, Don Petkov and Paul Schreuder for many helpful discussions.

References

Arbel, A. (1989). "Approximate articulation of preference and priority derivation", *European Journal of Operational Research*, **43**, 317–326.

Arbel, A. and Vargas, L.G. (1992), "The Analytic Hierarchy Process with interval judgements", *Multiple Criteria Decision Making : Proceedings of the 9th International Conference held in Fairfax, Virginia, 1990*, Eds. A.Goicoechea, L. Duckstein, and S.Zionts, 61–70.

Arbel, A. and Vargas, L.G. (1993), "Preference simulation and preference programming : robustness issues in priority derivation", *European Journal of Operational Research*, **69**, 200–209.

Crawford, G. and Williams, C. (1985), "A note on the analysis of subjective judgment matrices", *Journal of Mathematical Psychology*, **29**, 387–405.

Islei, G. and Lockett, A.G. (1988), "Judgemental modelling based on geometric least square", *European Journal of Operational Research*, **36**, 27–35.

Moreno-Jimenez, J.M. and Vargas, L.G. (1993), "A probabilistic study of preference structures in the Analytic Hierarchy Process with interval judgements", *Mathematical and Computational Modelling*, **17**, 73–81.

Olson, D.L. and Dorai, V.K. (1992), "Implementation of the centroid method of Solymosi and Dombi", *European Journal of Operational Research*, **60**, 117–129.

Saaty, T. L. (1980), *The Analytic Hierarchy Process*, McGraw-Hill, New York.

Saaty, T.L. and Vargas, L.G. (1984), "Comparison of eigenvalue, logarithmic least squares and least squares methods in estimating ratios", *Mathematical Modelling*, **5**, 309–324.

Saaty, T.L. and Vargas, L.G. (1987), "Uncertainty and rank order in the Analytic Hierarchy Process", *European Journal of Operational Research*, **32**, 107–117.

Salo, A. and Hämäläinen R.P. (1995), "Preference programming through approximate ratio comparisons", *European Journal of Operational Research*, **82**, 458–475.

Solymosi, T. and Dombi, J. (1986), "A method for determining the weights of criteria : the centralized weights", *European Journal of Operational Research*, **26**, 35–41.

The Stability of AHP Rankings in the Presence of Stochastic Paired Comparisons

Antonie Stam[1] and A. Pedro Duarte Silva[2]

[1] Department of Management, Terry College of Business, The University of Georgia, Athens, GA 30602, USA

[2] Universidade Católica Portuguesa, Centro Regional do Porto, Curso de Administração e Gestão de Empresas, Rua Diogo Botelho 1327, 4100 Porto, Portugal

Abstract. This paper develops a methodology for analyzing the stability of AHP rankings for decision problems in which the paired comparison judgments are stochastic. Multivariate statistical techniques are used to obtain both point estimates and confidence intervals of rank reversal probabilities. We show how simulation experiments can be used to assess the stability of the preference rankings under uncertainty. High likelihoods of rank reversal imply that the AHP rankings are unstable, and that additional analysis of the decision problem may be in order.

Keywords. Multi-criteria Decision Making, Decision Analysis, Analytic Hierarchy Process, Uncertainty, Preference Judgments

1 Introduction

The Analytic Hierarchy Process (AHP) (Saaty 1980) is a discrete multi-criteria methodology in which paired comparisons are used to derive relative priority weights (ratings) for the criteria and alternatives under consideration. The purpose of this paper is to study the nature of rank reversal in the presence of *stochastic* paired comparison judgments, offering a rigorous statistical approach to analyzing outranking and rank reversal properties of the AHP methodology under these conditions. This problem is unrelated to the well-documented controversy related to the rank reversal phenomenon in the case of decision problems with single-valued paired judgments (Dyer 1990; Saaty 1990; Schoner, Wedley and Choo 1992, 1993; Barzilai and Golani 1994). In fact, our method is not limited to the AHP and can be applied to other methods based on paired comparisons as well.

In the remainder of our paper, we distinguish between imprecise and stochastic judgments. Both cases imply judgment intervals, but whereas in the case of imprecise or fuzzy judgments the intervals reflect an inability on the part of the decision maker to express his/her relative preferences as a single value, stochastic judgments imply a probability distribution over the range of each judgment interval.

We believe that there exist many decision situations where the nature of the judgment intervals can be considered to be stochastic, justifying a probabilistic approach that uses standard statistical methodologies to study rank reversal likelihoods. For instance, when making an investment decision the actual net profit is unknown, so that any paired comparison between investment alternatives is clearly stochastic, in terms of achievement of the net profit criterion. The stochastic nature of paired judgments can reflect either subjective probabilities that a particular alternative better achieves a given goal, or objective probabilities that reflect uncertain consequences of selecting a particular alternative.

Some researchers have extended the single-valued AHP analysis to the case of imprecise paired preferences using subjective probabilities (Vargas 1982; Zahir 1991), others used fuzzy sets and interval judgments (Saaty and Vargas 1987; Arbel 1989; Arbel and Vargas 1992, 1993; Salo 1993; Salo and Hämäläinen 1992a, 1992b, 1995; Moreno-Jimenez and Vargas 1993). Specifically, Saaty and Vargas (1987) attempt to numerically approximate some of these measures through sampling experiments, but both their method and its interpretation have problems, among others because they try to estimate probabilistic quantities from non-probabilistic concepts.

We will focus on stochastic judgments. It is important to emphasize that the uncertainty associated with the intervals considered in our paper is inherent to the decision problem itself, and not to inconsistencies in the preference statements provided by the decision maker. While we represent the uncertainties in the paired comparisons by subjective probability distributions, it may be possible to extend aspects of our methodology to include imprecise (non-stochastic) judgments. However, in order to keep our focus clearly on the stochastic case, we will not discuss the case of imprecise (non-stochastic) paired judgments further.

2 Stochastic Judgments and Rank Reversal Probabilities

Denote the traditional single-valued $k \times k$ AHP matrix of single-valued paired preference ratios of k alternatives with respect to a criterion C by $\mathbf{A} = \{a_{ij}\}$. For simplicity, we restrict the discussion to a single criterion C. The generalization to multiple criterion matrices is provided in Stam and Duarte Silva (1997). We denote random variates in caps and realizations of random variables in lower case. In the case of stochastic judgments, we represent the paired judgments by the random variates M_{ij}, defined over the finite interval $\{[m_{ij}^L, m_{ij}^U]\}$, and denote the $k \times k$ matrix of stochastic intervals by $\mathbf{M} = [m_{ij}^L, m_{ij}^U]$. Since the M_{ij} are stochastic, the principal eigenvector W is random as well, and our sampling experiments are designed to derive probability statements about W. It can be shown that the range of values for each w_i is finite if the intervals in $\mathbf{M}$ are finite. Let the range of possible values of w_i associated with $\mathbf{M}$ be defined by $[w_i^L, w_i^U]$.

In our approach, the decision maker is asked to provide information that enables the construction of a probability distribution over the range of each judgment interval. The appropriate shape of the distributions over $[m_{ij}^L, m_{ij}^U]$ is problem-dependent and is to be elicited from the decision maker. Two reasonable candidates are (1) a discrete distribution with subjective values which reflect the decision maker's degrees of confidence associated with different paired comparison ratio levels; once normalized to sum to unity, these confidence levels can be interpreted as a subjective probability distribution over the range of each judgment interval, and (2) a continuous distribution which requires the decision maker to indicate the most optimistic, most pessimistic and most likely values for the preference ratios (for more details, see Stam and Duarte Silva 1997). Although many other probability distributions are possible, the amount of information required from the decision maker to build the above distributions is modest. Moreover, we note that our procedure is obviously not limited to uniformly distributed judgment intervals.

We assume that rank reversal between two alternatives i and j occurs if i would be preferred over j under perfect information, but is calculated to be less preferred based on relevant sample information (*i.e.*, $w_i < w_j$). Defining the true probability that the decision maker prefers alternative i over j by $\pi_{ij} = P(W_i > W_j)$, and assuming that π_{ij} approximately equals the probability that alternative i is preferred to j under complete information, the true probability of rank reversal Π_{ij} is given by (1),

$$\Pi_{ij} = 2\pi_{ij}(1 - \pi_{ij}). \tag{1}$$

Once the shape of the distributions over the judgment intervals of **M** has been specified, we use sampling techniques to obtain a representative sample of principal eigenvectors $w^1, ..., w^n$ from their respective sampled paired comparison matrices. This information is then used to estimate the Π_{ij} and to construct confidence intervals (CIs) for Π_{ij}, which provides the decision maker with a useful indication of the stability of the final AHP rankings. We develop two different stability measures, $\widehat{\Pi}_{ij}^1$ and $\widehat{\Pi}_{ij}^2$. These estimators are related to that proposed previously by Saaty and Vargas (1987) in the case of imprecise judgments ($\widehat{\Pi}_{ij}^S$). $\widehat{\Pi}_{ij}^1$ does not assume any distributional properties of **W**. $\widehat{\Pi}_{ij}^2$ is based on the assumption that **W** is multivariate normally distributed. We found this normality assumption to be reasonable in numerous simulation experiments (not reported in this paper, see Stam and Duarte Silva 1997).

2.1 Rank Reversal Probabilities Based on Sample Frequencies: $\widehat{\Pi}_{ij}^1$

$\widehat{\Pi}_{ij}^1$ is based on simple sample frequencies. Thus, π_{ij} in (1) can be estimated by P_{ij}, the relative sample frequency of the event that W_i exceeds W_j. Hence, we estimate Π_{ij} by $\widehat{\Pi}_{ij}^1$ in (2):

$$\hat{\Pi}_{ij}^1 = 2P_{ij}\,(1-P_{ij}). \tag{2}$$

After constructing an exact CI $[p_{ij}^L, p_{ij}^U]$ for π_{ij} (see Cooper and Pearson 1934), the end-points p_{ij}^L and p_{ij}^U of this interval can be used to construct the $(1-\alpha)$ CI for Π_{ij} shown in (3):

$$\left.\begin{array}{ll}
[2p_{ij}^L(1-p_{ij}^L),\ 2p_{ij}^U(1-p_{ij}^U)], & \text{if } p_{ij}^U \le 0.5, \\[6pt]
[2p_{ij}^L\,(1-p_{ij}^L),\ 0.5], & \text{if } p_{ij}^L,\ p_{ij} \le 0.5 \text{ and } p_{ij}^U \ge 0.5, \\[6pt]
[2p_{ij}^U\,(1-p_{ij}^U),\ 0.5], & \text{if } p_{ij}^U,\ p_{ij} \ge 0.5 \text{ and } p_{ij}^L \le 0.5, \\[6pt]
[2p_{ij}^U\,(1-p_{ij}^U),\ 2p_{ij}^L\,(1-p_{ij}^L)], & \text{if } p_{ij}^L \ge 0.5.
\end{array}\right\} \tag{3}$$

Note that the CI for Π_{ij} is not necessarily symmetric about the point estimate $\hat{\Pi}_{ij}^1$. For a more detailed derivation, see Stam and Duarte Silva (1997).

Besides its simplicity, an advantage of $\hat{\Pi}_{ij}^1$ is that it requires neither a specific assumption on the distribution of W, nor an assumption that the components of W are independent. Since each W_i is compared with the W_j measured in the same replication, $\hat{\Pi}_{ij}^1$ implicitly takes correlation into account. On the negative side, however, $\hat{\Pi}_{ij}^1$ does not provide information about the intensity of the preferences, as reflected by the relative values of W_i and W_j.

2.2 Rank Reversal Probabilities Based on Magnitude of Differences: $\hat{\Pi}_{ij}^2$

The derivation of $\hat{\Pi}_{ij}^2$ assumes that W is approximately normally distributed. $\hat{\Pi}_{ij}^2$ is derived directly from the magnitude of $D_{ij} = W_i - W_j$, the difference between the relative preference weights for alternatives i and j, and explicitly utilizes the variance-covariance structure of $W_1, ..., W_k$. If W is approximately multivariate normally distributed, then the same is true for D_{ij}. Specifically, the variance of D_{ij} equals $\sigma_{D_{ij}}^2 = \sigma_i^2 + \sigma_j^2 - 2\sigma_{ij}$, where σ_i^2 and σ_j^2 are the variances of W_i and W_j, respectively, and σ_{ij} is the covariance of W_i and W_j. The simulation study reported below shows that some of the W_i are in fact strongly correlated. This observation holds true for a sampling-based analysis of imprecise as well as stochastic judgments. The maximum likelihood estimator Q_{ij} of $\pi_{ij} = P(D_{ij} > 0)$ is given by $Q_{ij} \doteq P(Z > -(\overline{D}_{ij})/S_{D_{ij}})$, where Z is standard normally distributed, whereas $\overline{D}_{ij}$ and $S_{D_{ij}}$ are the sample mean and standard deviation of D_{ij}, respectively. Thus, $\hat{\Pi}_{ij}^2$ is determined as in (4):

$$\hat{\Pi}_{ij}^2 = 2\,Q_{ij}(1-Q_{ij}). \tag{4}$$

Using the Bonferroni method (Neter, Wasserman and Kutner 1985), we can derive a conservative simultaneous interval $[q_{ij}^L,\ q_{ij}^U]$ for π_{ij} from the individual $(1-\alpha/2)$ level CIs for $\mu_{D_{ij}}$ (the mean of D_{ij}) and $\sigma_{D_{ij}}$ (see Stam and Duarte Silva (1997) for details). A $(1-\alpha)$ CI for Π_{ij} is found in a way similar to (3), but now using q_{ij}^L, Q_{ij} and q_{ij}^U instead of p_{ij}^L, P_{ij} and p_{ij}^U, respectively.

Summarizing, the trade-off between $\widehat{\Pi}_{ij}^1$ and $\widehat{\Pi}_{ij}^2$ is that the former is of a simpler nature and can be used regardless of the distribution of W, whereas the latter may yield more accurate estimates if W is indeed approximately multivariate normally distributed. If the null hypothesis that W is normally distributed is rejected, then the use of $\widehat{\Pi}_{ij}^2$ is not recommended.

3 Simulation Experiments

We next illustrate our method using two simulation experiments (Experiments A and B), and compare the computational results for $\widehat{\Pi}_{ij}^1$ and $\widehat{\Pi}_{ij}^2$ with the corresponding estimates $\widehat{\Pi}_{ij}^S$ derived using the method of Saaty and Vargas (1987). The interval judgments in Saaty and Vargas are non-stochastic, and although their sampling experiment does yield a statistical analysis of rank reversal probabilities similar to ours, our comparison with their method is limited to the computational aspects only. The interpretation of these methods cannot be compared directly.

3.1 Experimental Design

In Experiment A, we generated 100 samples from the same matrix $\mathbf{M}^S$ of uniform stochastic judgment intervals that was used in the original study by Saaty and Vargas (1987), so that Experiment A corresponds directly with their study. Note that in the sampling procedure a logarithmic transformation is needed whenever a judgment interval of $\mathbf{M}$ is continuous, because of the ratio scale property (e.g., the interval [1/9,9] is asymmetric about 1).

It appears that, in practice, distributions over the interval judgments with a mode near the mean or median value may often be more realistic than uniform distributions. Hence, in Experiment B we again generated 100 samples from judgment intervals, but this time the variates were generated from the skewed discrete distributions in $\mathbf{C}$ in (5) over three specific paired judgment values, with probabilities proportional to the confidence levels. We selected asymmetric discrete distributions for Experiment B, because it offers an interesting comparison with the uniform logarithmic intervals in Experiment A. The entries of $\mathbf{C}$ give the paired comparison ratios used in Experiment B, followed within parentheses by their associated normalized confidence levels. The normalized confidence levels in $\mathbf{C}$ cover the same range of values as the matrix $\mathbf{M}^S$ used in Experiment A.

$$
\mathbf{C} =
\begin{array}{c}
\\ A1 \\ \\ \\ A2 \\ \\ \\ A3 \\ \\ \\ A4
\end{array}
\begin{bmatrix}
\overset{\text{A1}}{1(1.00)} & \overset{\text{A2}}{2(0.15)} & \overset{\text{A3}}{3(0.20)} & \overset{\text{A4}}{3(0.20)} \\
 & 3(0.60) & 4(0.50) & 4(0.50) \\
 & 4(0.25) & 5(0.30) & 5(0.30) \\
 & 1(1.00) & 1/2(0.15) & 2(0.30) \\
 & & 3/4(0.60) & 3(0.40) \\
 & & 1(0.25) & 5(0.30) \\
 & & 1(1.00) & 1/3(0.20) \\
 & & & 3/5(0.50) \\
 & & & 1(0.30) \\
 & & & 1(1.00)
\end{bmatrix}
\tag{5}
$$

For both Experiment A and Experiment B we determined point estimates $\widehat{\Pi}_{ij}^{1}$, $\widehat{\Pi}_{ij}^{2}$ and constructed 99 percent CIs for the rank reversal probabilities Π_{ij}. To distinguish between the two experiments, we denote the point estimates for Experiment A by $\widehat{\Pi}_{ij}^{A,1}$ and $\widehat{\Pi}_{ij}^{A,2}$, and those for Experiment B by $\widehat{\Pi}_{ij}^{B,1}$ and $\widehat{\Pi}_{ij}^{B,2}$. Whenever possible, we compare our results with Saaty and Vargas' rank reversal probability measures for Experiments A and B, represented by $\widehat{\Pi}_{ij}^{A,S}$ and $\widehat{\Pi}_{ij}^{B,S}$, respectively. For notational convenience, we will omit the superscripts A and B if the discussion applies to both experiments.

3.2 Results of Experiments A and B

A preliminary data analysis (not shown here) shows that overall the descriptive sample statistics of the W_i results for Experiments A and B are quite similar, probably due to the fact that the finite ranges of the discrete paired judgment distributions of Experiment B are identical to those in $\mathbf{M}^S$ used in Experiment A. The standard deviations of the weights in Experiment B were slightly larger than those in Experiment A. However, as we will see below, the estimated rank reversal probabilities for Experiment B are substantially different from those in Experiment A. Presumably, an estimator yields more stable and reliable results if the point estimates and CIs are not overly sensitive to the distributions and do not have a tendency to quickly jump to extreme values (0 and 1).

Table 1, which presents the number of times w_i exceeds w_j $(i < j)$ in Experiments A and B, shows that for the vast majority of samples the weight for alternative i exceeds that for alternative j $(i < j)$. For instance, w_1 exceeds w_2 for all 100 samples in both experiments, and w_3 is larger than w_4 in 94 out of 100 samples taken in Experiment A and in 69 of the 100 samples in Experiment B. Thus, most of the rank reversals involve alternatives 3 and 4, and these reversals are more pronounced in

Experiment B.

Table 1. Frequency of event $w_i > w_j$ among the _100_ eigenvectors simulated

		Experiment A						Experiment B		
			j					j		
		1	2	3	4		1	2	3	4
	1	–	100	100	100	1	–	100	100	100
i	2		–	99	100	2		–	98	98
	3			–	94	3			–	69
	4				–	4				–

Table 2. Sample correlation matrix for the w_i

		Experiment A					Experiment B		
	w_1	w_2	w_3	w_4		w_1	w_2	w_3	w_4
w_1	1	–.786	–.565	–.472	w_1	1	–.752	–.336	–.400
w_2		1	.149	.059	w_2		1	.088	–.140
w_3			1	.023	w_3			1	–.268
w_4				1	w_4				1

The sample correlation matrices of the W_i in Table 2 indicate that it would be erroneous to assume – as Saaty and Vargas (1987) did – that the principal eigenvector components are independent. For instance, in Experiment A the sample correlation between W_1 and W_2 was -0.786. Note that the correlation between the eigenvector components results strictly from the sampling procedure used, and in no way depends on whether the original judgment intervals were stochastic or imprecise (non-stochastic).

Kolmogorov-Smirnov tests for multivariate normality confirmed that for Experiment A W was approximately normally distributed, so that the rank reversal measures $\widehat{\Pi}_{ij}^{A,S}$ and $\widehat{\Pi}_{ij}^{A,2}$ (which both assume normality of W) indeed appear applicable. However, the validity of the normality hypothesis for Experiment B is doubtful, so that the use of $\widehat{\Pi}_{ij}^{1}$ (which is not based on distributional assumptions) is preferred for this experiment.

Table 3. Estimated rank reversal probabilities

	Experiment A			Experiment B		
Pair (i,j)	$\widehat{\Pi}_{ij}^{A,1}$	$\widehat{\Pi}_{ij}^{A,2}$	$\widehat{\Pi}_{ij}^{A,S}$	$\widehat{\Pi}_{ij}^{B,1}$	$\widehat{\Pi}_{ij}^{B,2}$	$\widehat{\Pi}_{ij}^{B,S}$
$(1,2)$	0.0	0.0	0.0	0.0	0.0	0.0
$(1,3)$	0.0	0.0	0.0	0.0	0.0	0.0
$(1,4)$	0.0	0.0	0.0	0.0	0.0	0.0
$(2,3)$	0.0198	0.0722	0.2156	0.0396	0.0446	0.2469
$(2,4)$	0.0	0.0032	0.0021	0.0396	0.0406	0.2325
$(3,4)$	0.1128	0.1486	0.5434	0.4278	0.4240	0.9087

Table 4. _99% confidence intervals for rank reversal probabilities_

	Experiment A		Experiment B	
Pair(i,j)	$\widehat{\Pi}_{ij}^{A,1}$	$\widehat{\Pi}_{ij}^{A,1}$	$\widehat{\Pi}_{ij}^{B,1}$	$\widehat{\Pi}_{ij}^{B,1}$
$(1,2)$	[0,0.10]	[0,0]	[0,0.10]	[0,0]
$(1,3)$	[0,0.10]	[0,0]	[0,0.10]	[0,0]
$(1,4)$	[0,0.10]	[0,0]	[0,0.10]	[0,0]
$(2,3)$	[0,0.13]	[0.01,0.15]	[0,0.16]	[0.01,0.19]
$(2,4)$	[0,0.10]	[0,0.02]	[0,0.16]	[0.01,0.18]
$(3,4)$	[0.03,0.25]	[0.037,0.29]	[0.32,0.49]	[0.34,0.50]

The point estimates and 99 percent CIs for $\widehat{\Pi}_{ij}$ for Experiments A and B are summarized in Tables 3 and 4. Saaty and Vargas (1987) do not discuss or calculate CIs for the rank reversal probabilities, so these are not included in Table 4. From Table 3 we see that in both Experiments A and B, each of the three measures yields a zero rank reversal probability between attribute 1 and any of the other attributes. For the attribute pairs not involving attribute 1, $\widehat{\Pi}_{ij}^1$ and $\widehat{\Pi}_{ij}^2$ are similar, but differ considerably from $\widehat{\Pi}_{ij}^S$. Typically, $\widehat{\Pi}_{ij}^2$ is between the corresponding $\widehat{\Pi}_{ij}^1$ and $\widehat{\Pi}_{ij}^S$ values, albeit much closer to $\widehat{\Pi}_{ij}^1$ than to $\widehat{\Pi}_{ij}^S$. In several cases, $\widehat{\Pi}_{ij}^{\cdot,S}$ differs significantly from the other two estimates. For instance, while $\widehat{\Pi}_{34}^{A,1} = 0.1128$ and $\widehat{\Pi}_{34}^{A,2} = 0.1486$, $\widehat{\Pi}_{34}^{A,S}$ is much higher at 0.5434.

Overall, Table 3 indicates that, not surprisingly, $\widehat{\Pi}_{ij}^S$ is less stable than $\widehat{\Pi}_{ij}^1$ and $\widehat{\Pi}_{ij}^2$, since the $\widehat{\Pi}_{ij}^S$ values varied considerably when the data conditions were only slightly changed (from Experiment A to Experiment B), while $\widehat{\Pi}_{ij}^1$ and $\widehat{\Pi}_{ij}^2$ appear less sensitive to the particular data condition. $\widehat{\Pi}_{ij}^S$ is less stable, and tends towards the extremes whenever the information regarding the rankings becomes somewhat conflicting. Again, we stress that the comparison of our stochastic measures with Saaty and Vargas' measure is limited to computational properties only, as the latter was designed for the case of imprecise (non-stochastic) paired judgments.

Summarizing, it appears that $\widehat{\Pi}_{ij}^1$ and $\widehat{\Pi}_{ij}^2$, the measures developed in this paper, yield attractive and robust estimates of the true rank reversal probabilities Π_{ij}, in the case of stochastic judgment intervals. Further details about the derivation of $\widehat{\Pi}_{ij}^1$ and $\widehat{\Pi}_{ij}^2$ and more extensive simulation experiments can be found in Stam and Duarte Silva (1997).

4 Concluding Remarks

We have developed and investigated two different measures of rank reversal probability in the AHP in the presence of stochastic paired comparisons, using multivariate statistical methods to derive both point estimates and tight confidence intervals. Two simulation experiments were used to show that our proposed measures yield robust and stable rank reversal probability estimates. Moreover, in a comparison with the stability measure proposed by Saaty and Vargas for the case of imprecise judgments, we found our measures to be more stable, thus providing a more accurate representation of how the uncertainty in the decision problem affects the rankings of the alternatives. It may be possible to extend our methodology to non-stochastic judgment intervals, for instance modifying our rank reversal measures in order to analyze rank reversal probabilities for fuzzy (non-stochastic) judgment intervals where not all points in the interval are considered equally by the decision maker. However, this extension is not trivial, because some of the statistical interpretations of our method would need to be carefully restated and perhaps adjusted. Nevertheless, this appears to be an interesting avenue of future research.

References

Arbel, A., "Approximate Articulation of Preference and Priority Derivation," *European Journal of Operational Research*, **43** (1989), 317–326.

Arbel, A. and L. G. Vargas, "The Analytic Hierarchy Process with Interval Judgments," in *Multiple Criteria Decision Making: Proceedings of the IXth International Conference: Theory and Applications in Business, Industry and Government*, A. Goicoechea, L. Duckstein and S. Zionts (Eds.), Springer-Verlag, New York, NY, 1992, 61–70.

Arbel, A. and L. G. Vargas, "Preference Simulation and Preference Programming: Robustness Issues in Priority Derivation," *European Journal of Operational Research*, **69** (1993), 200–209.

Barzilai, J. and B. Golani, "AHP Rank Reversal, Normalization and Aggregation," *INFOR*, **32**, 1994, 57–63.

Cooper, C. J. and E. S. Pearson, "The Use of Confidence or Fiducial Limits Illustrated in the Case of the Binomial," *Biometrika*, **26** (1934), 406–413.

Dyer, J. S., "Remarks on the Analytic Hierarchy Process," *Management Science*, **36**, 3 (1990), 249–258.

Moreno-Jimenez, J. M. and L. G. Vargas, "A Probabilistic Study of the Preference Structures in the Analytic Hierarchy Process with Interval Judgments," *Mathematical and Computer Modelling*, **17** (1993), 73–81.

Neter, J., W. Wasserman and M. H. Kutner, *Applied Linear Statistical Models*, Second Edition, Irwin, New York, NY 1985.

Saaty, T. L., *The Analytic Hierarchy Process*, McGraw-Hill, New York, NY, 1980.

Saaty, T. L., "An Exposition of the AHP in Reply to the Paper 'Remarks on the Analytic Hierarchy Process,'" *Management Science*, **36**, 3 (1990), 259–268.

Saaty, T. L. and L. G. Vargas, "Uncertainty and Rank Order in the Analytic Hierarchy Process," *European Journal of Operational Research*, **32** (1987), 107–117.

Salo, A. A., "Inconsistency Analysis by Approximately Specified Priorities," *Mathematical and Computer Modelling*, **17** (1993), 123–133.

Salo, A. A. and R. P. Hämäläinen, "Processing Interval Judgments in the Analytic Hierarchy Process," in *Multiple Criteria Decision Making: Proceedings of the IXth International Conference: Theory and Applications in Business, Industry and Government*, A. Goicoechea, L. Duckstein and S. Zionts (Eds.), Springer-Verlag, New York, NY, 1992, 359–372.

Salo, A. A. and R. P. Hämäläinen, "Preference Assessment by Imprecise Ration Statements," *Operations Research*, **40** (1992), 1053–1061.

Salo, A. A. and R. P. Hämäläinen, "Preference Programming Through Approximate Ratio Comparisons," *European Journal of Operational Research*, **82**, 1994, 458–475.

Schoner, B., W. C. Wedley and E. U. Choo, "A Rejoinder to Forman on AHP, With Emphasis on the Requirements of Composite Ratio Scales," *Decision Sciences*, **23**, 1992, 509–517.

Schoner, B., W. C. Wedley and E. U. Choo, "A Unified Approach to AHP with Linking Pins," *European Journal of Operational Research*, **64**, 1993, 384–392.

Stam and Duarte Silva, "Stochastic Judgments in the AHP: The Measurement of Rank Reversal Probabilities, *Decision Sciences*, **28**, 1997, 655–688.

Vargas, L. G., "Reciprocal Matrices with Random Coefficients," *Mathematical Modelling*, **3** (1982), 69–81.

Zahir, M. S., "Incorporating the Uncertainty of Decision Judgements in the Analytic Hierarchy Process," *European Journal of Operational Research*, **53** (1991), 206–216.

Convergence of Interactive Procedures of Multiobjective Optimization and Decision Support

Andrzej P. Wierzbicki

Institute of Telecommunications, Szachowa 1, 04-894 Warsaw, Poland
and the Institute of Control and Computation Engineering,
Warsaw University of Technology, Nowowiejska 15/19, 00-665 Warsaw, Poland[1]

Abstract. The paper presents an overview of issues of convergence of interactive procedures in multiobjective optimization and decision support. The issue of convergence itself depends on assumptions concerning the behavior of the decision maker – who, more specifically, is understood as a user of a decision support system. Known procedures with guaranteed convergence under classic assumptions are reviewed. Some effective procedures of accelerated practical convergence but without precise convergence proofs are recalled. An alternative approach to convergence based on an indifference threshold for increases of value functions or on outranking relations is proposed and illustrated by a new procedure called Outranking Trials.

Keywords: multiobjective optimization, decision support, interactive procedures, convergence

1 Introduction

An old and basic problem in interactive approaches to multiobjective optimization and decision support is the question how to organize interactive procedures which are convergent in some sense – or how to characterize and prove the convergence of an interactive procedure. A classic approach to this problem consists in emulating convergence analysis of mathematical programming algorithms – that is, in assuming that *the decision maker can be characterized by a value function* and the procedure helps her/him to maximize this function. Several approaches of this type are well known - such as the procedures of Geoffrion-Dyer-Feinberg (1972), of Zionts-Wallenius (1976), or of Korhonen-Laakso (1985).

However, other approaches assume that the essence of an interactive procedure is related to *learning, also by mistakes, thus changing value functions.* Therefore, an interactive procedure should concentrate more on exploring the set of attainable decision outcomes than on convergence. Again, there are several well known approaches of this type - starting with STEM (Benayoun *et*

---

[1] The research reported in this paper was partly supported by the grant No. 3P40301806 of the Committee for Scientific Research of Poland.

al, 1971), through *goal programming* (Charnes and Cooper, 1977), *reference point methods* (see *e.g.* Wierzbicki, 1980) and others.

Several questions arise from such a comparison. First, can we combine somehow an approach of the second type with one of the first type, to preserve the advantages of both types? The Korhonen-Laakso procedure can be viewed as such a combination; however, it will be shown here that some other combined procedures are also possible.

Secondly, how to account for uncertainty in such an interactive procedure? And what shall be meant by uncertainty in such a case? Full probabilistic description of decisions with uncertain outcomes, as in expected utility approaches, takes care of only one dimension of uncertainty. What if the decision maker is uncertain about her/his preferences – can we at all define the convergence of an interactive procedure in such a case?

Thirdly, what are possibly weakest assumptions necessary to define the convergence of an interactive procedure? Is the concept of a value or utility function necessary for this purpose, or can we use some weaker concepts of representing preferences?

The paper, after a short review of known approaches to the convergence of interactive procedures, addresses some aspects of the questions outlined above. However, before approaching these questions, we have to outline some general background.

We shall speak further about a decision maker in the sense of a *user* of a decision support system. The most frequent users of such systems are *analysts* that take advantage of computerized models to prepare analysis of various possible options for final political decision makers; or *modelers* that prepare and analyze computerized models describing a part of reality; or *designers* which have to construct a physical system – say, a computer network – while using the accumulated knowledge about such systems which can be converted into computerized models. When we reflect on the needs of analysts, modelers, designers, it becomes clear that a fast convergence toward the most preferred solution is of dubious value for them – they would rather like to be creative and thus do not know how to define the most preferred solution. Therefore, it is essential that a decision support system has properties supporting the exploration of possible outcomes, such as in reference point approaches.

After an exploration, the user might be interested in a convergent procedure of decision support; but we should not expect that she/he complies with the classical assumptions about the preferences of a decision maker. In order to define convergence, we might assume that the user has a value function – although we should not be surprised if she/he suddenly changes preferences and declares a finely tuned solution useless. Moreover, a typical user will instinctively evaluate objectives and value functions in ratio scales. Theoretically, this implies that his value function will be typically nonlinear and nonseparable – we can assume that value functions are typically linear or at least separable, but only if we consider these functions expressed in ordinal scale, see *e.g.* Rios, (1994).

Neither we can expect that the typical user will be interested in very small improvements of value. A preference structure with a threshold of indifference is best described by one of outranking relations of Roy, see *e.g.* Roy and Vincke (1981). In order to characterize convergence of various interactive procedures we shall assume that, after some time of exploration, the preferences of the user can be characterized by a value function; but we shall consider also the implications of an indifference threshold for this value function and shall also present a convergent procedure based on reference point exploration and on an outranking relation.

2 A multiobjective analysis problem

We shall discuss in the paper a general form of a multiobjective problem as follows. A *set of admissible decisions* X_0 in the decision space X is given – described by some inequality or equality constraints or other relations. We shall assume that the space X is topological and the set X_0 is compact. A space of decision outcomes or objectives (attributes, criteria) Y is defined; we shall assume $Y = I\!R^m$ *i.e.* limit our attention to the case of m criteria. Moreover, an *outcome mapping* $\mathbf{f} : X_0 \to Y$ is given; we shall assume that this mapping is continuous and thus the *set of attainable outcomes* $Y_0 = \mathbf{f}(X_0)$ is also compact.

The mathematical relations describing the set X_0 and the mapping $\mathbf{f}$ are called *the substantive model of a decision situation*; they describe the best knowledge about this situation we can rely on. The outcome relation is often defined *with uncertainty, i.e.* represented by a probabilistic or a fuzzy set model; but we shall not go into details of substantive models with uncertainty in this paper. Even if the substantive model might represent uncertainty, we assume that it is *structured, i.e.* mathematically well defined. If it is not, we call such a decision problem *unstructured*; although there exist techniques of dealing with unstructured information and knowledge (*e.g.* automatic learning by neural nets techniques), they are until now being developed and we shall not consider them here. In practice, most of decision problems are *semistructured:* out of the vast heritage of human knowledge, we must select such parts that are most pertinent to describe the model of given decision situation, then perform additional experiments *etc.*; the *art of model building*, necessary in all semistructured cases, relies on certain techniques but also on expert intuition, see *e.g.* Wierzbicki (1984, 1992a). Finally, a substantive model of a decision situation might have a *logical form* – with corresponding knowledge bases, inference engines *etc.* – or an *analytical form*; we shall limit our attention here to models of analytical form.

While a substantive model describes rather objective aspects and knowledge, each decision problem contains also much more subjective aspects, particularly concerning the preferences of a decision maker. The second part of a multiobjective analysis problem is thus a *preferential model* describing such preferences. A preferential model might have the form of a value function, *i.e.* a mapping $v : R^m \rightarrow R^1$, the higher values the better; or a utility function – a special form of a value function over the probabilistic distributions of decision outcomes – in the case of probabilistic representation of uncertainty. However, as discussed in the introduction, such a form of a preferential model is rather restrictive; we might use it in order to analyze the main subject of this paper, the convergence of interactive procedures, but it is better to consider also preferential models based on weaker assumptions.

One of them, basic for multiobjective analysis, is to represent the preferences of a decision maker by a partial order of Pareto type. We shall recall here some details of such representation. Such an order might be defined by selecting a positive (convex, proper, closed) cone D in outcome space and assuming the preference $\mathbf{y}^a \succ \mathbf{y}^b$ ($\mathbf{y}^a$ is preferred to $\mathbf{y}^b$) to be defined by the cone $\tilde{D} = D \setminus \{0\}$ as follows:

$$\mathbf{y}^a \succ \mathbf{y}^b \quad \Leftrightarrow \quad \mathbf{y}^a \in (\mathbf{y}^b + \tilde{D}) = (\mathbf{y}^b + D \setminus \{0\})$$
$$\mathbf{y}^a \succeq \mathbf{y}^b \quad \Leftrightarrow \quad \mathbf{y}^a \in \mathbf{y}^b + D \tag{1}$$

We shall consider here only the simplest case when $D = R_+^m$ which amounts to maximizing all objectives, but should note that an appropriate modification of the cone D takes care not only of the case when some objectives are minimized, but also of the case when some objectives are *stabilized*, *i.e.* maximized when below and minimized when above certain reference level – see *e.g.* Lewandowski and Wierzbicki (1989).

Given a positive cone D, the set of efficient decision outcomes relative to this cone (called Pareto-optimal, if $D = R_+^m$) is defined as follows:

$$\hat{Y}_0 = \{\hat{\mathbf{y}} \in Y_0 : Y_0 \cap (\hat{\mathbf{y}} + \tilde{D}) = \emptyset\} \tag{2}$$

The efficient decisions are defined as such that their outcomes are efficient, $\hat{\mathbf{x}} \in \hat{X}_0 \Leftrightarrow \mathbf{f}(\hat{\mathbf{x}}) \in \hat{Y}_0$. Beside the set of efficient decision outcomes, we shall use also the set of *weakly efficient* outcomes $\hat{Y}_0^w$ (and corresponding decisions) defined as in (2) but with $\tilde{D}$ substituted by the interior $IntD$, as well as the set of outcomes *properly efficient with a prior bound on trade-off coefficients* $\hat{Y}_0^{pe}$ defined as in (2) but with $\tilde{D}$ substituted by the interior $IntD_\varepsilon$, where D_ε can be defined in various ways, see Wierzbicki (1986, 1992b), while the most useful definition is:

$$D_\varepsilon = \{\mathbf{y} \in R^m : \max_{1 \leq i \leq m} y_i + \varepsilon \sum_{i=0}^m y_i \geq 0\} \tag{3}$$

where we assume that all objectives are dimensionless or have the same physical dimension (we shall later show how to weaken this assumption). While weakly efficient decisions and their outcomes are simpler to analyze mathematically but nonpractical, properly efficient decisions and outcomes with a prior bound on trade-off coefficients are eminently practical (in practice, decision makers do not distinguish cases of weakly efficient objectives and of objectives with very large trade-offs). It can be shown – see Wierzbicki (1992b), Kaliszewski (1995) – that, if $\hat{\mathbf{y}} \in \hat{Y}_0^{p\varepsilon}$, then the trade-off coefficients between objectives are bounded by the prior bound $M = 1 + \frac{1}{\varepsilon}$.

Maxima of a value function $v(\mathbf{y})$ correspond to efficient decision outcomes (or weakly efficient, or properly efficient with a prior bound) if this function is strictly monotone with respect to the cone $\tilde{D}$ (or $IntD$, or $IntD_\varepsilon$):

$$\mathbf{y}^a \in (\mathbf{y}^b + \tilde{D}) \Rightarrow v(\mathbf{y}^a) > v(\mathbf{y}^b) \tag{4}$$

When postulating that the decision maker has a value function, we shall assume such consistency of this function with the given partial order. We shall also use other functions – so called *scalarizing functions* $s(\mathbf{y}, \lambda)$ or $\sigma(\mathbf{y}, \bar{\mathbf{y}})$, where λ, $\bar{\mathbf{y}}$ are additional parameters, see e.g. Wierzbicki (1986) – which have also strict monotonicity property. Scalarizing functions might be considered as proxy value functions, but their actual role is to produce (when maximized) various efficient decisions and their outcomes and to control the selection of efficient outcomes by changing the parameters λ or $\bar{\mathbf{y}}$.

Another preferential model based on weaker assumptions than a value function is an outranking relation of the type proposed by Roy, see *e.g.* Vincke (1989). We assume then that the decision maker can tell, whether two decision outcomes $\mathbf{y}^a$ and $\mathbf{y}^b$ should be classified as incomparable, or one (strongly or weakly) preferred to another, or belonging to the same indifference class – while the most essential differences to the value function model are that not all decisions are necessarily comparable and and that there is a finite indifference threshold. We shall use later an outranking relation of this type together with a value function – thus, we shall assume that all decisions are comparable, but the finite indifference threshold implies that the decision maker does not distinguish small differences of value. We could thus define convergence of an interactive procedure only for the weaker model, an outranking relation, but to compare it with more classic approaches we must postulate that the outranking relation is consistent with a value function and an indifference threshold.

After this preliminaries, we can define now more precisely what is a *multiobjective analysis problem*. The word *problem* has here diverse meaning. First, given a *decision problem* (a *problem of real world*), we construct a model of the problem, consisting of two parts: substantive and preferential. Then, given a substantive model of a decision situation, we have a problem how to analyze the model – compared with the original decision problem, we might call this *analytical problem instance*, but we shall call it a *multiobjective analysis problem*. If we use a general preferential model based on Pareto-type order, multiobjective analysis means the analysis of the entire set of efficient outcomes and decisions. Analyzing a complicated, multidimensional set is itself a difficult issue: we must have at least tools for generating various specific, representative elements of this set. If we postulate a more detailed preferential model – an outranking relation, a value function – we might be interested either in finding the (or one of) most preferred decision and its outcomes, or again in generating a selected set of representative decisions and outcomes, possibly ranked and including the most preferred ones.

Practical experience in decision support shows that it is always good to use the more general, Pareto-type preferential model, possibly supplemented with more detailed ones, because the latter might easily change during the process of learning, inherent in decision support. Clearly, the more detailed types of preferential models should be consistent with the general one in the sense of monotonicity[2]. When we use a more detailed type of preferential model, we assume that the decision maker (the user of a decision support system) has already learned enough and thus the search for a most preferred decision is meaningful. The convergence of an interactive procedure of decision support is understood as the convergence of such search.

At this point, we could apply also the techniques used for *discrete* multi-objective (or multi-attribute) problems, when the set of admissible decisions X_0 is a discrete set, usually finite. Convergence in such a case is also finite. Thus, we might apply corresponding techniques of multi-attribute decision selection, say, between the ELECTRE, the PROMETHEE, the AHP (Analytical Hierarchy Process), see Roy and Vincke (1981), Brans and Vincke (1985), Saaty (1980), or any other known techniques – but not to produce a final answer, only in the sense of an iteration of an interactive process. If we actually have a *continuous* multiobjective problem, the set X_0 has infinite number of elements and continuous power, we could reduce the problem to a repetitive generation of a discrete, finite set of options and then to an application of some known technique of multiple attribute choice. In fact, some very attractive procedures of interactive decision support use similar ideas.

Another issue is the question of a parametric representation or covering the entire set of efficient decision outcomes.For example, parameterized compromise solutions might not necessarily cover all efficient set. This is not the case in a reference point approach: a full parametric representation of the set $\hat{Y}_0^{pe}$ can be obtained by maximizing the following order-consistent achievement function (which accounts also for various physical dimensions of objectives):

$$\sigma(\mathbf{y}, \bar{\mathbf{y}}) = \min_{1 \leq i \leq m} \frac{y_i - \bar{y}_i}{y_{i,up} - \bar{y}_i} + \varepsilon \sum_{i=1}^{m} \frac{y_i - \bar{y}_i}{y_{i,up} - \bar{y}_i} \tag{5}$$

with suitable changes of the reference point $\bar{\mathbf{y}}$; in the above function it is assumed that all objectives are maximized and $y_{i,up}$ denotes any upper bound (not less than utopia point component) for the objective y_i. Even if the set

---

[2]In practice, inconsistencies often occur. However, if the user of a decision support system discovers that an objective should no longer be maximized, the type of this objective can be changed to stabilized – hence the importance of dealing both with maximized or minimized and stabilized objectives.

Y_0 is not convex (or even if it is a discrete set), each maximal point of this function is efficient; moreover, we can attain this way any properly efficient outcome $\hat{\mathbf{y}}$ with trade-off coefficients scaled down by the deviations from the upper levels and bounded by:

$$t_{ij}(\hat{\mathbf{y}}) \le (1 + 1/\varepsilon)\,\frac{y_{i,up} - \hat{y}_i}{y_{j,up} - \hat{y}_j} \tag{6}$$

At any such point, we can define a reference point – by taking precisely $\bar{\mathbf{y}} = \hat{\mathbf{y}}$ – in such a way that the maximal point of $\sigma(\mathbf{y}, \bar{\mathbf{y}})$ coincides with $\hat{\mathbf{y}}$, see e.g. Wierzbicki (1992b). Equivalently, for any properly efficient $\hat{\mathbf{y}}$ we can choose a sufficiently small ε such that (6) is satisfied; then we can choose an aspiration point (equal *e.g.* to $\hat{\mathbf{y}}$, but any point on the line segment joining $\mathbf{y}_{up}$ and $\hat{\mathbf{y}}$ would do) together with the corresponding weighting coefficients such that the point $\hat{\mathbf{y}}$ is obtained by maximizing the achievement function. This property will be used in some interactive procedures of multiobjective optimization.

3   Interactive procedures with proven convergence

3.1   Geoffrion-Dyer-Feinberg procedure

The first of interactive procedures with guaranteed convergence was given by Geoffrion, Dyer and Feinberg (1972). For a detailed discussion of this procedure, see *e.g.* Steuer (1986); here we shall give only short comments. The procedure is applicable for general, nonlinear but differentiable substantive models; it is assumed that the decision maker has also nonlinear but differentiable value function[3]. The procedure is an extension of the Frank-Wolfe nonlinear optimization algorithm, where the interaction with human decision maker replaces the modules of function and gradient determination. The determination of the gradient of a value function is a complicated task, requiring many pairwise comparison questions to be answered by the decision maker. The procedure, though theoretically elegant, did not find many applications in multiobjective decision support. We have thus learned that, in order to be effective, an interactive procedure must concentrate more on the details of interaction with the decision maker, less on imitating mathematical programming algorithms.

However, this procedure is a prototype of other convergent procedures that exploit the concepts of convergence of mathematical programming algorithms. As it is known – see *e.g.* Luenberger (1973) – an unconstrained nonlinear programming algorithm based on directional searches converges under two general assumptions:

---

[3] Note that the differentiability of a value function implies that it is not an ordinal value function, but is measured on a stronger scale.

- the directions of search should give uniform improvement, that is the angle between these directions and the gradient should be uniformly (in the case of maximization) acute;
- the step-size coefficients determining the length of steps along the directions should not converge to zero too fast – which can be specified by various tests (of Goldstein, of Wolfe) or expressed by the assumption that, if they converge to zero, then the sequence of points generated by the algorithm converges to the maximum or at least a stationary point of the function.

The Frank-Wolfe algorithm produces directions of uniform improvement, and the second assumption sufficient for convergence is fulfilled in the Geoffrion-Dyer-Feinberg procedure because of the details of a specific line search used in the procedure.

3.2 Zionts-Wallenius procedure

The procedure of Zionts and Wallenius (1976) was designed for a much more limited class of models, but more attention was paid to the details of interaction with the decision maker. Substantive models were restricted to multiobjective linear programming models, where the set Y_0 is polyhedral; as a preferential model, the linear value function:

$$s(\mathbf{y}, \lambda) = \lambda^T \mathbf{y} = \sum_{i=1}^{m} \lambda_i y_i \qquad (7)$$

was chosen, but only as an approximation of any concave value function.

The procedure starts at any efficient extreme point of Y_0 – for example, from $\hat{\mathbf{y}}_{neu}^{(1)}$ corresponding to $\lambda_i = \frac{1}{m}, \forall i = 1, \ldots m$. The decision maker is asked whether she/he prefers:

- other extreme points of Y_0 that are adjacent to the current point but distinctly different from it, with a given difference threshold;
- trade-off vectors generated in an multiobjective linear programming algorithm and corresponding to such edges that lead to adjacent extreme points which are too close, below given difference threshold.

The adjacent extreme points are compared to the current one, but the trade-off vectors are evaluated in more general terms of liking them or not. Later experience has shown that decision makers might feel uncomfortable when answering such general questions. Since we actually ask the decision maker whether she/he likes improvements in a given direction, thus – if the adjacent extreme points are too close – an alternative way of asking the question is to present to her/him further, distinctly different points on the same straight line, not necessarily belonging to Y_0.

After such round of questions either a preference for an adjacent point or for a direction is established or all edges leading to adjacent extreme points of the current point are evaluated as nonprofitable. Conceptually, this would suffice to construct a convergent interactive procedure. Indeed, moving to the preferred extreme point results in an increase of the value function. Since the procedure moves among finite number of extreme points only, then the convergence (assuming compact Y_0) is actually finite – precise for linear value functions and only approximate for concave value functions that might have maxima on the facets, not at some extreme points of Y_0.

However, in the actual procedure of Zionts and Wallenius, the data obtained from the answers of the decision maker way are used additionally for updating the set of weighting coefficients $\Lambda^{(k)}$, where the upper index (k) denotes the iteration number. At the starting point, $\Lambda^{(0)} = \{\lambda \in \mathbb{R}^m : \lambda_i \geq 0, \sum_{i=1}^{m} \lambda_i = 1\}$ is assumed. Each increase of preference along a given direction, $\mathbf{y} \succ \mathbf{y}^{(k)}$, indicates that the set $\Lambda^{(k+1)}$ might be constructed from $\Lambda^{(k)}$ by adding the constraint $\lambda^T(\mathbf{y} - \mathbf{y}^{(k)}) \geq \delta$, where $\delta > 0$ is an accuracy threshold. Tradeoff vectors can be treated similarly as directions $\mathbf{y} - \mathbf{y}^{(k)}$. Errors in interaction might be corrected by deleting the oldest remembered constraints on λ.

Given an updated set $\Lambda^{(k+1)}$, the next extreme efficient point $\mathbf{y}^{(k+1)}$ is generated not as the preferred adjacent extreme point, but as a solution to the problem:

$$\max_{\mathbf{y} \in Y_0} \lambda^{(k+1)T} \mathbf{y} \tag{8}$$

where $\lambda^{(k+1)}$ is an (arbitrary) element of $\Lambda^{(k+1)}$.

Actually, a reasonable modification of Zionts-Wallenius procedure would be to let the decision maker choose between such $\mathbf{y}^{(k+1)}$ and the preferred adjacent extreme point. The original Zionts-Wallenius procedure, though it pays much attention to the interaction with the decision maker, is rather complicated in details; many further modifications, often simplifying this procedure, were proposed, see *e.g.* Zionts and Wallenius (1983). The main advantage of this procedure, as seen from todays perspective, was to use the properties of the set Y_0 and the thresholds of accuracy in interaction with the decision maker to produce a procedure with finite convergence.

3.3 Korhonen-Laakso procedure and Pareto Race

Korhonen and Laakso (1985) proposed a visual interactive method for controlling the selection of efficient outcomes and decisions, incorporated in a software tool called Pareto Race.

The essence of the Pareto Race method is a directional search of improvements of the satisfaction of the decision maker with the outcomes of efficient decisions (or of a value function). Starting from an efficient objective outcome point $\hat{\mathbf{y}}^{(k)}$, $k = 1$, the search is a repetition of following three phases:

- the determination of a direction $\mathbf{d}^{(k)} \in \mathbb{R}^m$ such that it would point towards some improvements of the satisfaction of the decision maker;

- the computation of several efficient objective outcomes and decisions corresponding to several reference points distributed along the direction;

- the choice of the step-length in the direction, *i.e.* a decision and objective outcome determined in previous phase that provides best improvement of the satisfaction of the decision maker; the resulting point becomes the next $\hat{\mathbf{y}}^{(k+1)}$.

Thus, the general idea of Korhonen-Laakso procedure is in a sense similar to that of Geoffrion-Dyer-Feinberg procedure, with some essential differences. The direction $\mathbf{d}^{(k)}$ is chosen in Pareto Race by the decision maker, but she/he might do it either arbitrarily (and take full responsibility for the convergence of the procedure) or be supported by an appropriate procedure. Such a procedure might consist in asking the decision maker, for each objective component y_i, how much this component should be – as compared to $\hat{y}_i$ – increased or decreased in relative terms (in % of the known range $[y_{i,lo}, y_{i,up}]$). This can be done *e.g.* in a prepared questionnaire form, asking the decision maker how much decreased or increased should be given objective y_i.

Such a direction $\mathbf{d}^{(k)}$ can be interpreted as a *reference direction*. Several reference points can be chosen along the direction *e.g.* as follows:

$$
\begin{aligned}
\bar{\mathbf{y}}^{(k,j)} &= \hat{\mathbf{y}}^{(k)} + \tau_j \mathbf{d}^{(k)}, \\
\tau_j &= 2^j, \; j = -3, -2, -1, 0, 1, 2, 3
\end{aligned}
\tag{9}
$$

which can be additionally projected on the ranges $[y_{i,lo}, y_{i,up}]$, if necessary. The efficient objective outcomes that are presented to the decision maker for evaluation are obtained in the Pareto Race method by a maximization of an order-consistent achievement function *e.g.* of the form (5):

$$
\hat{\mathbf{y}}^{(k,j)} = \underset{\mathbf{y} \in Q_0}{\operatorname{argmax}} \, \sigma(\mathbf{y}, \bar{\mathbf{y}}^{(k,j)})
\tag{10}
$$

These points can be considered as *projections*[4] of the directional changing reference points $\hat{\mathbf{y}}^{(k,j)}$ *on the properly efficient frontier* $\hat{Y}_0^{pe}$. Thus, another essential difference to the Geoffrion-Dyer-Feinberg procedure – and similarity to the Zionts-Wallenius procedure – is that Pareto Race moves among efficient points.

The choice of a step-length in the direction $\mathbf{d}^{(k)}$ can be prepared by displaying graphically these projections as profiles of changes of component objectives $\hat{y}_i^{(k,j)}$ along changing j. An important feature of the Pareto Race tool is the graphically supported interaction with the decision maker which uses

---

[4] It should be stressed, however, that these projections can be interpreted as results of a distance minimization only if $\mathbf{y}^{(k,j)}$ are "above to the north-east" of $\hat{Y}_0^{pe}$. If $\bar{\mathbf{y}}^{(k,j)} \in Y_0$ (which might occur *e.g.* in non-convex cases), then these projections represent rather a uniform improvement of all objective components than distance minimization, since an order-consistent achievement function is generally not a norm.

such profiles in order to select some j for which the improvement of all objective components is best balanced. This choice constitutes the third stage of the method and determines next starting point for a new iteration:

$$\hat{\mathbf{y}}^{(k+1)} = \hat{\mathbf{y}}^{(k,j)} \text{ for selected } j \tag{11}$$

If the decision maker states that she/he is not satisfied with any j – since neither of $\hat{\mathbf{y}}^{(k,j)}$ is for her/him better than $\hat{\mathbf{y}}^{(k)}$ – the decision support system should ask her/him to specify another direction $\mathbf{d}^{(k)}$, hopefully leading to an improvement, in which case the decision maker takes the responsibility for the convergence. The search stops if the decision maker states that she/he cannot suggest a direction of improvement. Korhonen and Laakso (1985) presented a convergence proof of this procedure under appropriate assumptions. Further details of convergence of such reference point or reference direction procedures were discussed by Bogetoft *et al.* (1988).

3.4 Stochastic approximation and quasi-gradient procedures

The interactive procedures described above have some corrective properties in cases of errors made by the decision makers, but do not take into account such errors systematically. There exist, however, a way of obtaining slow but certain convergence for directional search methods even if the evaluations of the decision maker are subject to random errors. This way is known from the old technique of *stochastic approximation,* or from the more general[5] technique of *stochastic quasi-gradient* of Ermolev – see e.g. Ermolev and Wets (1988). It consists in simply applying a harmonic sequence of fixed step-length coefficients for each iteration (k):

$$
\begin{aligned}
\tau_{(k)} &= \tau_{(1)}/k; \quad k = 1, 2 \ldots; \\
\bar{\mathbf{y}}^{(k+1)} &= \hat{\mathbf{y}}^{(k)} + \tau_{(k)}\mathbf{d}^{(k)}; \\
\hat{\mathbf{y}}^{(k+1)} &= \operatorname*{argmax}_{\mathbf{y} \in Q_0} \sigma(\mathbf{y}, \bar{\mathbf{y}}^{(k+1)})
\end{aligned}
\tag{12}
$$

The step to $\hat{\mathbf{y}}^{(k+1)}$ might be either made always (which results in a not monotone but still convergent variant of the procedure) or only if the decision maker accepts $\hat{\mathbf{y}}^{(k+1)}$ as better than $\hat{\mathbf{y}}^{(k)}$; if the step is not made, she/he has just to redefine the direction $\mathbf{d}^{(k)}$. Due to the fixed but harmonically decreasing coefficients, the procedure converges under the assumption that the directions specified by the decision maker approximate (with random errors but without bias) the stochastic quasi-gradient of her/his value function, see Ermolev and Wets (1988). Michalevich (1986) has used such a technique to propose an interactive procedure of decision support with proven convergence.

---

[5] Taking into account also nondifferentiable cases.

The essence of the convergence of such techniques consists in sufficiently slow convergence such that the stochastic errors of evaluations can be averaged. This accounts for the uncertainty about the preferences of the decision maker, if this uncertainty can be expressed in probabilistic terms. This idea is very attractive and powerful theoretically; however, practical experiments show that decision makers usually find such convergence too slow.

4 Procedures with accelerated convergence

In order to account for the value of the time used up by decision makers, different principles of interaction must be used. It is true that pairwise comparison is a relatively simple cognitive task and thus might be reliably used in interactive procedures; but most decision makers are able and would prefer to perform more complicated cognitive tasks if the time of procedure could be shorten this way. In a sense, decision makers are willing to work in a *parallel* fashion (as we can use today computers of remarkable processing power – and if this power is too small, we can always use computers with scalably parallel architecture).

There exists several approaches aimed at accelerating convergence of interactive procedures, all based on presenting to the decision maker a larger sample of decision options and their outcomes. We assume then that the decision maker is able to point to the most preferred outcome of such a sample. Theoretically, it is a difficult cognitive task; moreover, no convergence proofs were given for such procedures. Practically, however, such procedures save time. Therefore, we outline here briefly some of such procedures; then we shall propose a new procedure of accelerated convergence, together with convergence proof, indicating how the convergence of such approaches could be investigated.

4.1 Reference ball

One of the first ideas in reference point methods (see Wierzbicki, 1980) was to use – as an approximation of the preferred decision outcome – a central point $\hat{\mathbf{y}}^{(k)}$, which at $k = 1$ could be selected as a neutral point, and the most natural way of scattering the efficient points around it:

$$
\begin{aligned}
\delta_{(k)} &= 0.1/k, \quad k = 1, 2, \ldots \\
\bar{\mathbf{y}}^{(k,j)} &= \hat{\mathbf{y}}^{(k)} + \delta_{(k)}\mathbf{e}^{(j)}, \ \mathbf{e}^{(j)} = (0\,0\,\ldots\,1_{(j)}\,\ldots\,0\,0)^T \\
\hat{\mathbf{y}}^{(k,j)} &= \operatorname*{argmax}_{\mathbf{y}\in Q_0} \sigma(\mathbf{y}, \bar{\mathbf{y}}^{(k,j)})
\end{aligned}
\tag{13}
$$

The $m + 1$ efficient points $\hat{\mathbf{y}}^{(k)}$, $\hat{\mathbf{y}}^{(k,j)}$, $j = 1, \ldots m$ are presented then to the decision maker who selects the most preferred one as the next central point $\hat{\mathbf{y}}^{(k+1)}$. Because of the harmonic decrease of the radius of scattering, we can expect that the procedure might be convergent, even in the presence of random evaluation errors. However, no formal proof of convergence was given; moreover, practical trials have shown that this procedure is too slow for a typical decision maker.

On the other hand, this procedure illustrates the idea of a *reference ball:* given a current point $\hat{\mathbf{y}}^{(k)}$, scatter a given number of additional points in a ball of radius $\rho^{(k)}$ centered at the current point and use the additional points as reference points. If we take the Chebyshev norm to define the ball, it becomes just a box centered at $\hat{\mathbf{y}}^{(k)}$:

$$\tilde{B}(\hat{\mathbf{y}}^{(k)}, \rho^{(k)}) = \{\bar{\mathbf{y}} \in I\!\!R^m : |\bar{y}_i - \hat{y}_i^{(k)}| \leq \rho^{(k)}(y_{i,up} - y_{i,lo})\}, \quad \rho^{(k)} \in (0, 1) \quad (14)$$

where $y_{i,lo}$ is a lower bound for objective y_i and the symbol $\tilde{}$ in the denotation $\tilde{B}(\hat{\mathbf{y}}^{(k)}, \rho^{(k)})$ stresses that $\rho^{(k)}$ is only a relative radius of the box or ball. Suppose we choose some number P of randomly generated points $\bar{\mathbf{y}}^{(l,j)}$, $j = 1, \ldots P$ on the boundary of this box. To do this, it is sufficient to select randomly the index of the component i and then to generate randomly – *e.g.* with an uniform distribution – an aspiration component:

$$\bar{y}_i \in [\hat{y}_i^{(k)} - \rho^{(k)}(y_{i,up} - y_{i,lo}), \; \hat{y}_i^{(k)} + \rho^{(k)}(y_{i,up} - y_{i,lo})] \quad (15)$$

then project this component on the box boundary:

$$\begin{aligned}
\bar{y}_i^{(k,j)} &= \hat{y}_i^{(k)} + \rho^{(k)}(y_{i,up} - y_{i,lo}) \text{ if } \bar{y}_i \geq \hat{y}_i^{(k)} \\
\bar{y}_i^{(k,j)} &= \hat{y}_i^{(k)} - \rho^{(k)}(y_{i,up} - y_{i,lo}) \text{ if } \bar{y}_i < \hat{y}_i^{(k)}
\end{aligned} \quad (16)$$

For other components $i+1, \ldots, m, 1, \ldots i-1$ of the vector $\bar{\mathbf{y}}^{(k,j)}$, we repeat the generation as in (15), but not necessarily the projection as in (16) – if we repeat the projection, we obtain obtain a random sequence of corner points of the box.

Then we proceed to compute the corresponding properly efficient outcome points by repeated maximization of an order-consistent achievement function:

$$\hat{\mathbf{y}}^{(k,j)} = \underset{\mathbf{y} \in Q_0}{\operatorname{argmax}} \, \sigma(\mathbf{y}, \bar{\mathbf{y}}^{(k,j)}), \; j = 1, \ldots P \quad (17)$$

The resulting properly efficient outcomes $\hat{\mathbf{y}}^{(l,j)}$ will be not necessarily the most distant from $\hat{\mathbf{y}}^{(l)}$, even if we choose randomly aspiration points only in the box corners. Therefore, an additional filtering of these outcomes might be performed (if there are too many of them to be presented to the decision maker). Thus selected number of spread points $\hat{\mathbf{y}}^{(k,j)}$, $j = 1, \ldots P'$ and the center $\hat{\mathbf{y}}^{(k)}$ are presented to the decision maker who is supposed to select the most preferred one as the next $\hat{\mathbf{y}}^{(k+1)}$.

4.2 Contracted cone

A similar idea is employed in the *contracted cone* procedure of Steuer and Choo (1983). The original procedure was to define a contracting set of weighting coefficients which can be also interpreted as a cone of directions in objective space, starting at the upper bound or utopia point, going through the reference points and "illuminating" a part of the efficient surface. The decision maker would choose a point in the "illuminated" part, around which a new but contracted – "narrower" cone would be centered. The procedure was defined originally in terms of weighting coefficients and the Chebyshev norm defining the distance from the utopia point and interpreted in terms of contracting directional cones. Moreover, we can equivalently define this procedure in terms of a ball of reference points with a contracting radius and use the maximization of an order-consistent achievement function.

The convergence of such a procedure depends critically on the choice of the sequence $\rho^{(k)}$. We can choose either an exponential or a harmonic sequence:

$$\rho^{(k)} = 1/2^k \text{ or } \rho^{(k)} = 1/2k, \ k = 1, \dots \infty \tag{18}$$

Steuer and Choo proposed the exponential sequence. Although their procedure has found many successful applications, exponential sequences might result in a lack of convergence in special cases[6]. If we choose the harmonic sequence instead, it could be proved – as indicated earlier – that the procedure converges (in the expected value sense) to the maximum of the value function of the decision maker, constrained to Q_0, even if the decision maker makes occasional errors in selecting the most preferred point. However, the convergence with the harmonic sequence is rather slow.

4.3 Light beam search

Another similar idea was applied by Jaszkiewicz and Słowiński, see *e.g.* (1994), when illuminating a part of efficient surface in their Light Beam Search method. They noted, however, that the use of an order-consistent achievement scalarizing function instead of the Chebyshev metric method makes it possible to further modify the ideas of Steuer and Choo. They locate then "the light source" at any non-attainable reference point ("to the North-East" of efficient frontier, not necessary at the utopia point) and generate "the cone of light" with the help of additional weighting coefficients.

We should add here a further comment that, when using both aspiration and reservation levels instead of one reference point in an order-consistent achievement function, see *e.g.* Granat et al. (1994), we can use an aspiration point to locate the "source of light" and appropriate changes in reservation levels to generate the "cone of light". For example, if an aspiration point $\bar{y}^{(k)}$

---

[6] Consider the following single-dimensional example: maximize $y = -(x - 2)^2$ starting with $x^{(1)} = 0$ and with the sequence of limited steps $x^{(k+1)} - x^{(k)} = 1/2^k$, $k = 1, \dots \infty$. Obviously, $\lim_{k \to \infty} x^{(k)} = 1$ which falls short of $\hat{x} = 2$.

and a reservation point $\bar{\bar{\mathbf{y}}}^{(k)}$ were used to obtain an efficient point $\hat{\mathbf{y}}^{(k)}$ by maximizing an order-consistent achievement function $\sigma(\mathbf{y}, \bar{\mathbf{y}}^{(k)}, \bar{\bar{\mathbf{y}}}^{(k)})$, then a spread of reservation points around $\hat{\mathbf{y}}^{(k)})$ such as described above by (15), (16) will generate an appropriate "cone of light".

Another important aspect of the Light Beam Search method of Jaszkiewicz and Słowiński was the construction of an outranking relation in the sense of Roy, see *e.g.* Vincke (1989), in the illuminated part of efficient frontier. We shall not present here this feature of Light Beam Search method in detail; instead, we shall propose in next section another method which combines the construction of an outranking relation directly with the specification of reference levels. The proof of convergence of this method indicates how to approach convergence of methods such as Light Beam Search or other inter-active procedures with accelerated convergence that, until now, lack convergence proofs.

5 Outranking trials

The proposed method is as follows. Suppose the decision maker (the user of a decision support system) has performed as many experiments of multiob-jective optimization with the substantive model as she/he wished, but finally selected a preferential model in the form of a set of objectives with a speci-fied partial order; suppose a properly efficient outcome $\hat{\mathbf{y}}$ (say, summarizing the experience gained in the initial experiments) is also given.

Starting from this outcome, the decision maker asks for help in checking whether a more satisfactory decision outcome can be found and in finding the most satisfactory outcome; thus, the decision maker indicates that she/he has learned enough about the substantive model and has formed a preferen-tial model. Although this preferential model might be expressed by a value function, we assume that the decision maker does not care about small dif-ferences of value and does not want to compare decisions which result in too widely changing objectives. Thus, we shall use another preferential model of an outranking relation. Both the value function and the outranking rela-tion are assumed to be consistent with each other (which will be discussed in detail later) and with a partial order of objectives specified by the deci-sion maker. For simplicity of description, we assume that the partial order corresponds to the maximization of all objectives, although the method can be easily generalized for other cases.

Therefore, at this point of procedure, the decision maker is asked to specify some parameters of her/his outranking relation. We could use here *e.g.* the outranking relation of the type used in ELECTRE 3 or 4, see Vincke (1989). However, we shall modify these outranking relations to exploit the full pos-sibilities of reference point methods. The outranking relation proposed here is constructed as follows. For each component objective y_i, while knowing the range of changes of this objective and the point $\hat{\mathbf{y}}$, the decision maker is asked to specify four incremental threshold values – two obligatory, other two optional. If a range of reference levels (an aspiration level $y_{i,asp}$ and a reservation point $y_{i,res}$ such that $y_{i,res} < \hat{y}_i < y_{i,asp}$)) is given, then these thresholds can be also defined as corresponding parts of this range:

- the *veto threshold* $\Delta y_{av,i} > 0$, *e.g.* $\Delta y_{av,i} = \beta(\hat{y}_i - \bar{y}_{i,res})$, $\beta \in (0, 1]$;

- the *negative indifference threshold* $\Delta y_{ni,i} > 0$, $\Delta y_{ni,i} < \Delta y_{av,i}$ or $\Delta y_{ni,i} = \xi \Delta y_{av,i}$, $\xi \in (0, 1)$;

- the *component outranking threshold* $\Delta y_{co,i} > 0$, *e.g.* $\Delta y_{co,i} = \beta(\bar{y}_{i,asp} - \hat{y}_i)$, $\beta \in (0, 1]$;

- the *positive indifference threshold* $\Delta y_{pi,i} > 0$, $\Delta y_{pi,i} < \Delta y_{co,i}$ or $\Delta y_{pi,i} = \xi \Delta y_{co,i}$, $\xi \in (0, 1)$;

The number of objective components increased beyond positive indifference threshold will be compared with the number of components decreased beyond negative indifference threshold in order to determine specific outranking relations. All above threshold values might depend on the current point, but we shall not indicate this dependence explicitly, only let the decision maker correct them if needed.

We assume that all component objectives are measured on ratio scales; since the decision maker already learned about their ranges of change $y_{i,up} - y_{i,lo}$, all above thresholds might be also expressed in percentage terms of these ranges. We do not assume that any objective is much more important than another one, although moderate ratios of importance of objectives might be expressed implicitly by the selection of thresholds and their ratios. We assume here that the decision maker might specify separate negative and positive indifference threshold (although in ELECTRE the thresholds $\Delta y_{ni,i}$ and $\Delta y_{pi,i}$ are usually equal) because of known psychological studies – see *e.g.* Kahneman and Tversky (1982) – showing that we usually value differently a loss and a gain.

The specification of four outranking thresholds for each objective component might require too much effort from the decision maker. Therefore, we suggest that she/he specifies obligatory only two of them: the veto threshold $\Delta y_{av,i}$ and the component outranking threshold $\Delta y_{co,i}$ for each i; their suggested values might be just given percentage (*e.g.* 5%) of the objective ranges, or a given part β of the distances to the reservation and aspiration levels, as indicated above. The decision maker might define a number $\xi \in (0, 1)$ which defines the other (indifference) thresholds $\Delta y_{ni,i}$ and $\Delta y_{pi,i}$ generally, with the suggested value $\xi = 1/m$.

For any given pair of $\hat{\mathbf{y}}$ (which is interpreted as the current efficient point in objective space) and $\mathbf{y}$ (a compared point in objective space), the specified outranking thresholds might be used to define the following five index sets:

$$
\begin{aligned}
I_+ &= \{i \in \{1,\ldots m\} : y_i > \hat{y}_i + \Delta y_{pi,i}\} \\
I_- &= \{i \in \{1,\ldots m\} : y_i < \hat{y}_i - \Delta y_{ni,i}\} \\
I_0 &= \{1,\ldots m\} \setminus (I_+ \cup I_-) \\
I_c &= \{i \in \{1,\ldots m\} : y_i > \hat{y}_i + \Delta y_{co,i}\} \\
I_v &= \{i \in \{1,\ldots m\} : y_i < \hat{y}_i - \Delta y_{av,i}\}
\end{aligned}
\tag{19}
$$

Furthermore, the decision maker is asked to define an integer number m_0 in comparing the numbers of objectives increased and decreased beyond indifference thresholds. The *component C, normal P and weak Q outranking relations* between $\mathbf{y}$ and $\hat{\mathbf{y}}$ can be defined then as follows:

$$
\begin{aligned}
\mathbf{y} \, C \, \hat{\mathbf{y}} &\Leftrightarrow (I_c \neq \emptyset) \wedge (I_- = \emptyset) \\
\mathbf{y} \, P \, \hat{\mathbf{y}} &\Leftrightarrow (|I_+| - |I_-| \geq m_0) \wedge (I_v = \emptyset) \\
\mathbf{y} \, Q \, \hat{\mathbf{y}} &\Leftrightarrow (0 \leq |I_+| - |I_-| < m_0) \wedge (I_v = \emptyset)
\end{aligned}
\tag{20}
$$

where $|I_+|$, $|I_-|$ denote the numbers of elements of sets I_+, I_-. Thus, the normal outranking of $\hat{\mathbf{y}}$ by $\mathbf{y}$ means that there is no veto and that the number of objectives "in favor of $\mathbf{y}$" exceeds the number of objectives "in favor of $\hat{\mathbf{y}}$" at least by m_0. For small m, a natural choice is $m_0 = 1$. The weak outranking relation means that there is no veto and that the number of objectives "in favor of $\mathbf{y}$" is at least equal but does not exceed by m_0 the number of objectives "in favor of $\hat{\mathbf{y}}$". This weak outranking is interpreted, similarly as in ELECTRE, as a case when the decision maker might hesitate.

The component outranking is stronger than the normal one: one component improves beyond its component outranking threshold, greater than the positive indifference threshold, and others do not deteriorate beyond their negative indifference thresholds.

It is convenient to consider both the outranking relations P and Q as represented by one *parametric outranking relation* PQ_{m_1} with a given parameter $m_1 \in \{0, \ldots m - 1\}$:

$$
\mathbf{y} \, PQ_{m_1} \, \hat{\mathbf{y}} \Leftrightarrow (|I_+| - |I_-| = m_1) \wedge (I_v = \emptyset)
\tag{21}
$$

The outranking relations considered above are slightly more sensitive than those used in ELECTRE, because we will use them to produce decision options and their outcomes that *are suspected of outranking* the current decision and its outcome; *the final choice, whether the decision option really outranks the current one, is reserved for the decision maker* who plays a sovereign role in this procedure.

The essence of this procedure is the use of reference point methodology to produce points suspected of outranking the current one. This is based on a basic property of an order-consistent achievement scalarizing function of the type (5):

$$\bar{\mathbf{y}} \in Y_0 \Rightarrow \left\{ \begin{array}{r} \max_{\mathbf{y} \in Y_0} \sigma(\mathbf{y}, \bar{\mathbf{y}}) \geq 0; \\ \hat{\mathbf{y}} = \mathrm{argmax}_{\mathbf{y} \in Y_0} \sigma(\mathbf{y}, \bar{\mathbf{y}}) \geq \bar{\mathbf{y}} \end{array} \right\} \tag{22}$$

Thus, if $\max_{\mathbf{y} \in Y_0} \sigma(\mathbf{y}, \bar{\mathbf{y}}) < 0$, then $\bar{\mathbf{y}} \notin Y_0$. Therefore, in order to test for $\mathbf{y} \, \mathcal{C} \, \hat{\mathbf{y}}$, it is sufficient to test the attainability of the reference points $\bar{\mathbf{y}}^{(j)}$ with component reference levels:

$$\bar{y}_i^{(j)} = \left\{ \begin{array}{ll} \hat{y}_i + \Delta y_{co,i}, & i = j \\ \hat{y}_i - \Delta y_{ni,i} & i \neq j \end{array} \right\}, \, j \in \{1, \dots m\} \tag{23}$$

If the maximum of $\sigma(\mathbf{y}, \bar{\mathbf{y}}^{(j)})$ is nonnegative, we take the point maximizing this function as suspected for component outranking and present it to the decision maker for acceptance. If the maximum is negative, no component outranking point exists for the objective j. In this case, and also in the case when the decision maker does not accept a suggested point, we just check for next $j \in \{1, \dots m\}$; we must thus perform m maximizations of the order-consistent achievement function (5).

Instead of testing for normal and weak outranking, we shall actually test for parametric outranking $\mathcal{P}\mathcal{Q}_{m_1}$ with m_1 decreasing from $m - 1$ to 0. In order to do this, we define sets $I_+^{(j)}, I_-^{(j)}, I_0^{(j)}$ such that:

$$|I_+^{(j)}| - |I_-^{(j)}| = m_1 \tag{24}$$

where (j) denotes one of all possible subdivisions of the set $\{1, \dots, m\}$ into such sets. The number of such possible subdivisions, denoted here by $p(m, m_1)$, can be generated in a combinatorial way. This number denotes all possible partitions of the set $I = \{1, \dots m\}$ into three sets $I_+^{(j)}, I_-^{(j)}, I_0^{(j)}$, including the cases in which the sets $I_-^{(j)}, I_0^{(j)}$ are empty. For each of such partitions we can define a specific reference point (for each m_1, we start the count $j = 1, \dots p(m, m_1)$ anew):

$$\bar{y}_i^{(j)} = \left\{ \begin{array}{ll} \hat{y}_i + \Delta y_{pi,i}, & i \in I_+^{(j)} \\ \hat{y}_i - \Delta y_{av,i} & i \in I_-^{(j)} \\ \hat{y}_i - \Delta y_{ni,i} & i \in I_0^{(j)} \end{array} \right\}, \quad \begin{array}{l} i = 1, \dots m, \\ j = 1, \dots p(m, m_1) \end{array} \tag{25}$$

For each $j = 1, \dots p(m, m_1)$, we have now again to maximize the order-consistent achievement function $\sigma(\mathbf{y}, \bar{\mathbf{y}}^{(j)})$. Using again the basic property of such functions we note that if any of these maxima is greater or equal zero, then there might exist an attainable outcome outranking the point $\hat{\mathbf{y}}^{(k)}$ and such outcome can be chosen as the maximal point of this function. Conversely, if all these maxima are smaller than zero, then the reference points $\bar{\mathbf{y}}^{(j)})$ are not attainable for all $j = 1, \dots p(m, m_1)$ and there are no attainable points outranking the point $\hat{\mathbf{y}}$.

We can turn now to the full procedure of Outranking Trials which can be organized as follows:

1. At a given current outcome point $\hat{\mathbf{y}}^{(k)}$, where (k) denotes iteration number, check whether the decision maker would not like to *redefine the outranking thresholds.* If yes, correct these thresholds.

2. *Check for component outranking:* using reference point methodology, check whether there exist an admissible decision with outcomes $\mathbf{y}$ that component outrank the current point. If yes, present it for acceptance to the decision maker. If $\mathbf{y}$ is accepted, shift current outcome point $\hat{\mathbf{y}}^{(k+1)} = \mathbf{y}$ and start a new iteration. If $\mathbf{y}$ is not accepted, again (for other possible case of different objective components) check whether there exist another admissible decision with outcomes that component outrank the current point and, if yes, present it to the decision maker. Repeat until either a new point is accepted and new iteration started, or all possible cases of component outranking are tested negatively.

3. *Check for normal outranking:* proceed as above, but testing the existence and presenting to the decision maker possible cases of normal outranking by testing for parametric outranking with m_1 decreasing from $m-1$ to m_0, until a new iteration is started or all possible normal outranking cases are tested negatively.

4. *Check for weak outranking:* proceed as above for m_1 decreasing from $m_0 - 1$ to 0, testing the existence and presenting to the decision maker possible cases of weak outranking, until a new iteration is started or all possible weak outranking cases are tested negatively.

5. If all cases were tested negatively (either outranking decisions do not exist or the decision maker does not accept them as outranking), *stop;* the current point is most satisfactory according to the accepted outranking relation.

The use of three different outranking relations simplifies computations in the initial phases of the procedure: we start with checking the strongest relation and proceed to a weaker one first when the results of testing the stronger one are negative. When we come down to the weak outranking, we just test for decisions that might be only suspected of outranking. Thus, if there are too many of such suggestions, the decision maker has the right to stop the procedure at any point (on the understanding that further trials will produce only results that are less strongly outranking than former trials).

In order to accelerate convergence, we might also modify the above procedure of Outranking Trials as follows: when checking for outranking of a given type, points suspected of outranking are presented to the decision maker not separately, but jointly – for example, in the case of component outranking, for all $j \in \{1, \ldots m\}$ such that the maxima of $\sigma(\mathbf{y}, \bar{y}_i^{(j)})$ are nonnegative. The decision maker can either select any of these points or reject them all.

This modification is especially important when testing for normal and weak outranking, because there are many such tests. It can be seen that one iteration of Outranking Trials might be long, particularly if the decision maker wants to check all possible cases of outranking. However, the advantage of Outranking Trials is their assured and finite convergence.

In order to investigate convergence, we shall first note that the decision maker might decline to accept a possibly outranking option because she/he believes that this option does not increase her/his value function sufficiently. Thus, we can use the concept of a value function in order to analyze the convergence of Outranking Trials; we must only assume that there is a finite indifference threshold for a meaningful increase of the value function. Observe that the decision maker, because she/he decides sovereignly about all parameters and details of the procedure, can act consistently with her/his value function and with the indifference threshold, by defining sufficiently small outranking thresholds to produce all options suspected of outranking and accepting only those from such options that result in a meaningful increase of the value function, above the indifference threshold. Thus, we define the following condition of consistence of an outranking with a value function:

Outranking-value consistence. Outranking relations defined by (21) are consistent with a continuous, monotone value function $v(\mathbf{y})$ if there is a finite, positive indifference threshold Δv such that, if $v(\mathbf{y}) - v(\hat{\mathbf{y}}^{(k)}) \geq \Delta v$, then $\mathbf{y}$ outranks $\hat{\mathbf{y}}^{(k)}$ at least in the weak sense.

Assuming outranking-value consistence, the proof of the convergence of Outranking Trials is immediate. If the set Q_0 is compact, the continuous value function can only finitely increase in this set; since each iteration in which we find an outranking point increases this function at least by Δv, the procedure must stop after a finite number of iterations at a final point at which no outranking point can be found. We have shown earlier that if no such point can be found, then no attainable outranking point exists, due to the basic property of order-consistent achievement function used for seeking such points. Because of outranking-value consistence, no point exists which would increase the value function more than Δv when compared to the final point of the procedure.

We note that the use of a value function and of the outranking-value consistence property is necessary to investigate the convergence in classic terms; without such assumptions, we can naturally deduce the convergence in the more general terms of outranking relations.

We shall also note that the concept of convergence of Outranking Trials might be only *local*, since the use of veto thresholds implies a refusal to compare decisions with widely different objective components. However, if the assumed value function is concave and the set Y_0 is convex, local convergence is equivalent to the global one.

While it remains to be tested practically whether the Outranking Trials procedure is acceptable for a typical decision maker, this procedure provides at least an alternative theoretical approach to the issue of convergence of some interactive procedures of multiobjective optimization and decision support. For example, the convergence of Contracted Cone or Light Beam Search procedures might probably be investigated while using similar concepts as applied for Outranking Trials.

6 Conclusions

We have presented in this paper an overview of issues related to the investigation of convergence of interactive procedures of multiobjective optimization and decision support. While this subject has a long tradition, we have learned in many applications of decision support systems that the human decision maker cannot be treated as an automaton. It is the complexity of human behavior that necessitates alternative approaches to procedures of interactive decision support and their convergence.

References

[1] Bogetoft, P., Å. Hallefjord and M. Kok (1988) On the convergence of reference point methods in multiobjective programming. *European Journal of Operations Research* **34**, 56-68.

[2] Brans, J.P. and Ph. Vincke (1985) A preference ranking organization method. *Management Science* **31**, 647-656.

[3] Charnes, A. and W.W. Cooper (1977) Goal programming and multiple objective optimization. *J. Oper. Res. Soc.* **1** 39-54.

[4] Dinkelbach, W. and H. Isermann (1973) On decision making under multiple criteria and under incomplete information. In J.L. Cochrane, M. Zeleny (eds.) *Multiple Criteria Decision Making*. University of South Carolina Press, Columbia SC.

[5] Ermolev, Yu., and R. J-B. Wets (1988) *Numerical Techniques for Stochastic Optimization*. Springer Verlag, Berlin-Heidelberg.

[6] Geoffrion, A.M., J.S. Dyer and A. Feinberg (1972) An Interactive Approach for Multicriterion Optimization, with an Application to the Operation of an Academic Department. *Management Science* **19**, 357-368.

[7] Granat, J and A.P. Wierzbicki (1994) Interactive Specification of DSS User Preferences in Terms of Fuzzy Sets. Working Paper of the International Institute for Applied Systems Analysis, WP-94-29, Laxenburg, Austria.

[8] Granat, J., T. Kręglewski, J. Paczyński and A. Stachurski (1994) IAC-DIDAS-N++ Modular Modeling and Optimization System. Part I: Theoretical Foundations, Part II: Users Guide. Report of the Institute of Automatic Control, Warsaw University of Technology, March 1994, Warsaw, Poland.

[9] Jaszkiewicz, A. and R. Słowiński (1985) The LBS package - a microcomputer implementation of the Light Beam Search method for multi-objective nonlinear mathematical programming. Working Paper of the International Institute for Applied Systems Analysis, WP-94-07, Laxenburg, Austria.

[10] Kahneman, D. and A. Tversky (1982) The Psychology of Preferences. *Scientific American* **246** 160-173.

[11] Kaliszewski, I. (1994) *Quantitative Pareto Analysis by Cone Separation Techniques*. Kluwer Academic Publishers, Dordrecht.

[12] Kok, M. and F.A. Lootsma (1985) Pairwise comparison methods in multiple objective programming. *European Journal of Operational Research* **22**, 44-55.

[13] Korhonen, P. and J. Laakso (1985) A Visual Interactive Method for Solving the Multiple Criteria Problem. *European Journal of Operational Research* **24** 277-287.

[14] Lewandowski, A. and A.P. Wierzbicki (1989), eds. *Aspiration Based Decision Support Systems.* Lecture Notes in Economics and Mathematical Systems **331**, Springer-Verlag, Berlin-Heidelberg.

[15] Lootsma, F.A. (1993) Scale sensitivity in multiplicative AHP and SMART. *Journal of Multi-Criteria Decision Analysis* **2**, 87-110.

[16] Lootsma, F.A., T.W. Athan and P.Y. Papalambros (1985) Controlling the Search for a Compromise Solution in Multi-Objective Optimization. To appear in

[17] Luenberger, D.G. (1973) Introduction to Linear and Nonlinear Programming. Addison-Wesley, Reading, Ma.

[18] Michalevich, M.V. (1986) Stochastic approaches to interactive multicriteria optimization problems. WP-86-10, IIASA, Laxenburg.

[19] Ogryczak, W., K. Studzinski and K. Zorychta (1989) A generalized reference point approach to multiobjective transshipment problem with facility location. In A. Lewandowski and A. Wierzbicki, eds.: *Aspiration Based Decision Support Systems* Lecture Notes in Economics and Mathematical Systems, 331, Springer-Verlag, Berlin-Heidelberg.

[20] Rios, S. (1994) *Decision Theory and Decision Analysis: Trends and Challenges.* Kluwer Academic Publishers, Boston-Dordrecht.

[21] Roy, B. and Ph. Vincke (1981) Multicriteria Analysis: Survey and New Directions. *European Journal of Operational Research* **8**, 207-218.

[22] Saaty, T. (1980) *The Analytical Hierarchy Process.* McGraw-Hill, New York.

[23] Sawaragi, Y., H. Nakayama and T. Tanino (1985) *Theory of Multiobjective Optimization.* Academic Press, New York.

[24] Steuer, R.E. and E.V. Choo (1983) An interactive weighted Tchebycheff procedure for multiple objective programming. *Mathematical Programming* **26**, 326-344.

[25] Steuer, R.E. (1986) *Multiple Criteria Optimization: Theory, Computation, and Application.* J.Wiley & Sons, New York.

[26] Vincke, Ph. (1992) *Multicriteria Decision Aid.* J.Wiley, Chichester-New York.

[27] Wierzbicki, A.P. (1980) The use of reference objectives in multiobjective optimization. In G. Fandel, T. Gal (eds.): *Multiple Criteria Decision Making; Theory and Applications,* Lecture Notes in Economic and Mathematical Systems **177**, 468-486, Springer-Verlag, Berlin-Heidelberg.

[28] Wierzbicki, A.P. (1984) *Models and Sensitivity of Control Systems.* Elsevier-WNT, Amsterdam.

[29] Wierzbicki, A.P. (1986) On the completeness and constructiveness of parametric characterizations to vector optimization problems. *OR-Spektrum,* **8** 73-87.

[30] Wierzbicki, A.P. (1992a) The Role of Intuition and Creativity in Decision Making. WP-92-078, IIASA, Laxenburg.

[31] Wierzbicki, A.P. (1992b) Multiple Criteria Games: Theory and Applications. WP-92-079, IIASA, Laxenburg.

[32] Zeleny, M. (1974) A concept of compromise solutions and the method of the displaced ideal. *Comput. Oper. Res.* Vol **1**, 479-496.

[33] Zeleny, M. (1973) Compromise Programming. In J.L. Cochrane and M. Zeleny (eds.) *Multiple Criteria Decision Making.* University of Carolina Press, Columbia, SC.

[34] Zionts, S. and J. Wallenius (1976) An Interactive Programming Method for Solving the Multicriteria Problem. *Management Science* **22** 653-663.

[35] Zionts, S. and J. Wallenius (1983) An Interactive Multiple Objective Linear Programming Method for a Class of Underlying Nonlinear Utility Functions. *Management Science* **29** 519-529.

Fuzzy Parameters Multiobjective Linear Programming Problem: Solution Analysis

Arenas, M.[1], Bilbao, A.[1],
Rodríguez, M.V.[1], Jiménez, M.[2]

[1]Dto. Matemáticas. Univ.de Oviedo. e-mail:vrodri@.econo.uniovi.es
[2]Dto. Econ. Aplicada I. Univ. del País Vasco. e-mail:eupjilom@se.ehu.es

Abstract. In decision making problems, the estimate of the parameters defining the model is a difficult task. They are usually defined by the Decision Maker in an uncertain way or by means of language statements. Therefore, it is suitable to represent those parameters by fuzzy numbers, defined by their possibility distribution.

The fuzziness and/or inaccuracy of the parameters give rise to a problem whose solution will also be fuzzy as we saw in [1] and is defined by its possibility distribution, being the component of the same a fuzzy number.

In this paper we aim to find a decision vector which approximates as much as possible the fuzzy goals to the fuzzy solutions previously obtained. In order to solve this problem, we shall define an Interval Goal Programming Problem. Our proposal will be illustrated by an example.

Keywords: Multiobjective programming, decision making, fuzzy programming, possibility distribution, fuzzy number.

1 Introduction

In [1] we have obtained a solution $\tilde{z}^*$ of the following problem of linear multiobjective programming with fuzzy parameters:

$$\left. \begin{array}{ll} \text{maximize} & \tilde{Z} = (\tilde{c}_1 x, \tilde{c}_2 x, \ldots, \tilde{c}_k x) \\ \text{subject to} & x \in \mathcal{X}(\tilde{A}, \tilde{b}) = \{x \in I\!\!R^n \ / \ \tilde{A}x \leq \tilde{b}, \ x \geq 0\} \end{array} \right\} \text{(FP-MOLP)}$$

symbols with a 'tilde' on top, like $\tilde{c}_r$, $\tilde{A}$ and $\tilde{b}$ are all fuzzy parameters and we suppose that they are given by fuzzy numbers.

Fuzzy number is defined differently by different authors. The most frequently used definition is that a fuzzy number is a fuzzy set in the real line which is normal and bounded convex. The fuzzy numbers represent the continuous possibility distributions for fuzzy parameters and hence place a restriction on the possible values the variable may assume.

From the definition of a fuzzy number $\tilde{N}$, it is significant to note that the set of level α, N_α, can be represented by the closed interval which depends on the value of α: $N_\alpha = [n_\alpha^L, n_\alpha^R]$.

The parameters acting in the FP-MOLP are uncertain and/or inaccurate, so the solution will be. The proposed solving method in [1] gives fuzzy solutions in the objectives space defined by their distribution of possibilities.

Following the classical Bellman and Zadeh pattern for the fuzzy decision theory, we defined the possibility distribution $\Pi(\tilde{Z}^* = z^*)$ of the fuzzy set $\tilde{Z}^*$ that permits us to define a Pareto Optimal Solution (POS) of the fuzzy multiobjective problem:

$$\Pi(\tilde{z}^* = z^*) = \sup_{A,b,C} \{\Pi(A, b, C) / z^* \text{ is a POS of the MOLP}(A, b, C)\} \quad (1)$$

where MOLP(A, b, C) is the crisp linear multiobjective problem associated with the parameters A, b, C:

$$\left.\begin{array}{ll} \max & z = (c_1 x, c_2 x, \ldots, c_k x) \\ \text{s.t.} & x \in \mathcal{X}(A, b) = \{x \in \mathbb{R}^n / Ax \leq b, \ x \geq 0\} \end{array}\right\} \text{MOLP}(A, b, C)$$

The fuzzy solution of FP-MOLP in the set of objectives verified that every component was a fuzzy number. The method proposed in [1] also allowed us to have a fuzzy solution in the set of the decision vectors that, in general, we cannot state that it is a fuzzy number component by component. If the Decision Maker wants a more precise information with respect to the decision vector, then we can pose and solve a new problem. This is the objective of our work. We shall develop a model sustained on the initial solution, $\tilde{z}^*$ and it is based on Goal Programming. We want to obtain the values of the decision variables which determine, as accurately as possible the fuzzy solution $\tilde{z}^*$. Therefore, we try to solve the following problem:

Find a $x \in \mathcal{X}(\tilde{A}, \tilde{b}) = \{x \in \mathbb{R}^n / \tilde{A}x \leq \tilde{b}, \ x \geq 0\}$ such that:

$$\tilde{C}_r x_j \approx \tilde{z}_r^*, \forall r$$

where the relation $\approx$ is a relation between fuzzy numbers which will be read as "approximately equal". This is not any more an optimization problem as there is no objective to maximize or minimize but we have to find a vector which "approximates to" $\tilde{z}^*$ which acts as a target.

The relation $\approx$ will be defined sustained on the expected intervals [1] of fuzzy numbers $\tilde{C}_r x_j$ and $\tilde{z}_r^*, \forall r$. For each r we try that the expected

---

[1] Heilpern [2] defines the expected interval of a fuzzy number $\tilde{N}$, which will be noted $EI(\tilde{N})$, as:

$$EI(\tilde{N}) = \left[\int_0^1 n_\alpha^L \, d\alpha, \int_0^1 n_\alpha^R \, d\alpha\right]$$

where n_α^L and n_α^R are the upper and lower ends, respectively, of the α-cut of the fuzzy number.

interval of the fuzzy number which defines the r-th objective function is the nearest possible to the expected interval of the r-th component of the solution. We have to solve the following goal programming problem:

Find a $x \in \mathcal{X}(\tilde{A}, \tilde{b})$ such that:

$$EI \left(\sum_{j=1}^{n} \tilde{c}_{rj} x_j \right) \approx EI(\tilde{z}_r^*), \quad r = 1, \ldots, k \tag{2}$$

The uncertain and/or imprecise nature of the technological matrix and of the vector of resources which define the set of constraints of the model, implies that the feasibility of the decision vector, whatever the value of the restriction parameters within their respective supports, can only be guaranteed by considering the minimum feasible set corresponding to level $\alpha = 0$, i.e: $\mathcal{F} = \left\{ x \in I\!R^n \ / \ A_0^R x \leq b_0^L, \ x \geq 0 \right\}$ which is the intersection of all feasible sets possible. However, if the Decision Maker was ready to run some risks with respect to feasibility, he can establish the level of tolerance he is ready to assume in every constraint through a parameter $\beta_i \in [0,1]$, which may vary from some restrictions to the others. With the vector $\beta = (\beta_1, \ldots, \beta_m)$, the new feasible set with which we shall be working will be:

$$(\mathcal{F}_\beta) \begin{cases} \left[A_i^R - \beta_i (A_i^R - A_i^L) \right] x \leq b_i^L + \beta_i (b_i^R - b_i^L), \ i \in I \\ x \geq 0 \end{cases}$$

where $A_{i_{\alpha=0}} = [A_i^L, A_i^R]$ and $b_{i_{\alpha=0}} = [b_i^L, b_i^R]$, $\forall i \in I$ are the α-cut, $\alpha = 0$, of the coefficients of the technological matrix and of the vector of resources which define the set of constraints of the model.

Therefore, the problem to be analyzed will be to find a $x \in \mathcal{F}_\beta$ such that:

$$EI \left(\sum_{j=1}^{n} \tilde{c}_{rj} x_j \right) \approx EI(\tilde{z}_r^*), \quad r = 1, \ldots, k \tag{3}$$

We shall take as preferred the least wide interval, i.e. the least imprecise, and the relation $\approx$ in (3) will be interpreted as "approximately equal and less imprecise". In order to solve the model we introduce the following definitions and results:

Definition 1.1 *Given two intervals $A = [a^L, a^R]$ and $B = [b^L, b^R]$ of the real straight line, such that the width of the first is greater than the width of the second, we shall define the operation* **difference by extremes**, *which we shall denote by $\ominus$, as the interval: $A \ominus B = [a^L - b^L, a^R - b^R]$*

Definition 1.2 *Given an interval of the real straight line $A = [a^L, a^R]$ we shall define the **absolute value** of such interval in the following way :*

$$| [a^L, a^R] | = \begin{cases} [a^L, a^R] & \text{if} \quad a^L \geq 0 \\ [0, \max(-a^L, a^R)] & \text{if} \quad a^L < 0 < a^R \\ [-a^R, -a^L] & \text{if} \quad a^R \leq 0 \end{cases} \qquad (4)$$

Definition 1.3 *Given $A = [a^L, a^R]$ and $B = [b^L, b^R]$ with $b^R - b^L \leq a^R - a^L$, we shall define the **distance** between both intervals as a new interval obtained as the absolute value of its difference by extremes:* $D_{(A,B)} = |A \ominus B|$

From the equality between intervals and the definition of distance we have just formulated we have:

Theorem 1.1 *Given $A = [a^L, a^R]$ and $B = [b^L, b^R]$ intervals of the real straight line, it verifies:* $A = B \iff D_{(A,B)} = 0$

From this theorem the problem of the relation $\approx$ between the interval which defines the goals of the problem (3) can be transformed into one of minimizing the distance between them. By applying the definition of distance to the expected intervals, we have that:

$$D_r = \left| EI(\tilde{z}_r^*) \ominus EI\left(\sum_{j=1}^{n} \tilde{c}_{rj} x_j \right) \right| \qquad \forall r$$

and then

$$EI\left(\sum_{j=1}^{n} \tilde{c}_{rj} x_j \right) = EI(\tilde{z}_r^*) \qquad \iff \qquad D_r = 0 \quad \forall r$$

In order to work with the Goal Programming approach we are going to introduce in the model the positive and negative deviation variables for each goal. These variables quantify in terms of the extremes of the intervals how far is the solution of (3) from aspiration levels set by the Decision Maker, $EI(\tilde{z}_r^*)$, the negative deviation variables quantify the lack of achievement of an objective with respect to its level of aspiration, while the positive ones do the same with respect to its excess of achievement. Then, we can write:

$$EI \left(\sum_{j=1}^{n} \tilde{c}_{rj} x_j \right)^L + n_r^L - p_r^L = EI(\tilde{z}_r^*)^L,$$

$$\forall \, r \qquad (5)$$

$$EI \left(\sum_{j=1}^{n} \tilde{c}_{rj} x_j \right)^R + n_r^R - p_r^R = EI(\tilde{z}_r^*)^R,$$

so substituting in the expression of the total deviation for this goal, we have:

$$D_r = \left| \, [n_r^L - p_r^L, n_r^R - p_r^R] \, \right| \qquad (6)$$

and taking into account the definition of absolute value of an interval and the properties of deviation variables, we have the following proposition:

Proposition 1.2 $D_r = \left[D_r^L, D_r^R \right] = \left[n_r^L + p_r^R, \, \max \left(p_r^L, n_r^R \right) \right]$

Once the deviation variables have been introduced, the next step in the solving of a Goal Programming model will consist of determining the non-desired deviation variables. The following propositions will indicate the non-desired variables.

Proposition 1.3 *If* $D_r^R = 0$, *then goal interval coincides with the level of aspiration interval.*

Therefore, in order to obtain the equality aimed at, it would be sufficient to minimize the upper extreme of every distance interval. The process of minimizing the extremes of the distance intervals can be approached in different ways, giving rise every one of them to a variant of Goal Programming; the difference between them lies in the achievement function, remaining identical the working constraints. We shall use the lexicographic method.

In the lexicographic approach, the decisor sorts the objectives by grouping them in levels of importance called *priority levels*, so that the objectives corresponding to j-th level have absolute priority with respect to those of the (j+1)-th group. The introduction of such levels implies to state the Decision Maker's preferences but no to quantify them; it is a reorganization of the objectives so that those which cannot be waived go in the first places of a lexicographic order.

2 Solving of the problem

If $P_1, \ldots, P_l$ $(l \leq k)$ are the established priority levels and the j-th goal is in level P_i then, it will be denoted by $j \in P_i$. If several objectives share the same level of priority, their importance will be described by means of a weighting coefficient.

In order to solve the problem of homogeneity of goals sharing priority it is convenient to normalize the weights before being introduced in the achievement function. The normalization procedure to be used will consist of dividing every weighting coefficient by the expected value [2] of each aspiration level. So, we have: $\omega_r' = \dfrac{\omega_r}{|EV(\tilde{z}_r)|}$, $\forall r$. On the other hand, as we are trying to reduce uncertainty, it has to be verified: $n_r^L - p_r^L \leq n_r^R - p_r^R$, $\forall r$.

Also, taking into account the definition of the positive and negative deviation variables, it verifies: $n_r^L \times p_r^L = n_r^R \times p_r^R = 0$, $\forall r$.

The achievement function to be used is defined in terms of the lexicographic minimum of a ordered set of distances. As distances are expressed in function of the positive and negative deviation variables, they will be represented as:

$$\text{Lex min} \quad a(n, p) = [a_1(n, p), a_2(n, p), \ldots, a_l(n, p)] \tag{7}$$

Every component of the achievement function is a linear function of positive coefficients of the upper extremes of the distances. So the r-th component, $a_r(n, p)$, will be the following linear function:

$$a_r(n, p) = \sum_{s \in P_r} \omega_s' \max \left(p_s^L, n_s^R \right)$$

Let us assume that we want to achieve several goals simultaneously, i.e., we find that there are more than one objective in a certain level of priority. If we denote by $v_s = \max \left(p_s^L, n_s^R \right)$, we shall have to minimize the weighting addition of the v_s then the problem will be:

---

[2]Heilpern [2] defines the expected value of a fuzzy number $\tilde{N}$, which will be noted $EV(\tilde{N})$, as:

$$EV(\tilde{N}) = \frac{\int_0^1 n_\alpha^L \, d\alpha + \int_0^1 n_\alpha^R \, d\alpha}{2}$$

$$\min \sum_{s \in P_r} \omega'_s v_s$$

subject to:

$$p_s^L \leq v_s, \ n_s^R \leq v_s, \qquad\qquad s \in P_r$$

$$EI \left(\sum_{j=1}^{n} \tilde{c}_{sj} x_j \right)^L + n_s^L - p_s^L = EI(\tilde{z}_s^*)^L, \quad s \in P_r,$$

$$EI \left(\sum_{j=1}^{n} \tilde{c}_{sj} x_j \right)^R + n_s^R - p_s^R = EI(\tilde{z}_s^*)^R, \quad s \in P_r,$$

$$x \in \mathcal{F}_\beta$$

$$n_s^L - p_s^L \leq n_s^R - p_s^R, \qquad\qquad s \in P_r$$

$$n_s^L \times p_s^L = n_s^R \times p_s^R = 0, \qquad\qquad s \in P_r$$

$$n_s^L \geq 0, \ p_s^L \geq 0, \ n_s^R \geq 0, \ p_s^R \geq 0, \qquad s \in P_r$$

Let us assume that the decisor introduces $l \leq k$ levels of priority. The algorithm proposed in the following example to solve this problem through lexicographic goal programming is the linear sequential solving method. This algorithm has been implemented using the Matlab Library for Windows and its Toolbox of Optimization.

Example: Let us consider the linear multiobjective program with fuzzy parameters of [1]:

$$\max \tilde{Z} = (\tilde{c}_1 x, \tilde{c}_2 x, \tilde{c}_3 x)$$

subject to

$$\left. \begin{aligned} &\tilde{a}_{11} x_1 + 17 x_2 \leq 1400 \\ &3 x_1 + 9 x_2 + \tilde{a}_{23} x_3 \leq 1000 \\ &10 x_1 + \tilde{a}_{32} x_2 + 15 x_3 \leq 1750 \\ &\tilde{a}_{41} x_1 + 16 x_3 \leq 1325 \\ &\tilde{a}_{51} x_2 + 7 x_3 \leq 900 \\ &9.5 x_1 + \tilde{a}_{62} x_2 + 4 x_3 \leq 1075 \\ &x_j \geq 0, : j = 1, \ldots, n \end{aligned} \right\} \equiv x \in \mathcal{X}(\tilde{A}, \tilde{b})$$

where $\tilde{c}_1 x = \tilde{c}_{11} x_1 + 100 x_2 + 17.5 x_3$, $\tilde{c}_2 x = 92 x_1 + \tilde{c}_{22} x_2 + 50 x_3$ and $\tilde{c}_3 = \tilde{c}_{31} x_1 + 100 x_2 + 75 x_3$.

We assume that the fuzzy coefficients are characterized by triangular possibility distributions [3] as it is shown in the following table:

---

[3] In practice, input fuzzy data (of $\tilde{C}$, $\tilde{A}$, and $\tilde{b}$) are often assumed to be triangular or trapezoid functions. A triangular fuzzy number can be denoted as $\tilde{N} = (n_1, n_2, n_3)$ where n_2 is the central value, $n_2 - n_1$ is the left spreads, and $n_3 - n_2$ is the right spreads

	$\tilde{c}_{11}$	$\tilde{c}_{22}$	$\tilde{c}_{31}$	$\tilde{a}_{11}$	$\tilde{a}_{23}$	$\tilde{a}_{32}$	$\tilde{a}_{41}$	$\tilde{a}_{51}$	$\tilde{a}_{62}$
n_1	40	70	10	6	3	7	4	7	3.5
n_2	50	75	25	12	8	13	6	12	9.5
n_3	80	90	70	14	10	15	8	19	11.5

whose fuzzy optimal solution, $\tilde{z}^*$, obtained in [1] is a triangular fuzzy number component by component:

$$\tilde{z}_1^* = (6037, 7984, 12325)$$
$$\tilde{z}_2^* = (9784, 10057, 13984)$$
$$\tilde{z}_3^* = (7164, 9356, 13880)$$

We are going to find a decision vector which allow us to approximate as much as possible the fuzzy objectives to such a solution. The expected intervals of the solution, which will be the level of aspiration of this new problem, are:

$$EI(\tilde{Z}_1) = [6917.5, 9891]$$
$$EI(\tilde{Z}_2) = [9889.8, 11881]$$
$$EI(\tilde{Z}_3) = [8180.3, 11345]$$

and the expected intervals of the objectives:

$$EI(\tilde{c}_1 x) = [45x_1 + 100x_2 + 17.5x_3, 65x_1 + 100x_2 + 17.5x_3]$$
$$EI(\tilde{c}_2 x) = [92x_1 + 72.5x_2 + 50x_3, 92x_1 + 82.5x_2 + 50x_3]$$
$$EI(\tilde{c}_3 x) = [17.5x_1 + 100x_2 + 75x_3, 47.5x_1 + 100x_2 + 75x_3]$$

If the Decision Maker admits a risk vector β then the problem to be solved will be:

Find $x = (x_1, x_2, x_3) \in R^3$ such that:

$$EI(\tilde{c}_r x) \approx EI(\tilde{Z}_r)$$

$$\mathcal{F}_\beta = \begin{cases} (14 - \beta 8)x_1 + 17x_2 \leq 1400 \\ 3x_1 + 9x_2 + (10 - \beta 7)x_3 \leq 1000 \\ 10x_1 + (15 - \beta 8)x_2 + 15x_3 \leq 1750 \\ (8 - \beta 4)x_1 + 16x_3 \leq 1325 \\ (19 - \beta 12)x_2 + 7x_3 \leq 900 \\ 9.5x_1 + (11.5 - \beta 8)x_2 + 4x_3 \leq 1075 \\ x \geq 0 \end{cases}$$

by introducing the deviation variables in the problem and assuming that all objectives are considered at the same level of priority, we have that:

$$\min v_1 + v_2 + v_3$$
subject to:
$$v_s \geq p_s^L, \ v_s \geq n_s^R, \ s = 1,2,3$$
$$45x_1 + 100x_2 + 17.5x_3 + n_1^L - P_1^L = 6917.5$$
$$65x_1 + 100x_2 + 17.5x_3 + n_1^R - p_1^R = 9891$$
$$92x_1 + 72.5x_2 + 50x_3 + n_2^L - p_2^L = 9889.8$$
$$92x_1 + 82.5x_2 + 50x_3 + n_2^R - p_2^R = 11881$$
$$17.5x_1 + 100x_2 + 75x_3 + n_3^L - p_3^L = 8180.3$$
$$47.5x_1 + 100x_2 + 75x_3 + n_3^R - p_3^R = 11345$$

$$\mathcal{F}_\beta = \begin{cases} (14 - \beta 8)x_1 + 17x_2 \leq 1400 \\ 3x_1 + 9x_2 + (10 - \beta 7)x_3 \leq 1000 \\ 10x_1 + (15 - \beta 8)x_2 + 15x_3 \leq 1750 \\ (8 - \beta 4)x_1 + 16x_3 \leq 1325 \\ (19 - \beta 12)x_2 + 7x_3 \leq 900 \\ 9.5x_1 + (11.5 - \beta 8)x_2 + 4x_3 \leq 1075 \\ x \geq 0 \end{cases}$$

$$n_s^L \geq 0, \ p_s^L \geq 0, \ n_s^R \geq 0, \ p_s^R \geq 0, \ s = 1,2,3$$
$$n_s^L - p_s^L \leq n_s^R - p_s^R, \ s = 1,2,3$$
$$n_s^L \times p_s^L = n_s^R \times p_s^R = 0, \ s = 1,2,3$$

The solution $\tilde{z}^*$ and the solutions obtained for different values of β are:

	$\tilde{z}_1^*$	$\tilde{z}_2^*$	$\tilde{z}_3^*$
	(6061.43,7983.86,12324.702)	(9762.96,10057.32,13984.42)	(7196.22,9355.89,13880.21)
β	$\bar{c}_1 x$	$\bar{c}_2 x$	$\bar{c}_3 x$
0	(6061,6632,8343)	(9763,9909,10346)	(7196,8052,10619)
0.5	(7564,8124,9805)	(10544,10772,11454)	(8415,9256,11778)
1	(7576,8128,9784)	(10542,10771,11458)	(8521,9349,11832)

If we assume that the Decision Maker considers two priority levels, so that the first priority consists of the objectives second and third and the first is placed in the second priority, for $\beta_i = 0.5$, the result is:

$$\bar{c}_1 x = (7503, 8070, 9771)$$
$$\bar{c}_2 x = (10633, 10854, 11518)$$
$$\bar{c}_3 x = (8472, 9322, 11874)$$

3 Conclusions

We have found a decision vector which would allow us to approximate the goal functions with fuzzy coefficients of FP-MOLP to the fuzzy solution $\tilde{z}^*$ obtained in [1] for such program.

The described solving method is based on the Decision Maker's preferences and on the goal programming technics, being the expected intervals of the fuzzy solution the aspiration levels of the goal programming problem.

Taking into account that the α-cuts of fuzzy number are closed intervals of the real straight line, we might have developed other alternatives based on a finite set of α-cuts or have just worked with the expected intervals, only with the nucleus, with the support and the nucleus,, etc... The more information we introduce in the model, the greater will be the number of variables that we can work with, which obviously increases the complexity of the problem.

References

[1] Arenas Parra, M.; Bilbao Terol, A.; Jiménez, M.; Rodrí guez Uría. M.V., "A Fuzzy Approach to a Fuzzy Linear Goal Programming Problem". Lectures Notes in Economics and Mathematical Systems 448: Multiple Criteria Decision Making. pp: 255-264. Springer, 1997.

[2] HEILPERN, S. (1992): "The expected value of a fuzzy number", *Fuzzy Sets and Systems 47*, 81-86.

[3] IGNIZIO, J.P.(1976): *Goal Programming and Extensions* , Lexington Books.

[4] JIMÉNEZ M. (1994): *Modelos matemáticos aplicados a la toma de decisiones financieras en condiciones de incertidumbre* . Tesis Doctoral, Universidad del País Vasco, Bilbao.

The Static Approach To Quadratic Dynamic Goal Programming

Francisco Ruiz, Rafael Caballero, Trinidad Gómez, Mercedes González and Lourdes Rey.
Department of Applied Economics (Mathematics). University of Málaga.
Campus El Ejido s/n. 29071-Málaga. Spain.

Keywords: Goal Programming. Dynamic Optimization. Dynamic Target Values.

Abstract. This communication reports the development of a Goal Programming algorithm to solve quadratic dynamic goal programming models, from the static point of view. This approach is based on two main features:
a) The target values are also considered as functions of time, that is, there is a target value on the accumulated value of the objective functional for each period of time within the planning period. This allows the decision maker control the behavior of the functionals along the whole planning period, rather than only their final values.
b) The dynamic problem can be turned into a static one, where the decision variables are a vector formed by the values of the control variables in each period of time. This makes possible the use of a nonlinear Goal Programming package to solve the problem.
The algorithm has been implemented on a VAX computer, in FORTRAN language and with the aid of the NAG subroutine library. Results on some test problems are also reported.

1.- Introduction

Although some aspects of Dynamic Multiobjective Programming were originally treated in papers like Zadeh (1963), Nelson (1964), Chang (1966) and Da Cunha and Polak (1966a, 1966b), a further development of these ideas cannot be found until Lewandowski and Wierzbicki (1989), Wierzbicki (1991), Salukvadze (1979) and Li and Haimes (1989,1990). Nevertheless, the Dynamic Goal Programming problem was not taken in deep consideration until the works by Levary (1984), Caballero and González (1993), Trzaskalik (1996) and Caballero et al. (1996), where the concept of dynamic target values can be pointed out among the generalizations of the elements of the static problems.

In this paper, a linear quadratic dynamic Goal Programming model will be considered using the discrete time variable approach. The election of such a model is motivated by the fact that, among all convex programming models, the autonomous linear quadratic one is, after the linear model, the most commonly used one in the field of the economic modeling.

After stating the lexicographic Goal Programming problem in the next section, the problem will be turned into a quadratic static one in the third section, an algorithm for its resolution will be describes in the fourth section, and, finally, a result analysis is carried out in the fifth section.

2.- Statement of the problem

It will be assumed that the original problem is a dynamic multiple objective one, where the objective functionals are defined by linear quadratic functions of both the state ($\mathbf{x} \in R^n$) and control ($\mathbf{u} \in R^m$) variables, the time variables do not appear explicitly, and the dynamic system and the constraints are linear, that is,

$$\text{Min } J(u) = (J_1(u), J_2(u), ..., J_p(u))$$

where

$$J_i(u) = \sum_{k=0}^{N-1} \left\{ \frac{1}{2} \left[x(k)^T Q_i(k) x(k) + u(k)^T R_i(k) u(k) \right] + q_i(k)^T x(k) + r_i(k)^T u(k) \right\}$$
$$+ \frac{1}{2} x(N)^T S_i x(N) + s_i^T x(N)$$

$s.t.$ $x(k+1) = A(k)x(k) + B(k)u(k)$
$$x(0) = x_0$$
$$C(k)x(k) + D(k)u(k) \le b(k)$$
$$L \le u(k) \le U \qquad k = 0,..., N-1$$

where, for each i, the matrices $Q_i(k)$ and S_i are $n \times n$, and, at least, positive semidefinite, $R_i(k)$ is $m \times m$ and positive definite, and $A(k)$ is $n \times n$, $B(k)$ ix $n \times m$, $C(k)$ is $s \times n$ and $D(k)$ is $s \times m$.

Once the original problem has been formulated, our aim will be finding satisfying solutions through the use of Lexicographic Goal Programming. To this end, it is necessary to assign target values to each objective, and to group these goals in preference levels. The Lexicographic scheme has been chosen due to the fact that, from a strictly algorithmical point of view, Weighted Goal Programming is a particular case of Lexicographic Goal Programming, for the former can be regarded as a lexicographic problem with just one priority level. So, our implementation lets us treat Weighted Goal Programming problems as well.

With respect to the target values, it seems logical that, under a dynamic scheme, these target values are functions of time themselves (Caballero et al., 1996, Trzaskalik, 1996) and, therefore, in the discrete case, there is a target value for each period of time, that is $\alpha_i = (\alpha_i(0), \alpha_i(1), ..., \alpha_i(N-1))$ ($i = 1, ..., p$). These target values are imposed on the accumulated value of the corresponding objective functional up to the given period of time. This way, the decision maker can control the behavior of the objective functionals along the whole planning period, and not only their final value.

With respect to the preference levels among the goals, it will be assumed that several goals can share the same level, and thus, r priority levels will be defined. If the i-th goal belongs to the q-th priority level, this will be denoted by $i \in N_q$, and the decision maker can also, if he wishes, assign relative weights μ_i to the goals that share the same level.

Making use of this information, our aim will be to minimize the non desired deviations between the objective functionals and their target values. As it is assumed that the original problem was a minimization one, only the positive deviation variables, π, will have to be minimized. Thus, the problem corresponding to the q-th priority level, $\mathbf{P_q}$ takes the following form:

$$Min \quad R_q(\pi) = \sum_{i \in N_q} \mu_i \sum_{k=0}^{N-1} \left[\pi_i(k) \right]^{\beta}$$

$s.t. \quad x(k+1) = A(k)x(k) + B(k)u(k)$

$\qquad x(0) = x_0$

$\qquad C(k)x(k) + D(k)u(k) \le b(k)$

$\qquad L \le u(k) \le U$

$\qquad J_i(u,k) - \pi_i(k) \le \alpha_i(k) \qquad\qquad i \in N_q$

$\qquad \pi_i(k) \ge 0 \qquad\qquad\qquad\qquad i \in N_q$

$\qquad J_j(u,k) \le \alpha_j(k) \qquad\qquad j \in \left\{ N_1, ..., N_{q-1} \right\}$

for all $\quad k = 0,1,...,N-1$.

and where

$$J_i(u,k) = \sum_{\kappa=0}^{k} \left\{ \frac{1}{2} \left[x(\kappa)^T Q_i(\kappa)x(\kappa) + u(\kappa)^T R_i(\kappa)u(\kappa) \right] + q_i(\kappa)^T x(\kappa) + r_i(\kappa)^T u(\kappa) \right\} + $$
$$+ \frac{1}{2} x(k+1)^T S_i x(k+1) + s_i^T x(k+1)$$

Note that a parameter β has been used in the objective function of problem $\mathbf{P_q}$. As our problem is, in general, nonlinear, this parameter can, in some cases, accelerate the convergence of the procedure. Later on, in section 4, the use of this parameter β will be treated in more depth.

Although there are just r priority levels, r+1 levels will be taken into account in the algorithm, that is, one more will be added, called level 0, which corresponds to the determination of at least one feasible control for the original problem.

3.- Transformation into a static problem

The resolution of these problems could be carried out using techniques of dynamic optimization, but this could result into some practical difficulties as the number of variables and constraints increases. Fortunately, thanks to the especial structure of the problem, it is not necessary to solve the dynamic problems directly, because it is possible to transform them into static problems of the same kind, which only have the control variables as decision variables:

Theorem 1.
Each dynamic problem P_q can be explicitly transformed into a linear quadratic static problem.

Proof.
In order to prove this theorem, every element of the dynamic problem has to be turned into a static expression. To this end, it has to be taken into account that the difference equation system of our problem can be solved through the transition matrix as follows:

$$x(\kappa) = \Phi(\kappa,0)x_0 + \sum_{l=0}^{\kappa-1} \Phi(\kappa,l+1)B(l)u(l)$$

where

$$\Phi(\kappa,l) = A(\kappa-1)A(\kappa-2)....A(l)$$
$$\Phi(\kappa,\kappa) = I$$

This way, if the state variables are substituted by these expressions in each of the objective functionals and constraints, then they will all be expressed as functions of the mN-dimensional vector $\tilde{u}(N) = (u(0)^T, u(1)^T, \dots, u(N-1)^T)^T$. Let us now derive the analytic expression of each of the components of the original problem . For the linear constraints, we have

$$C(k)\left[\Phi(k,0)x_0 + \sum_{l=0}^{k-1}\Phi(k,l+1)B(l)u(l)\right] + D(k)u(k) \leq b(k) \quad k = 0,1,\dots,N-1$$

and then they can be expressed as $A\tilde{u}(N) \leq b$ where A is a matrix defined by blocks as follows :

$$
A = \begin{pmatrix}
D(0) & 0 & 0 & \cdots & 0 \\
C(1)B(0) & D(1) & 0 & \cdots & 0 \\
C(2)\Phi(2,1)B(0) & C(2)B(1) & D(2) & \cdots & 0 \\
\vdots & \vdots & \vdots & \ddots & \vdots \\
C(N-1)\Phi(N-1,1)B(0) & C(N-1)\Phi(N-1,2)B(1) & C(N-1)\Phi(N-1,3)B(2) & \cdots & D(N-1)
\end{pmatrix}
$$

and the right hand side is a vector, also defined by blocks :

$$b = \begin{pmatrix} b(0) - C(0)x_0 \\ \hline b(1) - C(1)\Phi(1,0)x_0 \\ \hline b(2) - C(2)\Phi(2,0)x_0 \\ \hline \vdots \\ \hline b(N-1) - C(N-1)\Phi(N-1,0)x_0 \end{pmatrix}$$

With respect to the goals, after carrying out the substitution, those corresponding to the current priority level N_q, will be expressed as:

$$\frac{1}{2}\tilde{u}(N)^T \mathbb{H}_{ik}\tilde{u}(N) + h_{ik}^T \tilde{u}(N) - \pi_i(k) \le \alpha_i(k) - \frac{1}{2}x_0^T Z_{ik}x_0 - z_{ik}^T x_0 \qquad i \in N_q$$

for every $k = 0,1,...,N\text{-}1$ (that is, there is one goal per period) and those corresponding to the subsequent priority levels take the form :

$$\frac{1}{2}\tilde{u}(N)^T \mathbb{H}_{jk}\tilde{u}(N) + h_{jk}^T \tilde{u}(N) \le \alpha_i(k) - \frac{1}{2}x_0^T Z_{jk}x_0 - z_{jk}^T x_0 \qquad j \in \{N_1,...,N_{q-1}\}$$

where, in both cases

$$\mathbb{H}_{ik} = \begin{pmatrix} H_{00} & \cdots & H_{0,k} & 0 & \cdots & 0 \\ \hline \vdots & \ddots & \vdots & \vdots & \ddots & \vdots \\ \hline H_{0,k}^T & \cdots & H_{k,k} & 0 & \cdots & 0 \\ \hline 0 & \cdots & 0 & 0 & \cdots & 0 \\ \hline \vdots & \ddots & \vdots & \vdots & \ddots & \vdots \\ \hline 0 & \cdots & 0 & 0 & \cdots & 0 \end{pmatrix}$$

and where $H_{\sigma,\tau}$ y is given by :

$$H_{\sigma\sigma} = R_i(\sigma) + \sum_{l=\sigma+1}^{k} \left[B(\sigma)^T \Phi(l,\sigma+1)^T Q_i(l)\Phi(l,\sigma+1)B(\sigma) \right] +$$

$$+ \sum_{l=\sigma+1}^{k+1} \left[B(\sigma)^T \Phi(l,\sigma+1)^T S_i \Phi(l,\sigma+1)B(\sigma) \right] \qquad for \quad \sigma = 0,1,...,k.$$

and the non-diagonal blocks :

$$H_{\sigma,\tau} = \sum_{l=\tau+1}^{k} \left[B(\sigma)^T \Phi(l,\sigma+1)^T Q_i(l)\Phi(l,\tau+1)B(\tau) \right] +$$

$$+ \sum_{l=\tau+1}^{k+1} \left[B(\sigma)^T \Phi(l,\sigma+1)^T S_i \Phi(l,\tau+1)B(\tau) \right] \qquad para \; \sigma < \tau$$

The linear terms of the quadratic functions are vectors of the form :

$$h_{i,k} = (h_0, h_1, ..., h_{N-1})$$

where each component h_σ takes the form :

$$h_\sigma = r_i^T(\sigma) + \sum_{l=\sigma+1}^{k} \left[x_0^T \Phi(l,0)^T Q_i(l) \Phi(l,\sigma+1) B(\sigma) + q_i(l)^T \Phi(l,\sigma+1) B(\sigma) \right] +$$

$$+ \sum_{l=\sigma+1}^{k+1} \left[x_0^T \Phi(l,0)^T S_i \Phi(l,\sigma+1) B(\sigma) + s_i^T \Phi(l,\sigma+1) B(\sigma) \right]$$

Finally, for the right hand side, we have:

$$\mathbb{Z}_{ik} = \sum_{l=0}^{k} \left[\Phi(l,0)^T Q_i(l) \Phi(l,0) \right] + \sum_{l=1}^{k+1} \left[\Phi(l,0)^T S_i \Phi(l,0) \right]$$

and

$$z_{ik} = \sum_{l=0}^{k} \left[q_i(l)^T \Phi(l,0) \right] + \sum_{l=1}^{k+1} \left[s_i^T \Phi(l,0) \right].$$

This way, the theorem is proved, for every expression of the original problem has been transformed into a explicit static function of the vector $\tilde{u}(N)$.

Thus, after the substitution, we are facing static linear-quadratic problems that can be solved using single objective non linear programming techniques.

4.- Resolution algorithm

The general scheme of the lexicographic goal programming algorithm is displayed in Figure 1. Let us point out some of its more important features :

1) Data Introduction.
All the data corresponding to a dynamic quadratic multiple objective problem can be introduced through a data file that includes : the dimensions and parameters of the problem, the data of the linear dynamic system, the constraints and the objective functionals.

With respect to the target values, they take the form of a bound in every period k for each objective functional, which can be introduced in running time through a double entry matrix $NA(i,k)$, where i corresponds to the i-th functional, and k corresponds to the period of time.

As for the priority levels, they are also introduced in running time. To this end, the user first gives the number r of priority levels. After that, the so called weighting matrix. This is a double entry matrix where the user has to introduce the weight assigned to each functional in each priority level.

Finally, let us observe that the algorithm calculates the payoff matrix and shows it to the user before he gives the target values, so that he has this additional information about the structure of the problem that can be very useful when giving such target values.

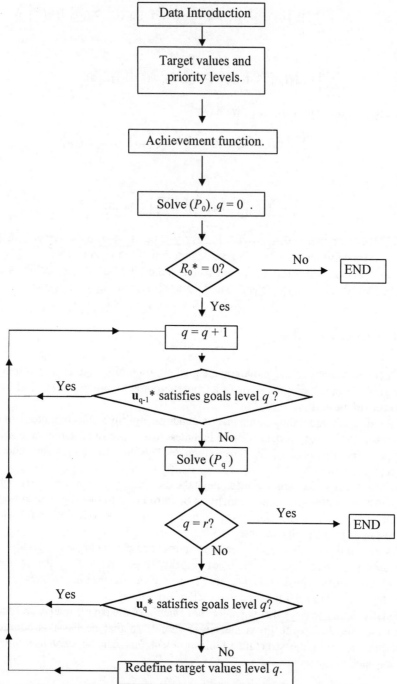

Figure 1. General scheme of the lexicographic Goal Programming Algorithm.

2) Options for the problems.

In all the intermediate problems that appear in this goal programming algorithm, the sum of some power (which will be denoted as β) of the deviation variables of the model is minimized, in such a way that the resulting functional is convex. Such value of :

a) For the level 0 problem, (feasibility level), the user can choose between minimizing a constant functional in the feasible set ($\beta = 0$) or minimizing the β powers ($\beta > 0$) of the accumulated sum of the deviation variables corresponding to the hard constraints of the problem.

b) For the intermediate problems corresponding to the subsequent priority levels, the power β, ($\beta > 0$), of the deviation variables must be introduced.

In the static case, the authors' experience in the resolution of high dimensional nonlinear problems (**Ruiz**, 1994) lets us advise to choose the constant option for level 0, and $\beta = 2$ for the subsequent levels (provided that a quadratic sequential algorithm is used to solve the nonlinear problems).

3) Resolution of the feasibility problem (level 0).

The following step of the algorithm consists in solving the problem corresponding to level 0, denoted as $\mathbf{P_0}$, devoted to find a feasible control trajectory for the hard constraints of the problem, if it exists. If the option $\beta = 0$ has been chosen, then no deviation variables are introduced, and the constant functional $R_0 \equiv 0$ is minimized over the feasible set of the original problem.

After solving the problem corresponding to level 0, if there exist feasible controls, then the algorithm will have found one of them $u^{(0)}*$, and we proceed to the following level.

If no such feasible control exists, then the algorithm ends, for no control exists that satisfies the hard constraints of the problem.

4) Loop per priority levels.

After solving the problem $\mathbf{P_0}$, the algorithm enters a loop per priority levels, which tends to solve the problem sequentially. Within this loop, the following steps are carried our for each level q:

- The algorithm tests whether the optimal solution corresponding to the previous level, $u^{(q-1)}*$, satisfies the goals of the current level. If so, the algorithm jumps to the following level without solving the problem corresponding to the current one.

- If not, then the constraints and bounds corresponding to problem $\mathbf{P_q}$ are built, including the goals that belong to level q, together with their deviation variables, as well as the goals already achieved in the previous levels.

- An initial estimate (u^0, π^0) of the optimal solution of the level is built, so that it is feasible, in the following way:

$$u^0 = u^{(q-1)} *$$

$$\pi_j^0(k) = \begin{cases} \sum_{\kappa=0}^{k} f_j(x(\kappa), u(\kappa), \kappa) + F_j(x(k+1)) - \alpha_j(k) > 0 \\ 0 \quad \text{in other case.} \end{cases}$$

This way, the initial estimate (u^0, π^0) is assured to be feasible, what in turn assures quicker running times.

- Problem $\mathbf{P_q}$ corresponding to the current level, is solved. The algorithm minimizes the weighted accumulated sum of the β powers of the deviation variables assigned to the goals of the level.

- The algorithm tests whether the optimal solution obtained, $u^{(q)}*$, satisfies the goals of the current level. If so, we proceed to the following level. If not, a redefinition of the target values is carried out,

$$\tilde{\alpha}_i(k) = \alpha_i(k) + \pi_i^*(k) \quad k = 0,1,..,N-1. \quad i \in N_q$$

so that the procedure is carried on for the following levels, after informing the user that the goals corresponding to the current level cannot be satisfied.

5.- Results analysis

Table 1 displays the results of the computational implementation of the goal programming algorithm. With the aim of observing the behavior of the algorithm, and in order to generate random problems whose dimensions are chosen by the user, a program has been developed, based on the scheme formulated by **Rosen** and **Suzuki** (1965).

The implementations have been carried out in FORTRAN language, with the aim of the N.A.G. mark *15* subroutine library, on a VAX computer with an Alpha coprocessor. The results obtained have been summarized in the following table that displays by columns the following information:

The first column shows the number of state variables of the dynamic problem, the second one displays the number of control variables, whose range is in relation with the number of state variables, while the third column indicates the number of objective functionals, which has ranged between two and five. The column entitled *Periods* displays the planning horizon, for which *5* or *10* periods have been considered. The next column shows the number of constraints, which includes pure state and control constraints and mixed ones, all of them in the same quantity, without taking into account simple bounds on the variables. Finally, the sixth column displays the processing times in seconds.

Also, each row shows the mean processing time of a set of five problems of the same dimensions.

State	Control	Obj. Funct.	Periods	Const.	C.P.U.
5	3	5	5	21	0.05
5	3	5	10	21	0.15
10	5	5	5	30	0.09
10	5	5	10	30	0.31
25	15	2	5	60	0.17
25	15	2	10	60	0.91
25	15	3	5	60	0.29
25	15	3	10	60	0.82
25	15	5	5	60	0.48
25	15	5	10	60	1.52

Table 1. C.P.U. times for several test problems solved by the lexicographic G.P. algorithm.

It must be observed that the number of periods causes an increase in the C.P.U. times, between two and four times their original values.

As the constraints that cause a higher complexity in the problem are the quadratic ones, any increase in the number of priority levels or of the periods cause an increase in the running time, although it also must be taken into account that such increases also affect the number of variables, for there is one per functional and period.

Finally, the number of problems to be solved grows as the number of objective functionals increases, but, generally, the satisfying set is also reduced and the global complexity of the problem can reduced because of the existence of redundant priority levels (see Romero, 1991).

Acknowledgements. The authors would like to express their gratitude to the anonymous referees for their helpful comments.

References:

Caballero, R., Gómez, T., González, M., Rey, L. and Ruiz, F. (1996). "Goal programming with dynamic goals". *Sent to Journal of Multi Criteria Decision Analysis.*

Caballero, R. and González, A. (1993). "Considerations on Dynamic Goal Programming" (in Spanish), *Estudios de Economía Aplicada,* 0, 121-145.

Chang, S.L. (1966). "General Theory of optimal process", *J. SIAM Control,* 4, 1.

Da Cunha, N.O. and Polak, E. (1966a). "Constrained minimization under vector-valued criteria in finite dimensional spaces", *Electronics Research Laboratory,* University of California, Berkeley, California, Memorandum ERL-188.

Da Cunha, N.O. and Polak, E. (1966b). "Constrained minimization under vector-valued criteria in linear topological spaces", *Electronics Research Laboratory,* University of California, Berkeley, California, Memorandum ERL-191.

Levary, R.R. (1984). "Dynamic programming models with goal objectives", *International Journal System Sciences,* 15, 3, 309-314.

Lewandowski, A. and Wierzbicki, A.P. (Eds) (1989). *Aspiration Based Decision Support Systems. Theory, Software and Applications.* Berlin. Springer-Verlag.

Li, D. and Haimes, Y.Y. (1989). "Multiobjective dynamic programming: The state of the art", *Control Theory and Advanced Technology,* 5, 4, 471-483.

Li, D. and Haimes, Y.Y. (1990). "New approach for nonseparable dynamic programming problems", *Journal of Optimization Theory and Applications,* 64, 2, 311-330.

N.A.G. (Numerical Algorithms Groups Limited) (1991). *The N.A.G. Fortran Library Introductory Guide, Mark 15.*

Nelson, W.L. (1964). "On the use of optimization theory for practical control system design", *IEEE Transactions on Automatic Control,* AC-9, 4.

Romero, C. (1991). Handbook of Critical Issues in Goal Programming. Oxford. Pergamon Press.

Rosen, J.B. and Suzuki, S. (1965). "Construction of nonlinear programming test problems", *Communications of ACM, 113.*

Ruiz, F. (1994). Convex Multiple Objective Programming. Foundations and Solution Determination Methods. Ph.D. Dissertation. University of Málaga, Spain. (In Spanish).

Salukvadze, M. (1979). *Vector- Valued Optimization Problems in Control Theory.* New York. Academic Press.

Trzaskalik, T. (1996). "Dynamic goal programming models", Presented to the *Second International Conference in Multi-objective Programming and Goal Programming, Málaga (Spain).* To appear in the proceedings volume.

Wierzbicki, A.P. (1991). "Dynamic Aspects of Multi-objective Optimization", en Lewandowski, A. and Volkovich, V. (eds). *Multiobjective Problems of Matematical Programming.* Berlin. Springer-Verlag.

Zadeh, L.A. (1963). "Optimality and Non-Scalar-Valued Performance Criteria", *IEEE Transactions on Automatic Control,* AC-8, 1.

Stochastic Simulation in Risk Analysis of Energy Trade

Makkonen Simo[1] and Lahdelma Risto[2]

[1] Process Vision Ltd, Ukonvaaja 2A FIN-02130 ESPOO, Finland
[2] Technical Research Centre of Finland, Energy Systems, P.O. Box 1606, FIN-02044 VTT, Finland

Abstract

Successful trade in the newly deregulated European electricity market requires both maximising the net profits and maintaining an optimal portfolio of energy sales and purchase contracts. Energy is acquired from multiple multi-tariff contracts, thermal and hydro power plants, cogeneration facilities and the spot market. Minimisation of the procurement costs results in a time dependent resource allocation problem which can be solved efficiently using decomposition techniques. To maximise the net profits in trade, the feasibility and profitability of different types of contracts can be analysed using for instance marginal cost analysis and scenarios.

Taking into account the uncertainties involved in load forecasts, availability and prices of future short term contracts and the spot market, we obtain a stochastic multi-criteria programming problem. The decision maker (DM) not only wants to maximise the expected profits, but also to minimise the risks. We apply Monte Carlo simulation (MCS) for analysing the risks involved in energy trade. By using parametric analysis on the MCS model the DM can then explore the dependencies between the profits and risks and choose an acceptable risk level. The mathematical models and algorithms have been implemented as part of the commercial EHTO NEXUS software, which is used by several energy companies in Finland.

Keywords. Monte Carlo simulation, risk analysis, energy management, energy market, optimisation

1 Introduction

The electricity market has recently been deregulated in many European countries and a large common market is likely to arise. Deregulation affects especially the daily operations of energy companies. The spot markets for electricity operate every weekday and the exploitation of the spot market requires continuous actions. When planning spot market operations and short term trade, the energy companies should simultaneously consider a large number of different factors.

1) The hourly availability and costs of energy from long term contracts and own production facilities are time dependent.
2) Cogeneration introduces dependencies between the supply and costs of electricity and heat.
3) Energy package contracts and hydro power introduce dynamic storage constraints to the energy system.
4) Optimisation of the unit commitment of power plants introduces dynamic constraints and cost function terms to the energy system.
5) Several uncertain factors exist in short and long term planning, such as load forecasts, availability of hydro power, the availability and price of the future short term contracts and the spot market.

The main objective of energy companies should be the maximisation of their expected net profits while maintaining an acceptable risk level. We identify four kinds of principal risks in trade.

1) Demand exceeds supply due to underestimation of demand or overestimation of supply (for instance hydro power) ⇒ the lacking energy must in general be purchased at a price much higher than normal.
2) Supply exceeds demand due to overestimation of demand or underestimation of supply ⇒ the sales income from the excess energy is lost, the excess is dumped on the spot market at a minimal price.
3) Over-expensive purchase due to overestimation of the future short term or spot market price level ⇒ unprofitable acquisition.
4) Too low sales price due to underestimation of the future short term or spot market price level ⇒ unprofitable sales.

Controlling these risks requires skilful management of the purchase and sales contract portfolio. This new setting requires that more attention be paid on the computational systems and sophisticated support tools to assist the decision making of energy trade.

In this paper we describe the EHTO concept, where simulation and optimisation is used to enhance the daily operations of energy companies. Cost optimisation is based on a deterministic mathematical model of the overall energy system with multiple multi-tariff contracts, cogeneration facilities, spot market trade and hydro power. The profitability of purchase and sales contracts can in the deterministic model be analysed using for instance marginal cost analysis and scenarios. Considering the uncertainties results in a stochastic multi-criteria programming problem, where the decision maker (DM) not only wants to maximise the expected profits, but also minimise the risks. Based on solving the overall energy system model, we apply Monte Carlo simulation (MCS) to assess the risks involved with a contract portfolio. By using parametric analysis on the MCS model we then compute the expected profits and risks as a function of the power limit of a particular contract. The parametric information is finally used in contract planning, where the DM can choose the power limit, leading to a preferred profit-risk combination from the Pareto-efficient frontier.

2 The EHTO Energy Trade Concept

The EHTO concept for energy trade modelling is based on an overall energy system model and efficient mathematical algorithms that are used systematically to fulfil different planning tasks (Makkonen S., Lahdelma R., 1996). The energy system model is configured graphically from object-oriented model components. A relational database stores the configured model, different sets of model parameters and the optimisation results. The same model is used in all problems (long and short term cost optimisation, purchase and sales planning, simulation, sensitivity and risk analysis) through a suitable selection of parameters and decision variables. In this paper, we focus on daily operations and short term decision making in energy trade. The model-based planning process with interactive graphical model configuration and parametrisation is illustrated in Figure 1.

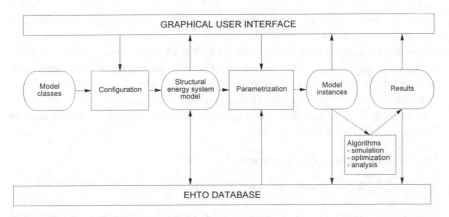

Figure 1. The model-based planning process.

3 Deterministic Models and Optimisation

In the following formulas, we will use the index $t \in T$ for the hours in the planning horizon, $i \in I$ for the subsystems (controllable supply and demand components), and $k \in K$ for the tariff components in multi-tariff contracts.

The objective of the deterministic energy system model is to maximise the net profit, that is, the sales income minus the purchase and production costs during the planning horizon subject to constraints

$$J(\alpha) = \max J(\alpha, x) \qquad (1)$$

where $J(\alpha)$ is the optimal net profit as a function of a parameter vector α and $J(\alpha,x)$ is the net profit, which depends both on the parameters and the vector of decision variables x. When the system is decomposed into subsystems, the overall model can be expressed as

$$\max J(\alpha,x) = \max \Sigma_{i \in I} J_i(\alpha_i,x_i)$$

s.t. (2)

$$H x = h,$$
$$G x \geq g,$$

where $J_i(\alpha_i,x_i)$ is the net profit of subsystem i and the vectors α_i and x_i are the parameters and decision variables specific to subsystems. The subsystems are connected to the overall energy system through hourly energy balances. The balance constraints are represented by the transmission matrix H and the vector h corresponding to aggregated demand forecasts. Power reserve constraints are included in the model similarly through matrix G and vector g. The net profit for the subsystems can be written as

$$J_i(\alpha_i,x_i) = \pm C_i(\alpha_i,x_i)$$

s.t. (3)

$$x_i \in X_i$$

where $\pm C_i (\alpha_i,x_i)$ is the income (+) or cost (-) from subsystem i, and the sets X_i represent subsystem-specific constraints. The different types of subsystems are defined as reusable model components using object-oriented techniques (Lahdelma, 1994). In the next sub-sections we describe only the trade components. A more complete description of the model components is found in Lahdelma, Makkonen (1996).

3.1 Long and Short Term Contracts

All 2^3 combinations of long and short term purchase and sales contracts for heat and electricity have a similar structure, consisting of $|K| \geq 1$ tariff components. Denoting by $T(i) \subseteq T$ the set of hours in the planning horizon when the contract is in effect, the contract model can be written as

$$C_i(\alpha_i,x_i) = c_0 + \Sigma_{k \in K} (c_k^{max} p_k^{max} + \Sigma_{t \in T(i)} c_k^t p_k^t), \qquad (4)$$

X_i:
$$\Sigma_{k \in K} p_k^t = p^t \qquad \forall t \in T(i), \qquad (5)$$
$$0 \leq p_k^t \leq p_k^{max} \qquad \forall k \in K, \forall t \in T(i),$$
$$p_k^t < p_k^{max} \Rightarrow p_{k+1}^t = ... = p_K^t = 0 \ \forall k \in K, \qquad \forall t \in T(i),$$

where c_0 is the *constant contract price*, c_k^{max} is the *order price* for *tariff power limit* p_k^{max}, and c_k^t is the hourly changing *energy price* for the traded *tariff energy* p_k^t. p^t is the total hourly traded energy. The last set of constraints specifies that the tariff components are to be used in a pre-determined order.

The main difference between purchase and sales contracts is that the energy company may decide how to use purchase contracts, but must provide energy according to customers' needs through sales contracts. Thus p_k^t and p^t are decision variables in purchase contracts, but parameters (forecasts) in sales contracts.

3.2 Spot Market Trade

Spot market trade can be handled as a special case of long and short term contracts. Spot market contracts consist of only one component and only the order price is defined. The set of effective hours T(i) consists typically of one or more consecutive hours. Spot market trade can thus be modelled as

$$C_i(\alpha_i, x_i) = c^{max} \, p^{max} \tag{6}$$

$$X_i: \qquad\qquad 0 \le p^t \le p^{max} \qquad\qquad \forall t \in T(i). \tag{7}$$

3.3 Solving the Deterministic Model

Through a suitable selection of parameters and decision variables $J(\alpha, x)$ can be used for multiple purposes. The Lagrangian decomposition and co-ordination technique (Nemhauser et al., 1989) was chosen for solving $J(\alpha, x)$, because it had been successfully used for similar kinds of problems (Baldick, 1995, Guan and Luh, 1992, Guan et al., 1995, Wang et al., 1995). Special algorithms have been tailored for solving each subproblem efficiently (Lahdelma et al., 1995). The decomposition approach also makes it easy to extend the system with new model components. (Lahdelma and Makkonen, 1996)

4 Stochastic Analysis

4.1 Sensitivity Analysis in Risk Analysis

Sensitivity analysis is a commonly used technique to evaluate the potential risks in long-term investment planning (Kliman, 1994). The same approach is also often applied when planning short-term contract portfolios. The sensitivity of the optimal overall costs $J(\alpha)$ with respect to a variation in an individual parameter α_j can be expressed as

$$[\, J(\alpha_1,...,\alpha_j + \Delta\alpha_j, \alpha_{j+1},...) - J(\alpha_1,...,\alpha_j, \alpha_{j+1},...)] \, / \Delta\alpha_j, \tag{8}$$

where $\Delta\alpha_j$ can be chosen to reflect the uncertainty in α_j. The above formula fails, however, to consider the joint effects of uncertainty in multiple parameters. To develop a more sophisticated risk analysis model with uncertainties, stochastic simulation techniques can be used (Spinney and Watkins, 1996).

4.2 Stochastic Marginal Cost Analysis for Planning Spot Market Trade

Spot market bids and asks can be planned based on the marginal cost of electricity

$$C_{marg}(p, \Delta p) = - [J(\alpha_1, ..., p+\Delta p, ...) - J(\alpha_1, ..., p, ...)] / \Delta p \qquad (9)$$

where p is the demand for electricity for one or more hours and Δp is a small increment or decrement. C_{marg} minus a required profit margin is then the highest profitable bid price and C_{marg} plus a required profit margin is the lowest profitable ask price. Parametric analysis could be used for computing a piecewise linear function of the highest profitable bids and lowest profitable asks. In Finland, however, the bids and asks must be entered into the spot market no later than two hours before the effective hour. The deterministic model fails to consider the uncertainties involved in demand forecasts making spot market trade prone to all four types of risk as described in Section 1.

To manage the risks we use a stochastic marginal cost analysis technique where we compute a stochastic cost function using a probability distribution for the power demand. Let us denote by $J_p(p)$ the net profit as a function of the stochastic power demand p and by f(p) the density function of p. The expected value for the profit is then obtained by integrating over the power demand:

$$\bar{J} = \int_0^\infty f(p) J_p(p) dp . \qquad (10)$$

Let us now purchase from the spot market an additional amount of power Δp with cost $C_\Delta(\Delta p)$. Negative values for Δp correspond to spot market sales. Spot market purchases (sales) reduce (increase) the original procurement which is equivalent to shifting the probability distribution to the left by an amount Δp. The influence of a spot purchase to the probability distribution is depicted in Figure 2.

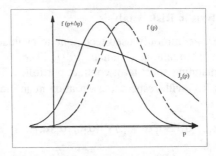

Figure 2. The shifted probability distribution after spot market purchase Δp.

The increase (decrease) in the profit $\Delta J(\Delta p)$ resulting from spot market purchases (sales) can be expressed as the difference between the spot market purchase costs C_Δ and the cost reduction in the original procurement.

$$\Delta J(\Delta p) = -C_\Delta(\Delta p) + \int_{-\Delta p}^{\infty} [f(p+\Delta p) - f(p)] J_p(p) dp \quad (11)$$

By setting the profit increase $\Delta J(\Delta p) = 0$ we can solve for the marginal cost function for spot market electricity.

$$C_\Delta(\Delta p) = \int_{-\Delta p}^{\infty} [f(p+\Delta p) - f(p)] J_p(p) dp. \quad (12)$$

Depending on the shape of the probability distribution we can obtain either an analytical or a numerical solution for the cost function.

4.3 Risk Analysis Based on Stochastic Simulation

The joint effects of uncertainty in multiple parameters can be analysed using Monte Carlo simulation. The uncertain parameters α_j are now stochastic variables. Their joint probability distribution is given as a density function $f(\alpha)$. In the special case, where the parameters are independent stochastic variables, the density function can be expressed as a product

$$f(\alpha) = \prod_j f_j(\alpha_j). \quad (13)$$

The probability distributions $f_j(.)$ are either specified by the DM or generated by statistical methods directly from observations using, for example, the bootstrap method. $J(\alpha,x)$ is then solved N times with different parameter values α^n drawn randomly from their probability distributions. As a result, distributions for x_j and $J(\alpha,x)$ are obtained. These distributions can be represented as histograms or using statistical measures. In this paper we assume a normal distribution for the profit and estimate the expected profit $\bar{J}(\alpha,x)$ and standard deviation $\sigma(J(\alpha,x))$ from the simulation results.

The DM wants to maximise the expected profit and minimise the risk. Different measures for the risk can be defined. Here we measure the risk by the 'worst case profit' $\rho(\alpha,x)$ which the DM wants to maximise. For simplicity, we define ρ to be the lower bound of the 68% confidence interval, that is

$$\rho(\alpha,x) = \bar{J}(\alpha,x) - \sigma(J(\alpha,x)). \quad (14)$$

4.4 Stochastic Multi-Criteria Programming Using Parametric Analysis

Given a fixed contract portfolio, the MCS model can be used for computing the expected profit $\bar{J}(\alpha,x)$ and worst case profit $\rho(\alpha,x)$. By using parametric analysis we can now compute the expected profit and worst case profit as function of the power limit p^{max} of a particular contract or spot market purchase. This is done by solving the MCS model iteratively with M different values for p^{max}. This results into a three-level stochastic multi-criteria optimisation model as illustrated in Figure 3.

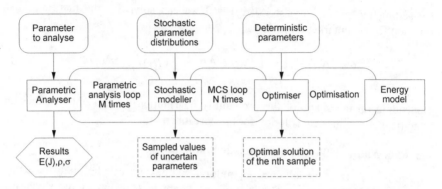

Figure 3. Schematic of the three-level stochastic multi-criteria programming framework.

4.5 Example

We consider a 336 hour (two week) planning horizon and the following components:
- a long-term purchase contract with three tariff components and two price zones (day, night),
- spot market with dynamically varying stochastic price,
- a fixed purchase contract to be determined by parametric analysis, and
- two customers with sales contracts and stochastic demands.

The deterministic optimisation problem is thus to maximise the operating profit over the planning horizon by using the long term purchase contract, the fixed purchase contract, and spot market trades. Probability distributions computed from past data are used for generating random market prices and power demands for each hour. Parametric analysis is used to explore the possibilities to purchase a fixed amount of electricity at a fixed price for each hour during the two week period. The problem is to determine the ideal power limit in the interval 0 to 200 MW.

Table 1. Expected value, standard deviation and worst case for profit.

Purchase component	0 MW	50 MW	100 MW	150 MW	200 MW
$\bar{J}$ (MFIM)	1.6	2.3	2.1	1.9	1.6
σ (MFIM)	1.5	1.4	0.7	0.4	0.3
ρ (MFIM)	0.1	0.9	1.4	1.5	1.3

The results of the parametric analysis are presented in Table 1 and Figure 4. Each power limit results into a different profit distribution as illustrated in Figure 4a. A more informative profit-risk graph is obtained by plotting the worst case profits against the expected profits as in Figure 4b. The points corresponding to different power limits are connected together by a spline. The DM can now choose her preferred profit-risk combination from this graph. A zero power limit is obviously not optimal, because when moving from zero towards 50 MW both the expected and worst case profits increase. A risk neutral DM would choose the 50 MW power limit corresponding to the maximum expected profit of MFIM 2.3. An extremely risk averse DM would choose the 150 MW power limit corresponding to the best worst case profit of MFIM 1.5. In the general case the DM would choose some point from the Pareto-efficient frontier between the 50 and 150 MW power limits.

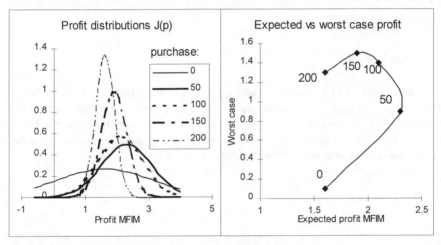

Figure 4a&b. Profit distributions and expected versus worst case profit.

5 Experiences and Discussion

The mathematical models and algorithms described in this paper have been implemented as part of the commercial EHTO NEXUS information system, which includes a measurement and contract database, an online data transfer system and several applications for system follow-up, reporting, planning of trading transactions, load forecasting and optimization of energy procurement and trade (Makkonen and Lahdelma, 1996). The EHTO optimization application, which is a basis for our MCDM concept, is currently used by 8 regional energy companies or energy purchasing consortiums, two educational institutes and one consulting company in Finland.

The EHTO optimisation package is used both for operative short-term planning and for long-term contract and investment planning. The basic concept illustrated in Figure 1 provides a useful tool for different kinds of planning tasks and the experiences of the users have been unconditionally affirmative. The graphical user interface and object-oriented model components are experienced as a key to wide use of complex algorithms in practical engineering work.

The MCDM concept based on stochastic simulation and a parametric analysis is a new extension to the EHTO concept. We have applied the MCDM model for operational planning of energy procurement. A further development will be to extend the planning horizon to cover a planning period of 1 to 3 years. One challenging task is to enhance the speed of the deterministic optimisation and to develop techniques for minimising the number of MCS iterations needed for obtaining sufficient accuracy. Another topic which requires further research is sensitivity and correlation analysis of the stochastic parameters because the shape of the Pareto-efficient frontier can be quite sensitive to the selection of the stochastic variables.

6 Conclusions

We have introduced a decision making procedure for energy trade based on a deterministic energy optimisation model, stochastic simulation and parametric analysis. The deterministic optimisation model defines a framework for more sophisticated risk analysis. The stochastic simulation provides a useful approach for decision making in energy trade under uncertainty.

7 References

Baldick R., 1995, "The Generalized Unit Commitment Problem", IEEE Transactions on Power Systems, Vol. 10, No. 1, February 1995.

Divekar U.M., Ribin E.S. "Optimal Design of Advanced Power Systems under Uncertainty", Proc. of the International Symposium of ECOS'96, Efficiency, Cost, Optimization, Simulation and Environmental Aspects of Energy Systems, June 25-27, Stockholm, Sweden, pp. 389-396.

Guan X., Luh P., 1992, "An optimization-based method for unit commitment", Electrical Power & Energy Systems, vol. 14, no. 1, February 1992.

Guan X., Luh P., Zhang L., 1995, "Non-linear Approximation Method in Lagrangian Relaxation-Based Algorithms for Hydrothermal Scheduling", IEEE Transaction on Power Systems, Vol. 10, No. 2., May 1995.

Kliman, Mel, "Competitive bidding for independent power, Developments in the USA", Energy Policy, January 1994, pp 41-56.

Lahdelma R., 1994, "An Object-Oriented Mathematical Modelling System", Acta Polytechnica Scandinavica, Mathematics and Computing in Engineering Series, No. 66, Helsinki, 77 p.

Lahdelma R., Makkonen S., 1996, "Interactive Graphical Object-Oriented Energy Modelling And Optimization", Proc. of the International Symposium of ECOS'96, Efficiency, Cost, Optimization, Simulation and Environmental Aspects of Energy Systems, June 25-27, Stockholm, Sweden, pp. 425-431.

Lahdelma R., Pennanen T., Ruuth S., 1995, "Comparison of Algorithms for Solving Large Network Flow Problems", Proc. 14th IASTED International Conference on Modelling, Identification and Control, ed. M.H. Hamza, February 20-22, IGLS, Austria, pp. 309-311.

Makkonen S., Lahdelma R., 1996 "A Concept for Simulation and Optimization of the Energy Trade", The International Conference: Simulation, Gaming, Training and Business Process Re-engineering in Operations, Riga, Latvia September 19-21.

Nemhauser G.L., Rinnooy Kan A.H.G., Todd M.J, (eds) 1989, "Handbooks in Operations Research and Management Science", Volume 1, Optimization, Elsevier Science Publishers B.V., Amsterdam.

Spinney P.J. and Watkins G.C., "Monte Carlo simulation technique and electric utility resource decisions", Energy Policy, Vol. 24, No 2, February 1996, pp. 155-164.

Wang S.J., Shahidehpour S.M., Kirschen D.S., Mokhtari S., Irisarri G.D., 1995, "Short term generation scheduling with transmission and environmental constraints using an augmented Lagrangian relaxation", IEEE Transaction on Power Systems, Vol. 10., No. 3, August 1995, pp. 1294-1301.

8 Biography

M.Sc. Simo Makkonen (born 1966) received his M.Sc. in Systems and Operations Research in 1990 at Helsinki University of Technology. He is currently the Managing Director of Process Vision Ltd. His research interests include systems analysis, artificial intelligence and international financing.

Dr. Risto Lahdelma was born in Helsinki, Finland, in 1961. He is currently Research Professor of Energy Markets at the Technical Research Centre of Finland. His research interests include mathematical modelling, optimisation tools and techniques, advanced software technology and knowledge engineering.

Part 4
Negotiation and Group Decision Support

Modeling Negotiations in Group Decision Support Systems

N.I. Karacapilidis[1], D. Papadias[2], and C.P. Pappis[3]

[1] Dept. of Computer Science, Swiss Federal Institute of Technology, IN Ecublens, 1015 Lausanne, Switzerland
[2] Dept. of Computer Science, The Hong Kong University of Science and Technology, Clear Water Bay, Hong Kong
[3] Dept. of Industrial Management, University of Piraeus, 80 Karaoli & Dimitriou, 185 34 Piraeus, Greece

Abstract. Group decision making processes are usually characterized by multiple goals and conflicting arguments, brought up by decision makers with different backgrounds and interests. This paper describes a computational model of negotiation and argumentation, by which participants can express their claims and judgements, aiming at informing or convincing. The model is able to handle inconsistent, qualitative and incomplete information in cases where one has to weigh multiple criteria for and against the selection of a certain course of action. It is implemented in Java, the aim being to deploy it on the World Wide Web. The basic objects in our terminology are positions, issues, arguments pro and con, and preference relations. The paper describes procedures for consistency checking, preference aggregation and conclusion of issues under discussion. The proposed model combines concepts from various well-established areas, such as Multiple Criteria Decision Making, nonmonotonic reasoning and cognitive science.

Keywords. Group Decision Support, Negotiation, Argumentation, Reasoning.

1 Introduction

The ability to comprehend and engage in arguments is essential for a variety of group decision making procedures, from political debates and court cases to every day's negotiations and human reasoning. In these environments, decision makers usually need appropriate systems for supporting argumentation, especially in cases where information is incomplete, uncertain and inconsistent, and conflicts of interest and opinion are common. Such a system should incorporate a variety of discourse acts, by which agents are able to pursue their individual objectives and resolve conflicts. It should also take into account that reasoning is defeasible; further information may cause another choice of action to be more preferable than what seems optimal at the moment. Finally, it should support both cooperative and non-cooperative interactions, and take into account that participants use different methods to express and evaluate knowledge and they are not proficient in formal models of logic (Karacapilidis 1995).

Several models of dialectical systems have been proposed (Gordon 1996; Prakken 1995). They consist of a set of abstract components, that specify the underlying *logic*, the *speech acts* allowed, and the *protocol* of rights and responsibilities of the participants. Formal models of argumentation have been built on various logics. For instance, Brewka (Brewka 1994b) reconstructed Rescher's theory of formal disputation (Rescher 1977), Prakken (Prakken 1993) used Reiter's default logic (Reiter 1980), and Gordon's work (Gordon 1993) was based on Geffner and Pearl's concepts of conditional entailment (Geffner & Pearl 1992). Whether it makes sense to use nonmonotonic, inductive or analogical logics is extensively discussed in (Prakken 1995).

Approaches for reasoning in the presence of inconsistency can be distinguished to these based on the notion of *maximal consistent subbase*, where coherence is restored by selecting consistent subbases, and those based on *argumentation principles*, where inconsistency is "accepted" by providing arguments for each conclusion (Cayrol 1995). Non-monotonic inference relations are basically used in the first category, defined by methods of generating *preferred* belief subbases and classical inference (Brewka 1989). The second category, inspired from work done on Artificial Intelligence and Law, views the validity and priority of belief bases as subject to debate. The role of decision makers here is to construct theories, that are robust enough to defend a statement. The argumentation concepts used in this paper result in a non-monotonic formalism, founded on argumentation principles.

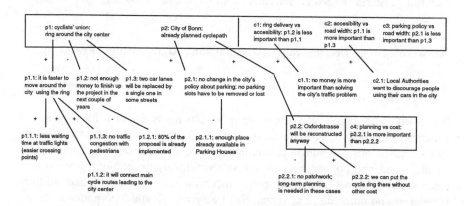

Figure 1: The discussion graph for the cyclists' path example.

This paper describes a computational model of argumentative discourse, by which agents can express their judgements and preferences during any kind of debate. We consider belief bases to be taken into account only if they have explicitly been stated by an agent. The model is able to handle ill-structured information in cases where one has to weigh reasons for and against the selection of a certain option. Throughout this paper, we refer to a real example, illustrated in Fig. 1. The discussion is about the planning of cyclepaths in the city of Bonn. The agents involved are members of the cyclists union and representatives from related City Hall departments. They express their claims and preferences for two alternatives, "a ring around the city center" and "the already planned cyclepath".

In the following section, we describe the argumentation-based framework our model is based on. We formally present the argumentation elements, proof standards and discourse acts involved. The latter relate to a variety of procedures for the propagation of the information, such as consistency checking, aggregation of preferences, and labeling of the argumentation elements according to different standards. We demonstrate the model using the above example. We then discuss related work aiming at extracting the pros and cons of our approach, and we conclude with directions for future work.

2 A Framework for Dialectical Argumentation

Our model extends the informal *Issue-Based Information System (IBIS)* model of argumentation (Kunz & Rittel 1970; Rittel & Webber 1973), later developed at MCC (Yakemovic & Conklin 1990; Conklin 1992). IBIS was first meant to support coordination and planning of political decision processes. Our approach maps a multi-agent decision making process to a *discussion graph* (tree).

2.1 Elements

Let L be a (finite) propositional language and $\mathcal{F}, \mathcal{D}$ sets of formulas of L. $\mathcal{F}$ represents the certain knowledge; it is assumed to be consistent and may be empty. Formulas in $\mathcal{D}$ may represent defeasible pieces of knowledge or beliefs (default rules); $\mathcal{D}$ is usually referred to as the belief base. Let also $\vdash$ denote classical entailment. The basic elements of our framework are *positions*, *constraints* (preferences), *arguments pro* and *con*, and *issues*.

Positions are the basic elements in our framework; they may contain data that have been brought up to declare alternative options, justify a claim, advocate the selection of a specific course of action, or avert one's interest from it (e.g., position p_1: "construct a ring around the city center"; position $p_{1.1}$: "it is faster to move around the city using the ring"; position $p_{1.3}$: "two car lanes will be replaced by a single one in some streets"). Each position has an *activation label* indicating its status (it can be *active* or *inactive*). The mechanisms for the calculation of these labels will be described in the sequel.

Definition 1: _A position is a pair_ _(p, act_[ab),_ _where_ _p_ _is a formula of_ _L_ _and_ _act_[ab_ _its activation label._

Constraints represent preference relations between two positions. The relations involved can be of the type _"more important than"_ and _"equally important to"_, denoted by the symbols ">" and "=", respectively. For instance, constraint c_2: "$p_{1.1}$ is _more important than_ $p_{1.3}$". Constraints, in addition to the activation label, have a _consistency label_ indicating whether the constraint is consistent or not.

Definition 2: _A constraint is a triple_ _(c, act_[ab, con_[ab),_ _where_ _c_ _is a formula_ _of_ _L_, _act_[ab_ _its activation label and_ _con_[ab_ _its consistency label. The formula_ _c is of the form_ _(p_i, p_j, r),_ _where_ _p_i,_ _p_j_ _are positions and_ _$r \in \{<, >, =\}$._

An _issue_ corresponds to a decision to be made, or a goal to be achieved. An issue can be interpreted as which alternative position to prefer, if any (e.g., issue I_1: "choose the most appropriate cyclepath in the city of Bonn"). Issues are brought up by agents and are open to dispute. The discussion graph may consist of numerous issues (in any case, one should be at the top). In their simplest form, issues consist of a set of positions $\mathcal{P}_i$ and a set of constraints C_i^1. An issue is resolved by selecting one of its alternative positions. More precisely:

Definition 3: _An issue_ _I_i_ _of the discussion is a triple_ _(g_i, $\mathcal{P}_i$, C_i),_ _where_ _g_i_ _is the_ _goal of the issue,_ _$g_i \in L$._

Arguments are assertions about the positions regarding their properties or attributes, which speak for or against them (multiple meanings of the term _argument_ are discussed in (Prakken 1995)). An argument links together two positions of the debate. The framework allows for _supporting arguments_ (e.g., $p_{1.1}$ is a supporting argument to p_1) or _counter-arguments_ (e.g., $p_{1.3}$ is a counter-argument to p_1). We consider that there are not unrefutable arguments and any combination of arguments can be applied; see also (Brewka 1989; Brewka 1994a). Various formal notions of an argument have been suggested in the literature. In (Cayrol 1995), an argument of $\mathcal{D}$ is a pair $(\mathcal{H}, h)$, where h is a formula of L (_conclusion_) and $\mathcal{H}$ a subbase of $\mathcal{D}$ (_support_) iff (i) $\mathcal{F} \cup \mathcal{H}$ is consistent, (ii) $\mathcal{F} \cup \mathcal{H} \vdash h$, and (iii) $\mathcal{H}$ is minimal (no strict subset of $\mathcal{H}$ satisfies (ii)). A similar notion of argument has been also given in (Benferhat & Dubois & Prade 1995). According to them, a consistent subbase $\mathcal{A}$ of L is said to be an argument to a rank i for a formula f if it satisfies the following conditions: (i) $\mathcal{A}$

---

[1] they may also contain subissues; due to space limitations, we don't describe them in detail here.

$\vdash f$, (ii) $\forall \alpha \in \mathcal{A}$, $\mathcal{A} - \{\alpha\} \dashv\vdash f$, and (iii) $i = \max \{ j \mid (f, j) \in \mathcal{A} \}$. The above notions are extensions of the one proposed in (Simari & Loui 1992). Our notion differs from them in that it allows the support base of an argument to be inconsistent with $\mathcal{F}$, and does not presume that is minimal. That is, a subset of the support base may also be another support base for the same conclusion. Formally:

Definition 4: An argument of $\mathcal{D}$ is a triple $(p_s , p_c , argLink)$, where p_s is a supporting position of L, p_c is a concluding position of L, and $argLink$ is the type of the argument link ($argLink \in$ {pro, con}). Moreover, for $p_s \in I_i$ and $p_c \in I_j$, it is $i \neq j$ (the positions an argument consists of should belong to different issues). If $argLink = $ pro, it is $\mathcal{F} \cup p_s \vdash p_c$; otherwise, $\mathcal{F} \cup p_s \neg\vdash p_c$.

Sometimes, we need to argue about the justification (warrant) for using the supporting position in favor of (or against) the concluding one. In these cases, an agent claiming a position p_c providing as data a position p_s has also to be prepared to provide another position, p_w, that justifies the desired link between the first two. For instance, one can claim that "Tweety flies", supporting it with the position "Tweety is a bird", and justifying his inference mechanism with the position "Birds fly". In our model situations like the above are treated as two separate arguments, one that links the supporting data with the claim, and another that links the justification with the claim.

As mentioned above, a constraint formula is of the form (p_i , p_j , r). Having defined issues and elements, we can speak about another restriction. In order for a formula to be valid, both p_i and p_j should be supporting positions in arguments that have their concluding positions in the same issue, the issue where the constraint belongs to. For instance, regarding constraint c_2, both $p_{1.1}$ and $p_{1.3}$ are supporting positions in arguments having p_1 as concluding position, and both c_2, p_1 belong to the same issue, I_1. Formally:

$$((p_i , p_j , r), act\_lab, con\_lab) \in I_i \equiv$$
$$\exists ((p_i , p_{c1} , argLink_1) \wedge (p_j , p_{c2} , argLink_2) \wedge (p_{c1} , p_{c2} \in I_i)). \tag{1}$$

The outcome of an argument is reflected to the activation label of a position. Similarly, since argumentation on constraints is also allowed, it has also to be reflected to the activation and consistency labels of a constraint. Keeping in mind that multiple levels of argumentation are desired, we need mechanisms able to determine the above outcomes taking into account the sub-tree below each position or constraint. A position/constraint with no argumentation underneath (e.g., a recently inserted position) is by default *active*. When there has been some discussion about it, its activation label is determined according to its proof standard.

2.2 Proof Standards

Different issues of a decision making process, even in the same course or debate, not necessarily need the same type of evidence. Quoting the well-used legal domain example, the arguments required to indict someone need not be as convincing as those needed to convict him (Farley & Freeman 1995). Therefore, a generic argumentation system requires different *proof standards* (work on Artificial Intelligence and Law uses the term *burdens of proof*). In the sequel, we describe those used and implemented in our system. We do not claim that the list is exhaustive (however, other standards, that match specific application needs, can be easily incorporated to the system), and the names used have exactly the same interpretation in the legal domain. In the rest of the paper, we use the term *alternative* whenever the context applies to either a *position* or *constraint*.

Definition 5: Scintilla of Evidence (SoE): An alternative with this standard is active, iff there is at least one supporting argument ($argLink = $ pro) that links it with an active position. Formally:

$$(p, active) \equiv \exists \ ((p_i, p, \text{pro}) \wedge (p_i, active)). \tag{2}$$

For instance, if $p_{2.2}$ in Fig. 1 had SoE as proof standard, it would be *active*, since it is supported by $p_{2.2.2}$ and the latter is *active* (there is no argumentation underneath).

Definition 6: Beyond Reasonable Doubt (BRD): An alternative with this standard is active iff there are no counter- arguments ($argLink = $ con) that link it with an active position[2]. Formally:

$$(p, active) \equiv \neg \exists \ ((p_i, p, \text{con}) \wedge (p_i, active)). \tag{3}$$

For instance, if p_1 in Fig. 1 had BRD as proof standard, it would be *active* iff both $p_{1.2}$ and $p_{1.3}$ were *inactive*.

Unlike the above two cases (where only the arguments pro and con are taken into account), constraints are used in the next proof standard in order to determine the activation label. A *scoring mechanism* (based on *topological sort*) calculates the weights of the positions involved. In the current implementation, weights fall within the range [*minimp .. maximp*], where *minimp (maximp)*

---

[2] it can be that either there are no counter-arguments at all, or all the counter-arguments link it with inactive positions (independently of the number of supporting arguments).

denotes minimum (maximum) importance for a position. In the absence of constraints, each position is given the average value of (*maximp* + *maximp*)/2. The scoring mechanism refines the weights when a new constraint is inserted. Every time a position is more important than another one (either in an implicit or explicit constraint), its weight is increased (and vice versa).

Definition 7: Preponderance of Evidence (PoE): An alternative with this standard is active iff the sum of weights of the active positions that are linked with supporting arguments is larger than that of the active positions that are linked with counter-arguments. Formally:

$$(p, active) \equiv \sum_{(p_i, p, \text{pro}) \wedge (p_i, active)} score(p_i) > \sum_{(p_j, p, \text{con}) \wedge (p_j, active)} score(p_j) \quad (4)$$

In our example, if $p_{2.2}$ had PoE as proof standard, it would be *inactive*, since $p_{2.2.1}$ is more important than $p_{2.2.2}$. The above standards can be applied to positions, constraints and issues. We assume that positions and constraints in an issue inherit its standard. We assume the following hierarchy of proof standards, according to the increasing importance of meeting them: $SoE << BRD << PoE$. In other words, the mere support of an argument to a position p is considered less important (needs lower burden of proof) than the defeat of all counterarguments to p (where agents have to provide extra argumentation in order to make them inactive), which in turn is considered less important than the predomination of the supporting arguments to p over the counterarguments to it (where preference relations, also debatable, should be brought up). In the labeling procedure presented below, the system can trace the discussion graph in a bottom-up approach, and calculate the most important standard that each element can "meet".

2.3 Discussion Graph: Acts and Labeling

Argumentation in our framework is performed through a variety of *discourse acts*, which may alter related parts of the discussion graph. The procedure of *labeling* the discussion graph results in an enumeration of all (affected) position and constraint labels. It involves topics of special interest, the most important being *constraint consistency checking* and *propagation*. The discourse acts implemented in our system and the procedures that they call are the following (we use a pseudo-Java type of code):

```
addIss( issue iss )/* opens a new issue */
{  iss = new issue();  }

addAltToIss( alternative alt, issue iss )   /* adds an alternative to an issue */
{ addElement( alt, iss );
   if (size( iss.alternatives ) >1 ) update( iss );}
```

addAltToPos(alternative *alt,* position *pos*)
      /* adds an altern. to a position; it first opens an issue */
{ *if* (*pos.inIssue* == *null*)
   { *iss* = new issue(*pos*);
     inIssue(alt) = *iss;*
     addElement(*alt, iss*);   }
   else addAltToIss(*alt, pos.inIssue*); }

linkPos(position *pos,* alternative *alt,* boolean *argLink*)
      /* links a position to an alternative */
{ *inIssue(pos)* = *null;*
  refersTo(pos) = *alt;*
  refersToLink(pos) = *argLink;*
  if (*argLink*) addElement(pos, *alt.positionsPro*);
    else addElement(pos, *alt.positionsCon*);
  update(alt); }

addConToIss(constraint *con,* issue *iss*)   /* adds a constraint to an issue */
{ *inIssue(con)* = *iss;*
  if (*not* alreadyExists(*con*))
  { addElement(*con, iss.constraints*);
    if (containsInactive(*con*)) *con.act_lab* = *false;*
    else
    { *con.act_lab* = *true;*
      if (*not* consistentWithPrevious(*con, iss*))
        con.con_lab = *false;*
      else
      { *con.con_lab* = *true;*
        update(iss); } } } }

addConToAlt(constraint *con,* alternative *alt*)
      /* adds a constraint to an alternative; it first opens an issue */
{ *iss* = new issue(*alt*);
  addConToIss(*con, iss*); }

Recall that constraints are of the form $(p_i,\ p_j,\ r)$. Whenever an agent is about to insert a new constraint *con,* alreadyExists(*con*) checks if both of its consistuent parts exist in another, already asserted constraint. That is, it checks whether both p_i and p_j are elements of another constraint con_i (the relation r is not checked at all; if r is the same then *con* is redundant, otherwise it is inconsistent with the existing discussion graph). In the case that alreadyExists(*con*) = *true,* the new-coming constraint is ignored. In the sequel, containsInactive(*con*) checks if *con* is valid, see (1), and whether both consistuent positions of *con* are *active*. If all happen, *con* is labeled *active*. Otherwise, if at least one of the p_i and p_j is *inactive*, *con* is labeled *inactive* (however, *con* is always inserted in the graph).

consistentWithPrevious(*con, iss*) involves a path consistency algorithm (Macworth & Freuder 1985), in order to determine if *con* contradicts with the other *active* (implicit or explicit) constraints of *iss*. If this is the case, con_label of *con* is *inconsistent*. During the discourse *con* will become *consistent* only if some of the conflicting constraints get inactivated. Note that path consistency, which is polynomial, suffices here because constraints involve only the relations > and =. If arbitrary disjunctions were allowed (e.g., ≠, ≥, etc.) exponential algorithms would be required for satisfiability.

Most of the discourse acts described above may affect parts of the graph. Insertion of a new position has to be propagated upwards; act_lab of its concluding position may change (upon its *proof standard*) which, in turn, may change the act_lab of its own concluding position, etc. The *propagation* procedure of our model reminds related work on Truth Maintenance Systems (Doyle 1979), but it is usually more complicated, due to the structure of the discussion graph. For instance, whenever a position p_i becomes *inactive*, constrains of the form (p_i, p_j, r) or (p_k, p_i, r) also become *inactive* (it is meaningless to keep *active* a constraint with *inactive* consistent part(s)). In general, propagation activate or inactivate affected positions and constraints and updates the issues they belong to either by taking into account a pre-selected *proof standard*, or by tracing the hierarchy of *proof standards* they can meet (according to their increasing importance). The update procedures for positions and issues are as follows (the update(constraint *con*) is straightforward from the first one):

update(position *pos* **)**
{ *if* (*pos.inIssue* ≠ *null*) update(*pos.inIssue*);
  else
  { *temp* == findStatus(*pos*);
    if (*temp* ≠ *pos.act_lab*)
    { *if* (*temp*) activate(*pos*);
     else inactivate(*pos*);
     if (*pos.refersTo* ≠ *null*)
       update(*pos.refersTo*); } } }

update(issue *iss* **)**
{ *scoresGraph* = new graph(*iss*);
  calcScores(*scoresGraph*);
  temp = max *score(iss.alternatives)*;
  tempStatus = $act\_lab(k)$ | $k \in iss.alternatives \wedge score(k) = temp$;

```
for ( i=1; i<size(iss.alternatives); i++ )
{ if ( temp == score(iss.alternatives(i)) )
      act_lab(i) = activate( i );
  else act_lab(i) = inactivate( i );
  if ( inIssue(i) ≠ null ) update( inIssue(i) );
  else if ( i.refersTo ≠ null ) update( i.refersTo ); } }
```

In the example of Fig. 1, we assume that the proof standard for the two issues (denoted by rectangles) is PoE, whereas for any other position and constraint is SoE. Positions $p_{2.1}$ and $p_{2.2}$ become *inactive* because of $p_{2.1.1}$ and c_4, respectively. Therefore, both supporting arguments for the selection of p_2 have been defeated. Constraint c_3 will also become *inactive*, because it includes the *inactive* position $p_{2.1}$. Regarding the other alternative position, p_1, $p_{1.1}$ and $p_{1.3}$ are *active*, while $p_{1.2}$ becomes *inactive* due to $p_{1.2.1}$. Constraint c_1 is also *inactive*, due to $p_{1.2}$. Therefore, the only active constraint in I_1 is c_2, which makes p_1 *active*. Note that p_1 is able to meet both SoE and PoE standards, while p_2 none of them.

The model presented above is already implemented in Java, the aim being to deploy it on the Web. Aiming at developing an efficient system of argumentative discourse for multi-agent decision making on the Internet, we need to address practical requirements such as availability of the system on all prominent operating systems and hardware platforms. Java is ideal for this purpose. An appropriate set of user interfaces has also been implemented, which allows decision makers to participate in a variety of discourses and exploit the features of the model described in this paper.

3 Discussion

In this section we focus on previous, well-tried concepts and theories, that have addressed the problems of practical and substantial reasoning, attempting to account for how humans combine various types (i.e., deductive, defeasible, inductive, probabilistic) of reasoning. We aim at extracting the pros and cons of our approach, and motivations for further work.

Toulmin's early work (Toulmin 1958) suggested an argument schema that distinguishes a variety of components in order to clarify the different types of assertions made in the construction of an argument. According to him, logic is a set of norms regulating practical discourse. Our approach is in accordance with it, in that it encompasses all norms required for regulating such discourses, and not just the subset concerning logical necessity and contradiction, while is also able to support derivability relations. Focusing on constructing theories and norms able to regulate a debate, we try to get away from classical non-monotonic logic approaches to gain more on efficiency and simplicity. Fig. 2 illustrates how his structure of arguments can be mapped to ours. We view any type of his framework components (i.e., *datum, warrant, backing, claim* and *rebuttal*) as positions linked with the appropriate type of argument. Toulmin's *qualifier* is similar to our *score* concept. Main weaknesses of Toulmin's theory are the lack of the appropriate formalism for addressing multiple issues and ordering competing arguments, and its orientation towards only cooperative-type argumentation. Instead, our framework is applicable to any kind of adversarial or cooperative group decision making process, and it provides a means of ordering arguments and positions in an issue.

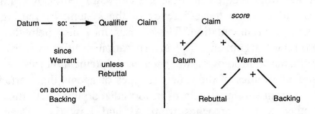

Figure 2: Structure of an argument according to Toulmin's (left) and the proposed (right) model.

Pollock's *OSCAR* model of defeasible reasoning (Pollock 1988) was one of the first attempts to base defeasible reasoning on arguments, influencing later work; see for example (Simari & Loui 1992; Geffner & Pearl 1992). His model does not deal with resolving disputes, but with prescribing the set of beliefs a single rational agent should hold, under certain simplifying assumptions. According to him, the initial *epistemic basis* of an agent consists of a set of positions, either supported by perception or recalled from memory that are not necessarily believed, a set of *defeasible inference rules* and a set of *nondefeasible inference rules*. Belief on positions stands until defeated by new reasons, disregarding the original ones. Pollock also distinguishes between *warranted* and *justified belief*; we address it using the *proof standards*. A weakness of *OSCAR*, addressed in our framework, is that there are no reasoning schemata provided for the validity of inference rules, or their relative weight and priority. For an extensive criticism on Toulmin's and Pollock's theories, see (Gordon 1993).

Rescher, in his theory of *formal disputation* (Rescher 1977), considers disputation to be a three-party game, taking place with a *proponent*, asserting a certain position, an *opponent*, able to challenge proponent's position and a *determiner* which decides whether the proponent's position was successfully defended or not. Our model plays the last role, mediating the discussion among multiple agents. A more formal reconstruction of Rescher's theory is presented in (Brewka 1994b), based on Reiter's Default Logic and, more especially, on its SDL variant (SDL has also been developed by Brewka, and it is a variant of classical DL, allowing for inclusion of the *specificity principle*). Brewka's work clarifies Rescher's concepts and goes ahead defining *elementary* and *legal moves* during a dispute, as well as *winning situations*. Nevertheless, both approaches are limited in that the players have no chance to disagree about defaults.

Our model provides the required discourse acts in order for agents to adhere to the ten basic principles (known as the "Ten Commandments") of a critical discussion, introduced by Eemeren and Grootendorst (Eemeren & Grootendorst 1992).

We intend to integrate our model with case-based reasoning methods and decision theory techniques such as multi-attribute utility theory. Sycara's *PERSUADER* was implemented in these lines (Sycara 1987). Keeping track of previous cases will increase the efficiency of the system, by reusing previous instances of similar argumentation. These cases may also help the agents to construct more robust arguments. The incorporation of utilities in our model is rather straightforward. Moreover, we need to implement norms or *protocols* about the rights and duties of the different types of agents that participate in a discussion. Ideas from similar structures in formalized public activities should be exploited together with approaches from AI and Law (i.e., Deontic Logic, Argumentation Theory) and Distributed AI. Protocols should take into account the roles of participants and the evolution of the decision making procedure. Finally, future work should consider various logic-based models for group decision making and pure negotiation techniques, in order to extract the pros and cons of our approach. For instance, *NEGO* (Kersten 1985) is a well-tried model, based on the generalized theory of negotiations formulation (Kersten and Szapiro 1986), that supports an interactive process of individual proposal formulation and negotiation, and allows decision makers to change their strategies, form coalitions, and compromise on the issues under consideration.

4 Concluding Remarks

Use of information technology may assist multi-agent decision making in various ways (Karacapilidis and Pappis 1997). An important goal is to provide easy access to the current knowledge, at any time. This would be greatly facilitated if all relevant data and documents are made available and maintained in electronic form, in a well-structured and organized way. Our system's primary task is to provide direct computer support for the argumentation, negotiation and mediation process in these environments. A computer network can be used as a medium for such a process, in which documents are indexed according to their role and function, using the presented model of argumentation. The model presented in this paper combines concepts from various well-established areas such as Decision Theory, Non-Monotonic Reasoning, Constraint Satisfaction and Cognitive Modeling. We intend it to act as an assistant and advisor, by recommending solutions and leaving the final enforcement of decisions and actions to agents.

Acknowledgements: This work was partially funded by the European Commission through the GeoMed project (IE 2037), when the first two authors were with GMD. We would like to thank Gerhard Brewka, Thomas Gordon and Hans Voss for helpful discussions and comments on various topics of the paper; also Klaas Sikkel and David Kerr for bringing up the example used in the paper. Nikos Karacapilidis is financed by the Commission of the European Communities through the ERCIM Fellowship Programme.

References

Benferhat, S., Dubois, D., Prade, H. 1995. How to infer from inconsistent beliefs without revising? In Proceedings of the 14th IJCAI, 1449-1455.

Brewka, G. 1989. Preferred Subtheories: An extended logical framework for default reasoning. In Proceedings of the 11th IJCAI, 1043-1048.

Brewka, G. 1994a. Reasoning about Priorities in Default Logic. In Proceedings of the 12th AAAI, 940-945.

Brewka, G. 1994b. A Reconstruction of Rescher's Theory of Formal Disputation Based on Default Logic. In Working Notes of the 12th AAAI Workshop on Computational Dialectics, 15-27.

Cayrol, C. 1995. On the Relation between Argumentation and Non-monotonic Coherence-Based Entailment. In Proceedings of the 14th IJCAI, 1443-1448.

Conklin E.J. 1992. Capturing Organizational Memory. In D. Coleman (ed.) *Groupware '92*, 133-137.

Doyle, J. 1979. A Truth Maintenance System. *Artificial Intelligence* 12(3):231-272.

Dung, P.M. 1993. On the acceptability of arguments and its fundamental role in non-monotonic reasoning and logic programming. In Proceedings of the 13th IJCAI, 852-857.

Eemeren, F.H. van and Grootendorst, R. 1992. *Argumentation, communication, and fallacies. A pragma-dialectical perspective*. Hillsdale, NJ: Lawrence Erlbaum Associates.

Elvang-Goransson, M., Fox, J., Krause P. 1993. Acceptability of arguments as "logical uncertainty". In LNCS 747, 85-90. Berlin: Springer-Verlag.

Farley, A.M., Freeman, K. 1995. Burden of Proof in Legal Argumentation. In Proceedings of the 5th International Conference on AI & Law, 156-164.

Geffner, H., Pearl, J. 1992. Conditional Entailment: Bridging two Approaches to Default Reasoning, *Artificial Intelligence* 53(2-3):209-244.

Gordon, T. 1996. Computational Dialectics. In P. Hoschka (ed.), *Computers as Assistants - A New Generation of Support Systems*, 186-203. Hillsdale, NJ: Lawrence Erlbaum Associates.

Gordon, T. 1993. The Pleadings Game: An Artificial Intelligence Model of Procedural Justice. Ph.D. diss., Fachbereich Informatik, Technische Hochschule Darmstadt.

Hurwitz, R., Mallery, J.C. 1995. The Open Meeting: A Web-Based System for Conferencing and Collaboration. In Proceedings of the 4th International WWW Conference.

Karacapilidis, N.I. 1995. Planning under Uncertainty: A Qualitative Approach. In C. Pinto-Ferreira and N.J. Mamede (eds.), *Progress in Artificial Intelligence*, LNAI 990, 285-296. Berlin: Springer-Verlag.

Karacapilidis, N.I., Pappis, C.P. 1997. A framework for group decision support systems: Combining AI tools and OR techniques. *European Journal of Operational Research* 103: 373-388.

Kersten, G.E. 1985. NEGO-group decision support system. *Information and Management* 8: 237-246.

Kersten, G.E., Szapiro, T. 1986. Generalized approach to modeling negotiations. *European Journal of Operational Research* 26: 124-142.

Kunz, W., Rittel, H.W.J. 1970. Issues as Elements of Information Systems. Working Paper 0131, Institut fuer Grundlagen der Plannung, Universitaet Stuttgart.

Mackworth, A., Freuder, E. 1985. The Complexity of some Polynomial Network Consistency Algorithms for Constraint Satisfaction Problems", *Artificial Intelligence* 25:65-74.

Pollock, J. 1988. Defeasible Reasoning. *Cognitive Science* 11:481-518.

Prakken, H. 1995. From Logic to Dialectics in Legal Argument. In Proceedings of the 5th International Conference on AI and Law, 165-174.

Prakken, H. 1993. Logical Tools for Modelling Legal Argument. Ph.D. diss., Free University of Amsterdam.

Reiter, R. 1980. A Logic for Default Reasoning. *Artificial Intelligence* 13:81-132.

Rescher, N. 1977. *Dialectics: A Controversy-Oriented Approach to the Theory of Knowledge*. Albany:State University of New York Press.

Rittel, H.W.J., Webber, M.M. 1973. Dilemmas in a General Theory of Planning. *Policy Sciences* 4:155-169.

Simari, G.R., Loui, R.P. 1992. A Mathematical Treatment of Defeasible Reasoning and its Implementation. *Artificial Intelligence* 53(2-3):125-157.

Sycara, K. 1987. Resolving Adversarial Conflicts: An Approach Integrating Case-Based and Analytic Methods. Ph.D. diss., School of Information and Computer Science, Georgia Institute of Technology.

Toulmin, S.E. 1958. *The Uses of Argument*. Cambridge University Press.

Yakemovic, K.C.B., Conklin, E.J. 1990. Report on a Development Project Use of an Issue-Based Information System. In Proceedings of CSCW 90, 105-118.

Modelling Multi-Issue Negotiation

May Tajima and Niall M. Fraser

Department of Management Sciences, University of Waterloo, Waterloo, Ontario, Canada, N2L 3G1

Abstract. Multi-issue (criteria) negotiation belongs to a domain of relatively new research area which deals with both multiple criteria decision making (MCDM) and multiple participant decision making (MPDM). Its trade-off process is modelled using *logrolling* in order to achieve *integrative* solutions. Logrolling is the exchange of loss in some issues for gain in others resulting in mutual gain. Since the trade-offs among issues are also the trade-offs among negotiating parties, the solution approach using logrolling attempts to deal with MCDM and MPDM simultaneously. This study considers two-issue two-party negotiation. It is shown that logrolling is always possible except for one special case. Logrolling does not suggest a unique solution to multi-issue negotiation, but it can provide an efficient frontier which is shown to be linear or piecewise linear when the parties' value functions are linear. Characteristics of the Nash bargaining solution are also determined.

Keywords. Negotiation, multiple criteria decision making, integrative bargaining, logrolling, trade-offs

1 Multi-Issue Negotiation Methodology

In politics, business, and other international and interpersonal relations, people negotiate everyday and make agreements that involve many issues. Decision making in multi-issue negotiation is difficult. The difficulty comes from two sources: having multiple issues (criteria) and having multiple decision makers. This makes multi-issue negotiation belong to a domain of relatively new research area which deals with both multiple criteria decision making (MCDM) and multiple participant decision making (MPDM) (such as game theory). A multi-issue negotiation methodology can help negotiators make better decisions by analyzing and dealing with the complexity caused by multiple issues and multiple decision makers.

The perspective in multi-issue negotiation methodology can range from strictly competitive to cooperative. This study takes the *integrative* approach which tries to find a high benefit solution to all parties by reconciling, or integrating, the parties' interests (Pruitt, 1981). One of the methods for achieving integrative solutions mentioned by Pruitt (1983) is called *logrolling*. Logrolling is the key concept of this study.

Logrolling in general means the exchanging of assistance or favours. The term originated from a former American custom of neighbours assisting one another in rolling logs into a pile for burning. It also means the trading of votes by legislators (Webster On-Line Dictionary, 1996). One area of research in logrolling is voting (Bernholz, 1973; Miller, 1977; Tullock, 1959; and others). The other main research area is in psychology of bargaining behaviour (Froman and Cohen, 1970; Roloff and Jordan, 1991; Thompson, 1990; and others). This study defines logrolling similarly to that of the psychological studies which discuss logrolling as one of the bargaining strategies, but the analysis is quantitative and it focuses on the trade-off process.

In the following sections, logrolling and the related concepts are defined and discussed for *two-issue two-party* negotiation. Then the efficient frontier and the Nash bargaining solution are examined.

2 Logrolling

Logrolling in two-issue two-party negotiation is the exchange of loss in one issue for gain in the other which results in an increase of the overall value for *both* parties. For example, two parties are negotiating over a large quantity of apples and bananas. Originally, party 1 has more apples and party 2 has more bananas. However, party 1 loves bananas and party 2 prefers apples over bananas. In this case, each party can increase the overall value by trading the less preferred fruit for the more preferred, provided that the trade-off ratio is satisfactory. This exchange of fruits is an example of logrolling since it results in mutual gain.

In this study, the type of bargaining is assumed to be distributive over each issue, and maximization is the objective for both parties. Let i denote the issue which increases and j denote the issue which decreases in an exchange for party 1. The *exchange rate* between issues i and j is defined to be the ratio of the increase in i to the decrease in j, denoted by S_{1i}/S_{1j}, $S_{1i}, S_{1j} > 0$. Correspondingly, the exchange rate for party 2 is S_{2j}/S_{2i}, $S_{2i}, S_{2j} > 0$, and $S_{1i} = S_{2i}$ and $S_{1j} = S_{2j}$. In other words, for example, $S_{1i} = S_{2i}$ implies that the amount party 1 gains in issue i is equal to the amount party 2 loses in issue i. S_{1i}, S_{1j}, S_{2i}, and S_{2j} are assumed to be in the original units of measurement. Based on this notion of exchange rate, the formal definition of logrolling is given as follows.

Definition. *Logrolling* is the exchange based on the exchange rates, S_{1i}/S_{1j} for party 1 and S_{2j}/S_{2i} for party 2, which satisfies

$$V_1(x_{1i} + S_{1i}, x_{1j} - S_{1j}) > V_1(x_{1i}, x_{1j}) \text{ and } V_2(x_{2i} - S_{2i}, x_{2j} + S_{2j}) > V_2(x_{2i}, x_{2j}).$$

$V_k(X_k)$, $X_k \in \Re^2$, is the value function for party k, and (x_{ki}, x_{kj}) is the starting point for party k, $k = 1, 2$. The value function is assumed to exist for each party, and is assumed to be linear for this study.

3 Logrolling Range

By definition, logrolling provides mutual gain. The key to identifying mutual gain, and therefore integrative solutions, is recognition of the exchange rates that permit logrolling. The range of such exchange rates can be determined using the parties' criteria weights and is called the *logrolling range*. The justification and the existence of the logrolling range are based on the concept underlying the weight assessment technique using trade-offs (Keeney and Raiffa, 1976).

Definition. Suppose S_{1i}/S_{1j} and S_{2j}/S_{2i} are party 1 and party 2's exchange rates. w_i and w_j are party 1's weights for issues i and j, and z_i and z_j are party 2's weights for i and j.

(i) Two parties have different priorities if and only if $w_i>w_j$ and $z_i<z_j$. In this case, the logrolling ranges for S_{1i}/S_{1j} and S_{2j}/S_{2i} are as follows.

$$\frac{w_j}{w_i} < \frac{S_{1i}}{S_{1j}} < \frac{z_j}{z_i} \quad \text{and} \quad \frac{z_i}{z_j} < \frac{S_{2j}}{S_{2i}} < \frac{w_i}{w_j}$$

These ranges exist if both parties gain in their more important issue and lose in the less important issue.

(ii) Two parties have the same priority but at different preference intensity if and only if $(w_i/w_j)\neq(z_i/z_j)$ where $w_i>w_j$ and $z_i>z_j$. In this case, the logrolling ranges for S_{1i}/S_{1j} and S_{2j}/S_{2i} are as follows.

$$\text{If } (w_i \,/\, w_j) > (z_i \,/\, z_j), \quad \frac{w_j}{w_i} < \frac{S_{1i}}{S_{1j}} < \frac{z_j}{z_i} \quad \text{and} \quad \frac{z_i}{z_j} < \frac{S_{2j}}{S_{2i}} < \frac{w_i}{w_j}.$$

$$\text{If } (w_i \,/\, w_j) < (z_i \,/\, z_j), \quad \frac{w_i}{w_j} < \frac{S_{1j}}{S_{1i}} < \frac{z_i}{z_j} \quad \text{and} \quad \frac{z_j}{z_i} < \frac{S_{2i}}{S_{2j}} < \frac{w_j}{w_i}.$$

These ranges exist if the party having the higher (lower) preference intensity gains in issue i (j) and loses in issue j (i).

The only case in which logrolling range cannot exist is when two parties have the same priority at the same preference intensity, that is, $(w_i/w_j)=(z_i/z_j)$. In other words, dealing with a negotiator who is different from oneself actually facilitates the discovery of mutual gain. Even if the parties prefer the same issue, as long as there is a difference in how much they prefer, logrolling can still apply. Hence, according to the definition of the logrolling range, logrolling, a method to search for mutual gain, is always possible except only for one special case.

4 Two-Issue Logrolling Procedure

The potential mutual gain is indicated by the logrolling range and is exploited by the actual exchange of issues. The negotiation procedure using logrolling is described in Figure 1. The procedure either identifies the starting point as Pareto optimal or produces a Pareto optimal solution as the end result.

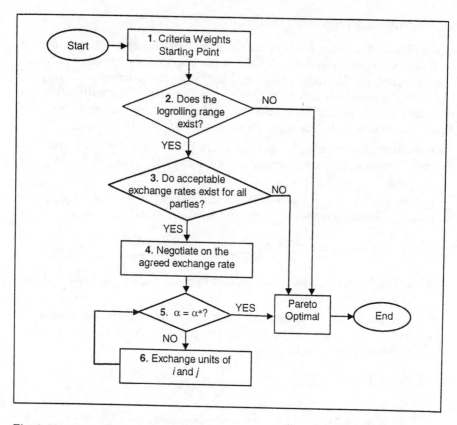

Fig. 1. Negotiation procedure using logrolling

Notes on the terminology used in Figure 1:

Step 3. The exchange rate is *acceptable* for party k if it yields $\Delta V_k > \mathcal{A}_k$ when it is used in logrolling. ΔV_k is the difference in the value before and after logrolling and $\mathcal{A}_k$ is the activation cost of negotiation, $k = 1, 2$.

Step 4. Agreed exchange rate is the exchange rate which is acceptable to all parties and agreed by all parties to be used in logrolling. The agreed exchange rate is assumed to remain constant over one session of logrolling.

Steps 5 and 6. α is a multiplying factor, $\alpha \geq 0$, which is used in expressing the total amount of change in issues. Let s_{1i}/s_{1j} be party 1's agreed exchange rate with particular values. Then αs_{1i} and αs_{1j} be the total amount of change in issues i and j, respectively, for party 1. This way, the exchange rate remains constant $(\alpha s_{1i}/\alpha s_{1j} = s_{1i}/s_{1j})$, and the value after logrolling is expressed as $V_1(x_{1i} + \alpha s_{1i}, x_{1j} - \alpha s_{1j})$. It can be shown that $V_1(x_{1i} + \alpha s_{1i}, x_{1j} - \alpha s_{1j})$ obtains the maximum when α takes its maximum, α^*. The value of α^* can be determined from how the exchange ends. The exchange stops when one of the following conditions is met.

(i) Issue i reaches maximum for party 1: $\bar{x}_i = x_{1i} + \alpha s_{1i} \Leftrightarrow \alpha = (\bar{x}_i - x_{1i})/s_{1i}$

(ii) Issue j reaches minimum for party 1: $\underline{x}_j = x_{1j} - \alpha s_{1j} \Leftrightarrow \alpha = (x_{1j} - \underline{x}_j)/s_{1j}$

where $(\bar{x}_i, \bar{x}_j)$ represents a set of maximum values and $(\underline{x}_i, \underline{x}_j)$ represents a set of minimum values for the two issues. Then, $\alpha^* = min\{(\bar{x}_i - x_{1i})/s_{1i}, (x_{1j} - \underline{x}_j)/s_{1j}\}$. The total amount of change in i is $min\{\alpha^* s_{1i}, (\bar{x}_i - x_{1i})\}$, and the total amount of change in j is $max\{\alpha^* s_{1j}, (x_{1j} - \underline{x}_j)\}$.

5 Efficient Frontier for Two-Issue Logrolling

In actual negotiations, the negotiating parties must negotiate over an agreed exchange rate. At this point, multi-issue negotiation is actually a negotiation over an agreed exchange rate. Once chosen, it determines the trade-off between the issues which is also the trade-off between the parties. Therefore, the solution approach using logrolling attempts to deal with MCDM and MPDM simultaneously. This is worth mentioning because many studies that deal with MCDM-MPDM problems, such as multi-issue negotiation, take an approach that decomposes the two components. If an agreed exchange rate is not specified, logrolling does not suggest a unique solution. However, one can alternatively study the efficient frontier over the logrolling range in order to further analyze the possible solutions.

5.1 Linear Value Functions

In order to study the efficient frontier, the parties' values as functions of the exchange rate are first identified. Considering party 1 first, $V_1(x_{1i} + \alpha S_{1i}, x_{1j} - \alpha S_{1j})$ is the end result of logrolling using the exchange rate S_{1i}/S_{1j}. $V_1(x_{1i} + \alpha S_{1i}, x_{1j} - \alpha S_{1j})$ is further specified to be $V_1(x_{1i} + \alpha S_{1i}, \underline{x}_j)$ if logrolling stops when issue j reaches minimum for party 1 or $V_1(\bar{x}_i, x_{1j} - \alpha S_{1j})$ if logrolling stops when issue i reaches maximum for party 1.

In $V_1(x_{1i} + \alpha S_{1i}, \underline{x}_j)$, αS_{1i} indicates the amount of increase in issue i as the exchange rate S_{1i}/S_{1j} increases. Note that αS_{1i} is the only changing quantity in $V_1(x_{1i} + \alpha S_{1i}, \underline{x}_j)$. Assume that issues i and j are independent of each other. Then $V_1(x_{1i} + \alpha S_{1i}, \underline{x}_j)$ can be treated as a function of αS_{1i} over the logrolling range. Now let us find the slope of $V_1(x_{1i} + \alpha S_{1i}, \underline{x}_j)$ by letting αS_{1i} increase by a small amount ε, $\varepsilon > 0$.

$$\frac{V_1(x_{1i} + \alpha S_{1i} + \varepsilon, \underline{x}_j) - V_1(x_{1i} + \alpha S_{1i}, \underline{x}_j)}{(\alpha s_i + \varepsilon) - \alpha s_i}$$

$$= \frac{V_1[(x_{1i} + \alpha S_{1i} + \varepsilon, \underline{x}_j) - (x_{1i} + \alpha S_{1i}, \underline{x}_j)]}{\varepsilon} \quad \text{(by } V_1(X_1) \text{ linear)}$$

$$= \frac{V_1(\varepsilon, 0)}{\varepsilon}$$

Since the slope is constant, it is concluded that $V_1(x_{1i} + \alpha S_{1i}, \underline{x}_j)$ is linear over the logrolling range.

Following the same procedure above, $V_1(\overline{x}_i, x_{1j} - \alpha S_{1j})$ can be shown to be linear. The slope of $V_1(\overline{x}_i, x_{1j} - \alpha S_{1j})$ by letting $-\alpha S_{1j}$ increase by a small amount δ, $\delta > 0$, is $V_1(0,\delta)/\delta$. The same procedure also applies to party 2's value functions.

5.2 Linear Efficient Frontier

Sometimes logrolling ends only by reaching issue j's minimum (case 1) and sometimes ends only by reaching issue i's maximum (case 2) over the logrolling range. The combination case is possible, but let us discuss cases 1 and 2 first.

In case 1, party 1's value function is $V_1(x_{1i} + \alpha S_{1i}, \underline{x}_j)$ and it is linearly increasing over the logrolling range. Party 2's value function is $V_2(x_{2i} - \alpha S_{2i}, \overline{x}_j)$, and it can be shown linearly decreasing over the logrolling range. Then as V_1 increases, corresponding V_2 decreases at a fixed rate. Therefore, the efficient frontier is linear.

In case 2, replace party 1's value function with $V_1(\overline{x}_i, x_{1j} - \alpha S_{1j})$ and party 2's value function with $V_2(\underline{x}_i, x_{2j} + \alpha S_{2j})$ in the previous paragraph. The same argument applies and it can be shown that the efficient frontier is linear.

5.3 Piecewise Linear Efficient Frontier

In the combination case, as the exchange rate increases, party 1's value function switches from $V_1(x_{1i} + \alpha S_{1i}, \underline{x}_j)$ to $V_1(\overline{x}_i, x_{1j} - \alpha S_{1j})$. This implies that the slope changes from $V_1(\varepsilon,0)/\varepsilon$ to $V_1(0,\delta)/\delta$. Here, the units of ε and δ may not be the same because the units of issues i and j can be different. However, it can be shown that scaling does not affect the basic expression of the nominators having one issue equal to 0 and the other equal to a positive score. So the slope of $V_1(x_{1i} + \alpha S_{1i}, \underline{x}_j)$ can be written as $V_1(\varepsilon',0)$ and the slope of $V_1(\overline{x}_i, x_{1j} - \alpha S_{1j})$ as $V_1(0,\delta')$. Now these expressions can be compared. By assuming issues i and j are independent, one of these cases holds.

$$V_1(\varepsilon', 0) > V_1(0, \delta') \quad \text{if issue } i \text{ has higher priority than issue } j. \tag{1}$$
$$V_1(\varepsilon', 0) < V_1(0, \delta') \quad \text{if issue } j \text{ has higher priority than issue } i. \tag{2}$$

In either case, the difference between the two slopes creates a corner point, or *break point*, at where the two functions intersect. Note that the break point occurs only once over the logrolling range and occurs simultaneously for two parties. The existence of the break point indicates that $V_1(x_{1i} + \alpha S_{1i}, x_{1j} - \alpha S_{1j})$ is piecewise linear over the logrolling range since it is composed of two linear functions, $V_1(x_{1i} + \alpha S_{1i}, \underline{x}_j)$ and $V_1(\overline{x}_i, x_{1j} - \alpha S_{1j})$, which have different slopes.

5.4 General Shape of the Piecewise Linear Efficient Frontier

Two cases are further considered regarding the shape of the piecewise linear efficient frontier: the case in which two parties have the different priorities (case 3) and the case in which the parties have the same priority but at different preference intensity (case 4).

Case 3. Suppose that issue i has higher priority than issue j for party 1 and suppose the opposite for party 2. Then the slopes of $V_1(x_{1i} + \alpha S_{1i}, \underline{x}_j)$ and $V_1(\overline{x}_i, x_{1j} - \alpha S_{1j})$ have relationship (3) and the slopes of $V_2(x_{2i} - \alpha S_{1i}, \overline{x}_j)$ and $V_2(\underline{x}_i, x_{2j} + \alpha S_{1j})$ have relationship (4) over the logrolling range based on relationships (1) and (2).

$$V_1(\varepsilon', 0) > V_1(0, \delta') \text{ where } V_1(\varepsilon', 0) \text{ and } V_1(0, \delta') > 0. \tag{3}$$
$$V_2(\varepsilon'', 0) > V_2(0, \delta'') \text{ where } V_2(\varepsilon'', 0) \text{ and } V_2(0, \delta'') < 0. \tag{4}$$

The slope of the efficient frontier before the break point is $[V_2(\varepsilon'',0)/V_1(\varepsilon',0)]<0$, while the slope of the efficient frontier after the break point is $[V_2(0,\delta'')/V_1(0,\delta')]<0$. From relationships (3) and (4), it can be seen that

$$0 > \frac{V_2(\varepsilon'',0)}{V_1(\varepsilon',0)} > \frac{V_2(0,\delta'')}{V_1(0,\delta')} \ .$$

Figure 2 illustrates the above result.

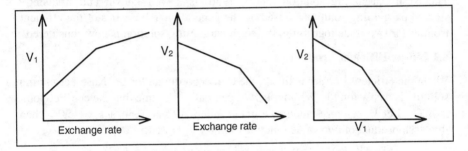

Fig. 2. Value functions and the efficient frontier for case 3

Case 4. Suppose that issue i has higher priority than issue j for both parties. Also suppose that party 1 has higher preference intensity than party 2. Then the slopes of $V_1(x_{1i} + \alpha S_{1i}, \underline{x}_j)$ and $V_1(\overline{x}_i, x_{1j} - \alpha S_{1j})$ have relationship (5) and the slopes of $V_2(x_{2i} - \alpha S_{1i}, \overline{x}_j)$ and $V_2(\underline{x}_i, x_{2j} + \alpha S_{1j})$ have relationship (6).

$$V_1(\varepsilon',0)>V_1(0,\delta') \text{ where } V_1(\varepsilon',0), V_1(0,\delta')>0, \text{ and } V_1(0,\delta')=1/a \cdot V_1(\varepsilon',0) \tag{5}$$
$$V_2(\varepsilon'',0)<V_2(0,\delta'') \text{ where } V_2(\varepsilon'',0), V_2(0,\delta'')<0, \text{ and } V_2(0,\delta'')=1/b \cdot V_2(\varepsilon'',0) \tag{6}$$
In (5) and (6), $a,b > 0$ and $a>b$ which implies $a/b > 1$.

The slope of the efficient frontier before the break point, $[V_2(\varepsilon'',0)/V_1(\varepsilon',0)]<0$, and the slope of the efficient frontier after the break point, $[V_2(0,\delta'')/V_1(0,\delta')]<0$, have the following relationship based on (3) and (4) above.

$$0 > \frac{V_2(\varepsilon'',0)}{V_1(\varepsilon',0)} > \left(\frac{V_2(0,\delta'')}{V_1(0,\delta')} = \frac{1/b}{1/a} \cdot \frac{V_2(\varepsilon'',0)}{V_1(\varepsilon',0)} = a/b \cdot \frac{V_2(\varepsilon'',0)}{V_1(\varepsilon',0)} \right)$$

Figure 3 summarizes the above result.

184

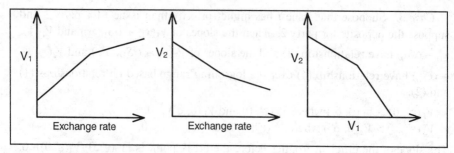

Fig. 3. Value functions and the efficient frontier for case 4

In summary, the efficient forntier for two-issue logrolling is shown to be linear or piecewise linear which is concave.

6 Nash Bargaining Solution

The Nash bargaining solution (Nash, 1950) is a well known and frequently studied bargaining solution. Based on the general characteristics of the efficient frontier for two-issue logrolling, the Nash bargaining solution is now considered.

6.1 Linear Efficient Frontier

When the efficient frontier is linear, the exact coordinate for the Nash bargaining solution is easily found. Without loss of generality, assume that the conflict point is $(0,0)$. Let V_2 be represented by a linear function of v_1, $-mv_1 + k$, $m,k>0$. Then the Nash product of two value functions V_1 and V_2 is written as:

$$(V_1 \cdot V_2)(v_1) = v_1 \cdot (-mv_1 + k) = -mv_1^2 + kv_1, \; m, \, k > 0.$$

The Nash product is a concave quadratic function, and therefore, a unique global maximum exists and it is found at $(v_1, v_2) = (k/(2m), k/2)$.

6.2 Piecewise Linear Efficient Frontier

For an efficient frontier which is piecewise linear and is composed of two linear functions, let us denote the two functions as F_1 and F_2. Let $P_1=(p_1,V_2(p_1))$ and $P_2=(p_2,V_2(p_2))$ be the Nash bargaining solutions for F_1 and F_2, respectively. Let $P_3=(p_3,V_2(p_3))$ be the break point of the efficient frontier. Based on whether P_1 and P_2 are located on the efficient frontier, four different combinations exist.

(i) P_1 is on the efficient frontier and not P_2.
(ii) P_2 is on the efficient frontier and not P_1.
(iii) Neither P_1 nor P_2 is on the efficient frontier.
(iv) Both P_1 and P_2 are on the efficient frontier.

Cases (i)-(iii) are illustrated in Figure 4. Case (iv) can be shown to be impossible.

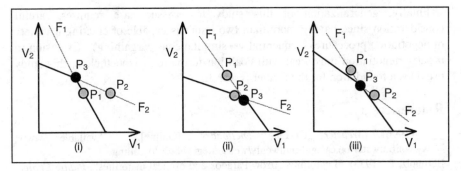

Fig. 4. Nash solution for logrolling: three possible cases

Let us examine case (i) closely. First recall that the Nash product is a concave quadratic function when the efficient frontier is linear. Let $(p^*,V_2(p^*))$ be the Nash bargaining solution. Then the Nash product has a positive slope at a point, $(p,V_2(p))$, if $p<p^*$ and has a negative slope if $p>p^*$. This implies the following which is referred to as the *quadratic property* of $V_1 \cdot V_2$.

For any $p' < p'' < p^*$, $(V_1 \cdot V_2)(p') < (V_1 \cdot V_2)(p'') < (V_1 \cdot V_2)(p^*)$.
For any $p^* < p' < p''$, $(V_1 \cdot V_2)(p^*) > (V_1 \cdot V_2)(p') > (V_1 \cdot V_2)(p'')$.

On F_2, among the points on the efficient frontier, P_3 maximizes $V_1 \cdot V_2$ according to the quadratic property of $V_1 \cdot V_2$. Note that P_3 is on F_1. On F_1, however, P_1 maximizes $V_1 \cdot V_2$ since it is the Nash solution for F_1. Therefore, P_1 must be the Nash bargaining solution since it maximizes $V_1 \cdot V_2$ on the efficient frontier.

Using the same argument, it can be shown that if either P_1 or P_2 is on the efficient frontier, but not both, then it is the Nash bargaining solution. If neither P_1 nor P_2 is on the efficient frontier, then the break point P_3, the distinct characteristic point of the piecewise linear efficient frontier, is the Nash bargaining solution.

7 Conclusion

In this study, logrolling and the related concepts are introduced and discussed for two-issue two-party negotiation. The solution approach using logrolling is distinctive in two ways: it deals with MCDM and MPDM simultaneously and it searches for integrative solutions via mutual gain. The potential mutual gain is identified by the logrolling range and is exploited through the negotiation procedure. Commonly, the difference in opinion or preference is seen as an obstacle in reaching an agreement between the parties. An interesting implication from the definition of the logrolling range is that the difference in parties' priority or preference intensity actually facilitates the search for mutual gain. Although logrolling does not suggest a unique solution to multi-issue negotiation, it can provide an efficient frontier for further analysis such as identifying the Nash bargaining solution.

Finally, generalization of this study to m-issue case requires careful consideration since having more than two issues is capable of creating a variety of negotiation procedures (sequential vs. simultaneous bargaining). Extension to n-party negotiation must deal with coalition formation. Nonetheless, this study provides a foundation for the further research.

References

___ (1996), *Webster On-Line Dictionary*, [Online]. Available www: www.lib.uwaterloo.ca/cgi-bin/uwonly/webster.cgi?word=logrolling

Bernholz, P. (1973), "Logrolling, Arrow Paradox and cyclical majorities", *Public Choice*, 15, 87-95.

Froman, L.A., Jr. and M.D. Cohen (1970), "Compromise and logroll: comparing the efficiency of two bargaining processes", *Behavioral Science*, 15, 180-183.

Keeney, R. and H. Raiffa (1976), *Decisions with Multiple Objectives: Preferences and Value Tradeoffs*, John Wiley & Sons.

Miller, N.R. (1977), "Logrolling, vote trading, and the paradox of voting: a game-theoretical overview", *Public Choice*, 30(0), 51-75.

Nash, J.F. (1950), "The bargaining problem", *Econometrica*, 18, 155-162.

Pruitt, D.G. (1983), "Achieving integrative agreements", in M.H. Bazerman and R.J. Lewicki (eds.), *Negotiating in Organizations*, Sage Publications, Beverly Hills, CA, 35-50.

Pruitt, D.G. (1981), *Negotiating Behavior*, Academic Press, New York, NY.

Roloff, M.E. and J.M. Jordan (1991), "The influence of effort, experience, and persistence on the elements of bargaining plans", *Communication Research*, 18(3), 306-332.

Thompson, L. (1990), "Negotiation behavior and outcomes: empirical evidence and theoretical issues", *Psychological Bulletin*, 108(3), 515-532.

Tullock, G. (1959), "Problems of majority voting", *Journal of Political Economy*, 67-571-579.

Influence in Group Decision Making

Geoff Lockett[1], Peter Naudé[2,] and Gerd Islei[3]

[1] Leeds University, Leeds, LS2 9JT, United Kingdom
[2] Manchester Business School, Booth Street West, Manchester, M15 6PB, UK
[3] Templeton College, Kennington, Oxford, OX1 5NY, United Kingdom

Abstract. Group decision making is becoming a common activity in many industrial settings, and a number of methods for assisting are regularly proposed. These may be theoretical, such as most of the MCDM approaches, or technologically based computer systems. However nearly all of them do not consider the individual influences that are brought to bear on the group decision. This paper presents a case study where this aspect is investigated using a method of psychological profiling. The experimental results show that the make-up of the group of decision makers may have a dramatic impact on the final decision. Although only a single small case study, the outcome suggests that group characteristics may have a direct impact on the use of MCDM models. It may be a fruitful avenue of future research in group decision making.

Keywords. Judgmental Modelling, Group Decision Making, Psychological Profiling

1 Introduction

Group decision making is a growing area of research and application for many organisations. It is particularly important now that IT/IS has become pervasive, and use of technology is being made by many companies to enhance their decision making capabilities. But an underlying assumption behind most of the methods and associated systems is that all the individuals concerned have equal influence. Therefore cultural, organisational and psychological factors are often neglected.

There have been many advances in the application of IT/IS to help decision making. One of the main themes for research has been in the general area of Decision Support Systems (DSS), and there is little doubt of the impact of some very simple systems such as spread sheets on most organisations. Similarly the effects of the utilisation of many types of system are increasingly evident e.g. data base systems. However most of the applications are on structured problems, which were not the original focus of DSS, and the supporting of less formalised decision making has not received as much attention. There have been some attempts to tackle the problem from a series of differing perspectives, and with varying degrees of success. One of the important areas is that of multi criteria decision making (MCDM). The international MCDM conferences give a flavour of what is discussed (Lockett and Islei, 1989; Goicoechea et al, 1991). Although numerous theoretical approaches have been developed, true reported applications are rare, and it is not uncommon for much of the literature to be devoted to solving mathematical puzzles rather than the problem of implementation. Although there are some excellent examples of successful practice (e.g. Saaty 1980; Islei et al, 1991), most of the published work avoids attempting to deal with the issues surrounding the wider organisational environment. Some authors do attempt to put the models in context (e.g. French, 1984; Belton, 1993; Hoch and Schkade, 1996), but they are few and far between and discussions of failure are almost non-existent. This may mean that there are very few attempts at application, but that is unlikely.

In addition to this lack of focus on the organisational and behavioural issues, most of the models so far produced have been devoted to the case of the single decision maker. Even when groups have been involved, the models have generally assumed that the group will act as a single decision maker, and that an 'averaging' of opinions will occur. Some approaches to overcome this problem have been made (see for example Islei et al, 1991: Lockett and Naudé, 1996), but this has not been a strong area for MCDM research. This paper aims to develop a better understanding of the way people behave when using decision making tools in a group situation.

A complementary area has been the implementation of many forms of group systems. Through the seminal work of Nunamaker et al (1988), it is generally accepted that group work will become a very important aspect of organisational systems in the next decade. However, even after all the apparent success, there are some implicit assumptions behind the systems that have yet to be tested. These were dealt with earlier and will not be repeated, but it is important to note that the systems do not take into account any of the work on group behaviour undertaken in the social sciences. So far there is little research on the topic, and most of the comments are anecdotal i.e. that there is some resistance from the important decision makers to the use of bulletin boards. Nevertheless it is highly likely that a much better understanding of human behaviour in this type of decision and organisational environment will be necessary if the full potential of such systems is to be realised.

This paper looks at one important characteristic i.e. group composition and uses a method of psychological profiling to investigate its effect on a case study of a real decision. It is based on an application of judgmental modelling in the production of a forecast. A group of managers from a variety of types of organisations took part, but they were all from a similar technical background. The group members used the models to structure their decision, and then in an informal manner negotiated final solutions. In order to investigate some of the problem areas discussed above, psychological profiles were taken of each of the participants, and this information was used in the experimental design. The study is described in full in the next section.

2 The Experiment

The main purpose of the exercise was to produce a forecast for the future of pharmaceutical dosage forms for the year 2000 and beyond. All nine participants were members of the Institute of Chemical Engineers, and are part of the Pharmaceutical, Toiletries and Cosmetics Group. In this paper they will be referred to as individuals A to I. They wanted to produce forecasts that would be published by the Institute and which would be the basis for an informed discussion. The participants had some knowledge of the methods to be adopted, and were all computer literate.

As well as using the exercise for the solution to the particular problem, the authors used the occasion, with the group's knowledge, to also investigate the effect of the type of individual on the group negotiation process. Therefore prior to the meeting the individuals each completed a psychological profiling questionnaire. These were based on the work of Belbin (1981; 1993), which is founded on the concept of team roles. There is some discussion in the literature about the best method of profiling and other solutions have been developed [for the interested reader see Kirton, 1989; Furnham et al, 1993; Dulewicz, 1995; and Allison and Hayes, 1996]. However for this exploratory exercise, given the ease of implementation and its recognition by most managers, the Belbin methodology was used. The different team roles are presented in Table 1, and the profiles are shown in Figure 1 (note that boxes indicate a high score on the attribute, circles a low score). On the day when all the participants met, this research aspect was not to the fore and all of them concentrated on the primary task of producing a good forecast.

Table 1 A Brief Description of the Different Team Roles

The Company Worker: Conservative, dutiful, and predictable, little self-interest. Good organising ability, hard working, but inflexible and unresponsive to new ideas.
Chairman: Adept at drawing on group resources. Calm, self-confident, controlled. Welcomes all potential contributors on merit. Strong sense of objectives.
Shaper: Driving, action-based leader. Changes, challenges, disturbs the group. Highly strung, outgoing, and dynamic.
Plant: Introverted, but unorthodox. Has the imagination and intellect to solve difficult problems. Inclined to disregard practical ideas or protocol.
Resource Investigator: Outgoing, extroverted, inquisitive. Capacity for exploring anything new, and responding to challenges. Liable to lose interest.
Monitor-Evaluator: Serious minded, prudent, and slow in making decisions. Makes shrewd judgements, but lacks inspiration or ability to motivate others.
Team Worker: Socially oriented. Ability to respond to people and to situations, and to promote team spirit. Indecisiveness during moments of crisis.
Completer-Finisher: Ability to finalise thoroughly. Often in short supply. Painstaking, orderly, anxious. Worries about small things, and reluctant to delegate.

	Completer-Finisher	Company Worker	Chairman	Shaper	Plant	Monitor Evaluator	Team Worker	Resource Investigator
A	6	7	10	14	(3)	9	(3)	[18]
B	6	11	10	11	(3)	6	15	10
C	[11]	[23]	(6)	(5)	10	(0)	11	(4)
D	5	11	11	15	7	7	(5)	9
E	[14]	14	8	10	9	(5)	(3)	7
F	(0)	14	8	10	8	11	(6)	[13]
G	[12]	15	(4)	14	9	9	(3)	(4)
H	5	(0)	[16]	[36]	(0)	9	(0)	(4)
I	8	12	7	[18]	6	(2)	(5)	[12]

NOTE: [ ] = high score

( ) = low score

Figure 1 The Individuals' Belbin Profiles

Using the results of the questionnaire, the group of nine was divided into three sub groups:- 1 (D,H,I), 2 (C,E,G), and 3 (A,B,F). The aim was to allocate the individuals in sub-groups so as to ensure that, as far as possible, those individuals within a sub-group would have similar types of profiles, while also maximising the difference between the three sub-groups so that they would be distinct. Given the real nature of the problem it is not possible to have perfect groupings, and therefore some latitude had to be exercised. The groups' characteristics and expected behaviour, from the Belbin theory, are presented in Figure 2. These three groups first produced problem structures that were very similar. The whole group then developed a model of the problem based on a hierarchical approach which was assessed using a judgmental model through an interactive computer based package JAS (Islei and Lockett, 1988; Lockett and

Islei, 1989). It has been widely used on a variety of problems and is very user friendly and unobtrusive (Islei et al, 1991). The outline structure of the problem is shown in Figure 3.

Group 1 (D, H, I)

A bit of a mixture, but mostly Shapers
Driving, action-based leader. Changes, challenges, disturbs the group.
Highly strung, outgoing, and dynamic.

Expected:
Shapers: Action based, like to challenge and disturb
One dominant individual (H)
Expect lack of agreement

Group 2 (C, E, G)

The Completer-Finshers

Ability to finalise thoroughly. Often in short supply. Painstaking, orderly, anxious. Worries about small things, and reluctant to delegate.

Tendency towards Company Worker and Plants as well

Expected:
Fairly cohesive
Wish to finalise thoroughly
Will agree

Group 3 (A, B, F)

The Resource Investigators
Outgoing, extroverted, inquisitive. Capacity for exploring anything new, and responding to challenges. Liable to lose interest.
No other particular characteristics

Expected:

Find the exercise interesting
No great ability to finalise, or to focus on detail
Will reach consensus easily due to lack of attention to detail
Expect relatively large swings in scores

Figure 2 The Groups' Characteristics and Expected Behaviour

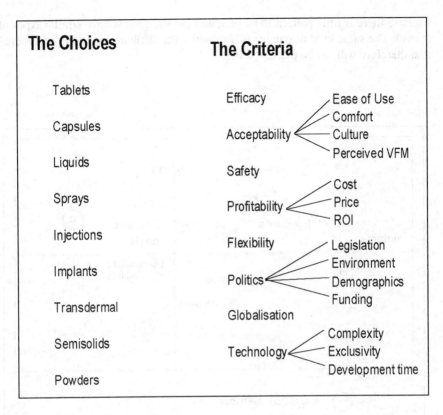

Figure 3 The Agreed Problem Structure

3 The Results

As stated previously, the three groups first discussed the problem structure independently, and after a short brainstorming session they collectively produced the problem structure shown in Figure 3. There was very little disagreement on this part of the exercise and the group members moved very quickly into a judgmental analysis of the model they had generated. This was done by each of the participants individually without any discussion about their perceptions. The results for their final choices are shown in Figure 4. The diagram has been derived using correspondence analysis (Greenacre, 1984), a perceptual mapping technique (Greenacre and Hastie, 1987; Hoffman and Franke, 1986). Essentially, the individuals/choices that are plotted close together are similar across the profiles of the different choices/individuals. Hence individuals C and G have very similar perceptions of the choices, whereas E and H vary substantially. Our interest is not in the values themselves, but in how the participants compare to each other, and such types of maps clearly identify how the emphases differ. Each of the individual results is identified by a letter A to I, and the groups to which the individuals were subsequently allocated is also identified. At this stage of the exercise there is little pattern to be seen and the sub-groups have similar types of spread. The same kind of pattern is obtained if the attribute scores are considered and therefore will not be presented here.

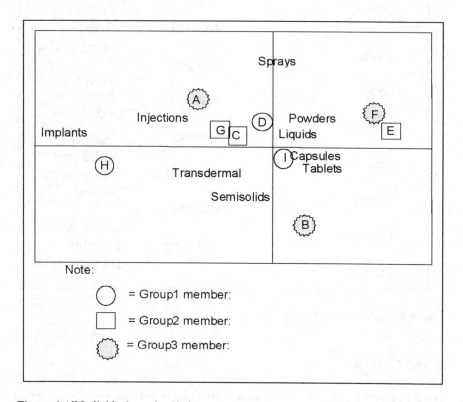

Figure 4 All Individuals on the Choices

These results were presented to the whole group for discussion to see if the members thought that they could form the basis for a forecast. After gaining a better understanding of their differences and the meaning of the numerical scores the second forecast was attempted.

This was undertaken by putting the participants into the three sub-groups presented earlier. They did not have knowledge of the method of group formation, and this was not revealed to them until well after the whole exercise had been completed and the results fully analysed. For the second round of judgmental modeling they used the same model structure. Some discussion took place about possible revision, but it was quickly agreed that it was an adequate representation of their beliefs. Each sub-group of three people then went away and completed a second exercise where they negotiated the group perceptions. The data input was done on a group basis and each input figure was obtained after negotiation between the group members. This finally produced three aggregate forecasts. The aim was to see if, during this negotiation phase, the psychological make-up of the group members had had any impact on the results compared to their earlier individual scores. Of course some variation between the two exercises would be expected, but the profiles may not have any significant effect on the outcome. In Figure 2 presented earlier, some of the expected behaviour of the sub groups was documented, and they can be summarised as :-

Group 1:	Expect lack of agreement
Group 2:	Will probably agree
Group 3:	Expect relatively large swings in scores.

These led directly from the Belbin theory and the way the sub groups were composed from the types of profile that were contained within the overall group. The results of the scores on the choices for the three sub-groups are shown in Figure 5, and can be readily compared to the above predictions. In each case we identify the individual's prior score (i.e. the score generated as an individual), and then also show the outcome of the three groups' negotiated position at the end of the second scoring exercise (these are labeled G1, G2, G3) [We can see that the inclusion of data points G1, G2, and G3 has resulted in the plot being presented as an almost exact mirror-image of Figure 4].

In the case of Group 1, it can be seen that they produce a result which is outside the configuration of their first answers. No averaging has taken place, and the dominant individual (H) has clearly 'pulled' the other two group members towards his position.

Group 2 has produced a much more interesting outcome. In this case the final negotiated outcome is a balance between their original preferences. They have in effect averaged, and negotiated from a fixed position. The result is much more stable than the other groups.

Group 3 have produced a final agreed result which is the most different from their first attempts. It shows the largest movement from the previous average of their exercise, and instead of moving between the three preliminary outcomes, the result shows a large swing away from two of the groups first ideas. Individual F seems to have persuaded A and B to completely change their opinions.

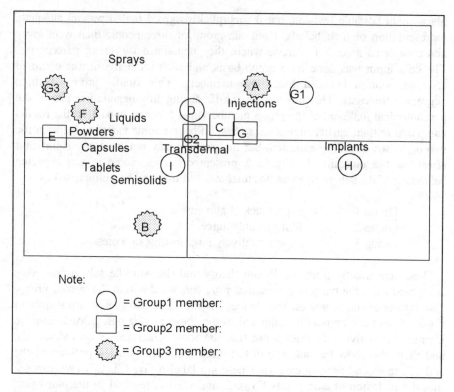

Figure 5 The Final Outcome

These final judgments were the result of considerable negotiation between the group members over a period of time. They were obtained after the individuals had just performed the same type of assessment separately. It was therefore fresh in their minds, and they had been presented with a detailed breakdown of their own analysis and had been shown the aggregate results of the whole group. The exercise was also conducted on a problem with which they were familiar and they were very interested in the outcome. It was generally thought that the final agreed judgements would in effect be averaged. Apart from G2 this is clearly not the case. Even with this small data set the results suggest that the psychological make up of the group participants can be an important factor in the eventual group decision. One cannot assume that averaging will take place. Of course, because of the small case set, the results should be interpreted with care. However even with this cautionary note they do suggest that group composition may have a crucial impact on the outcome of group negotiations. The methodology used for psychological profiling was a very simple instrument, and was chosen to be unobtrusive and not to inhibit the task at hand. Although used for research purposes, it did not interfere with the reality of the empirical exercise. Even though there was no control on the large group composition, and therefore theoretically sampled sub-groups could not be constructed, the groupings did allow the possible effects of profiles on group behaviour to be demonstrated.

4 Discussion

In this paper a case study has been presented which has used the psychological profiles of group members as a variable in the composition of the group. It has been done in order to explore how individual psychological profiles influence the outcome of negotiations within a group. The vehicle for this task has been the production of a technology forecast. Although only a small case study, and using a form of judgmental modelling as a tool for perception collection, these preliminary results provide us with insights into how the groups come to a conclusion during negotiation for a real life decision. They suggest that the psychological make-up may be one of the most important factors influencing the final outcome i.e. certain group members have strong influences on other members of the group. Again this small exercise has to be put in context, but at least it suggests that the subject needs to be the focus of much more research.

References

Allison, C. W., and Hayes, J. (1996) Measuring the Cognitive Styles of Managers and Professionals, *Journal of Management Studies*, 33:1, pp. 119-135

Belbin, R. M. (1981) *Management Teams: Why They Succeed or Fail*, London, Heinemann

Belbin, R. M. (1993) *Team Roles at Work*, Oxford, Butterworth-Heinemann

Belton, V. (1993) Project Planning and Prioritizing in the Social Services, *Journal of Operational Research*, (44:2), pp. 115-124, 1993

Dulewicz, V. (1995) A Validation of Belbin's Team Roles from 16PF and OPQ Using Bosses' Ratings of Competence, *Journal of Occupational and Organizational Psychology*, 68, pp. 81-99

French, S. (1984)Interactive multi-objective programming: its aims, applications, and demands, *Journal of the Operational Research Society*, 35, pp. 827-834

Furnham, A., Steele, H., and Pendleton, D. (1993) A psychometric assessment of the Belbin Team-Role Self Perception Inventory, *Journal of Occupational and Organizational Psychology*, 66, pp. 245-257

Goicoechea, A., Duckstein, L., and Zionts, S. (Eds.) (1991) *Multiple Criteria Decision Making*, Proceedings of the Ninth International Conference: Theory and Applications in Business, Industry, and Government, Springer-Verlag, Berlin

Greenacre, M. J. (1984) *Theory and Applications of Correspondence Analysis*, Academic Press, London

------- and Hastie, T. (1987) The Geometric Interpretation of Correspondence Analysis, *Journal of the American Statistical Association*, (82:398), pp. 437-447

Hoch, S.J. and Schkade, D.A. (1996) A Psychological Approach to Decision Support Systems, *Management Science*, 42:1, pp. 51-64

Hoffman, D. L. and Franke, G. R. (1986) Correspondence Analysis: Graphical Representation of Categorical Data in Marketing Research, *Journal of Marketing Research*, 23 (Aug), pp. 213-227

Islei, G and Lockett, A. G. (1988) Judgemental Modelling Based on Geometric Least Squares, *European Journal of Operational Research*, (36:1), pp. 27-35

--------, --------, Cox, B., Gisbourne, S. and Stratford, M. (1991) Modelling Strategic Decision Making and Performance Measurements at ICI Pharmaceuticals, *Interfaces*, (21:6), pp. 4-22

Kirton, M. J. (1989)Adaptors and Innovators at Work, in, M. J. Kirton (ed.) *Adaptors and Innovators; Styles of creativity and problem solving*, pp. 56-78. London: Routledge

Lockett, A.G. and Islei, G. (Eds.) (1989) *Improving Decision Making in Organisations*, Proceedings of the Eighth International Conference on MCDM, Springer-Verlag, Berlin

Lockett, A.G. and Naudé, P. (1998) The Stability of Judgemental Modelling: An Application in the Social Services, *Group Decision Making and Negotiation (Special Issue on Multiple Criteria Decision Making Models in Group Decision Support* (Forthcoming – vol. 7:1)

Nunamaker, J.F., Applegate, L.M., and Konsynski, B.R. (1988) Computer-Aided Deliberation: Model Management and Group Decision Support, *Operations Research*, (36:6), pp. 826-848

Saaty, T. L. (1980) *The Analytic Hierarchy Process*, New York, McGraw-Hill

Monotonicity of Power Indices[1]

František Turnovec

Center for Economic Research and Graduate Education, Charles University,
& Economics Institute of the Academy of Sciences of the Czech Republic,
Politických vězňů 7, 111 21 Prague 1, Czech Republic

Abstract: The paper investigates general properties of power indices, measuring the voting power in committees. Concepts of local and global monotonicity of power indices are introduced. Shapley-Shubik, Banzhaf-Coleman, and Holler-Packel indices are analyzed and it is proved that while Shapley-Shubik index satisfies both local and global monotonicity property, Banzhaf-Coleman satisfies only local monotonicity without being globally monotonic and Holler-Packel index satisfies neither local nor global monotonicity.

Keywords: Anonymity, committee, dummy member, marginality, monotonicity, power index, power axioms, quota, voting, weights, winning configuration

1 Introduction

Measuring of voting power in committee systems was introduced by Shapley and Shubik [1954] more than 50 years ago. Recently we can observe growing interest to different aspects of analysis of distribution and concentration of power in committees, related to development of new democratic structures in countries in transition as well as to institutional reform in European Union. Both methodological and applied problems of analysis of power are being intensively discussed during last few years (see e.g. Gambarelli [1990], [1992], Holler and Li [1995], Holubiec and Mercik [1994], Roth [1988], Turnovec [1995, 1996], Widgrén [1994, 1995]).

This paper focuses on methodological issues of power analysis. General properties of different measures of power are investigated. A model of a committee in terms of quotas and allocations of weights of committee members is used instead of traditional model of cooperative simple games. A subset of the set of all committee members is called a winning voting configuration if the total

---

[1] This research was undertaken with support from the European Commission's Phare ACE Programme, project No. P96-6252-R and cosponsored by the Grant Agency of the Czech Republic, project No. 402/98/1470. The author benefited significantly from the comments of Manfred Holler from University of Hamburg on the earlier draft of the paper.

weight of all its members is at least equal to the quota. The power index is defined as a mapping from the space of all committees into the unit simplex, representing a reasonable expectation of the share of voting power given by ability to contribute to formation of winning voting configurations.

Using Allingham [1975] axiomatic characterization of power, amended by an intuitively appealing axiom of so called global monotonicity, the most widely known power indices are confronted with the five axioms of power: dummy (a member of a committee who can contribute nothing to any winning voting configuration has no power), anonymity (power does not depend on committee members' names or numbers), symmetry (members with equal contribution to any voting configuration have the same power), local monotonicity (a member with greater weight cannot have less power than a member with smaller weight), and global monotonicity (if the weight of one member is increasing while the weights of all other members are decreasing or staying the same, the power of the "growing weight" member will at least not decrease).

The most widely known and used power indices, Shapley-Shubik index [1954], Banzhaf-Coleman index [1965, 1971] and Holler-Packel index [1978, 1983], are analyzed and it is proved that while Shapley-Shubik index satisfies all five power axioms, Banzhaf-Coleman satisfies dummy, anonymity, symmetry and local monotonicity axioms without being globally monotonic and Holler-Packel index satisfies only dummy, anonymity and symmetry axioms without being locally monotonic and globally monotonic.

The paper has no ambition to answer the straightforward question: which index is right? It rather indicates that there is something missing in power analysis: a unified approach to the problems of modelling and evaluating of voting and decisional power.

2 Power in Committees

Let $N = \{1, ..., n\}$ be the set of members (players, parties) and ω_i ($i = 1, ..., n$) be the (real, non-negative) weight of the i-th member such that

$$\sum_{i \in N} \omega_i = 1, \ \omega_i \geq 0$$

(e.g. the share of votes of party i, or the ownership of i as a proportion of the total number of shares, etc.). Let γ be a real number such that $0 \leq \gamma \leq 1$.
The (n+1)-tuple

$$[\gamma, \omega] = [\gamma, \omega_1, \omega_2, ..., \omega_n]$$

such that

$$\sum_{i=1}^{n} \omega_i = 1, \ \omega_i \geq 0, \ 0 \leq \gamma \leq 1$$

we shall call a committee of the size n = card N with quota γ and allocation of weights

$$\omega = (\omega_1, \omega_2, ..., \omega_n)$$

(by *card* S we denote the cardinality of the set S, for empty set *card* $\varnothing = 0$).

Any non-empty subset $S \subset N$ we shall call a voting configuration. Given an allocation ω and a quota γ, we shall say that $S \subset N$ is a winning voting configuration, if

$$\sum_{i \in S} \omega_i \geq \gamma$$

and a losing voting configuration, if

$$\sum_{i \in S} \omega_i < \gamma$$

(i.e. the configuration S is winning, if it has a required majority, otherwise it is losing).

Let

$$\Gamma = \left\{ (\gamma, \omega) \in \mathbb{R}_{n+1} : \sum_{i=1}^{n} \omega_i = 1, \ \omega_i \geq 0, \ 0 \leq \gamma \leq 1 \right\}$$

be the space of all committees of the size n and

$$E = \left\{ e \in \mathbb{R}_n : \sum_{i \in N} e_i = 1, \ e_i \geq 0 \ (i=1,...,n) \right\}$$

be the unit simplex.

A *power index* is a vector valued function

$$\pi : \Gamma \rightarrow E$$

that maps the space Γ of all committees into the unit simplex E. A power index represents a reasonable expectation of the share of decisional power among the various members of a committee, given by ability to contribute to formation of winning voting configurations. We shall denote by $\pi_i(\gamma, \omega)$ the share of power that the index π grants to the i-th member of a committee with weight allocation ω and quota γ. Such a share is called a *power index of the i-th member*.

2 Axiomatization of Power

A member $i \in N$ of the committee $[\gamma, \omega]$ is said to be **dummy** if he cannot benefit any voting configuration by joining it, i.e. the player i is dummy if

$$\sum_{k \in S} \omega_i \geq \gamma \Rightarrow \sum_{k \in S-\{i\}} \omega_i \geq \gamma$$

for any winning configuration $S \subset N$ such that $i \in S$.

Example 1
Let $[\gamma; \omega] = [0.7; 0.4, 0.4, 0.2]$, then we have only one winning configuration $\{1, 2, 3\}$ with the member 3:

$$\omega_1 + \omega_2 + \omega_3 = 1 > \gamma$$

and we know that

$$\omega_1 + \omega_2 = 0.8 > \gamma$$

then player 3 is dummy.

Two distinct members i and j of a committee $[\gamma, \omega]$ are called **symmetric** if their benefit to any voting configuration is the same, that is, for any S such that $i, j \notin S$

$$\sum_{k \in S \cup \{i\}} \omega_k \geq \gamma \quad \leftrightarrow \quad \sum_{k \in S \cup \{j\}} \omega_k \geq \gamma$$

Obviously, if for two members i and j of a committee $[\gamma, \omega]$ holds $\omega_i = \omega_j$, then i and j are symmetric.

Example 2
Let $[\gamma; \omega] = [0.7; 0.5, 0.3, 0.2]$. There is only one configuration not containing members 2 and 3, which is $\{1\}$. We can see that

$$\omega_1 + \omega_2 = 0.8 > \gamma, \quad \omega_1 + \omega_3 = 0.7 = \gamma$$

In this case the members 2 and 3 are symmetric.

Let $[\gamma, \omega]$ be a committee with the set of members N and

$$\sigma : N \rightarrow N$$

be a permutation mapping. Then the committee

$$[\gamma, \sigma\omega]$$

we shall call a **permutation** of the committee $[\gamma, \omega]$ and $\sigma(i)$ is the new number of the member with original number i.

Example 3
Let $[\gamma; \omega] = [0.7; 0.5, 0.3, 0.2]$ and $\sigma(N) = (3, 1, 2)$, then $\sigma(1) = 3$, $\sigma(2) = 1$, $\sigma(3) = 2$, and $[\gamma, \sigma\omega] = [0.7; 0.2, 0.3, 0.5]$.

Let $\pi_i(\gamma, \omega)$ be a measure of power of a member i in a committee with quota γ and allocation ω, then (assuming no additional information about the structure of the committee and specific voting rules) it is natural to expect that some minimal intuitively acceptable properties should be satisfied by a reasonable π. The following axiomatic characterization of power indices (in slightly different form) was introduced by Allingham [1975]:

Axiom D (dummy): Let [γ; ω] is a committee and i is dummy, then

$$\pi_i(\gamma;\omega) = 0$$

Dummy member has no power.

Axiom A (anonymity): Let [γ; ω] is a committee and [γ; $\sigma\omega$] its permutation, then

$$\pi_{\sigma(i)}(\gamma,\sigma\omega) = \pi_i(\gamma,\omega]$$

The power is a property of committee and not of players names and numbers.

Axiom S (symmetry): Let [γ; ω] is a committee and i and j (i $\neq$ j) are symmetric, then

$$\pi_i(\gamma,\omega) = \pi_j(\gamma,\omega)$$

The power of symmetric members is the same.

Axiom LM (local monotonicity): Let [γ; ω] be a committee and $\omega_i > \omega_j$, then

$$\pi_i(\gamma,\omega) \geq \pi_j(\gamma,\omega)$$

The member with greater weight cannot have less power than the member with smaller weight.

Turnovec [1994] suggested the additional fifth axiom:

Axiom GM (global monotonicity): Let [γ, α] and [γ, β] be two different committees of the same size such that $\alpha_k > \beta_k$ for one k $\in$ N and $\alpha_i \leq \beta_i$ for all i $\neq$ k, then

$$\pi_k(\gamma,\alpha) \geq \pi_k(\gamma,\beta)$$

If the weight of one member is increasing and the weights of all other members are decreasing or staying the same, then the power of the "growing weight" member will at least not decrease.

3 Marginality

We shall term a member i of a committee [γ, ω] to be **marginal** (essential, critical, decisive) with respect to a configuration S $\subseteq$ N, i $\in$ S, if

$$\sum_{k\in S} \omega_k \geq \gamma \quad and \quad \sum_{k\in S\setminus\{i\}} \omega_k < \gamma$$

A voting configuration S $\subseteq$ N such that at least one member of the committee is marginal with respect to S will be called a critical winning configuration (CWC).

Let us denote by $C(\gamma, \omega)$ the set of all CWC in the committee with the quota γ and allocation ω. By $C_i(\gamma, \omega)$ we shall denote the set of all CWC the member $i \in N$ is marginal with respect to, and by $C_{is}(\gamma, \omega)$ the set of all CWC of the size s (by size we mean cardinality of CWC, $1 \leq s \leq n$) the member $i \in N$ is marginal with respect to. Then

$$C_i(\gamma,\omega) = \bigcup_{s=1}^{n} C_{is}(\gamma,\omega)$$

$$C(\gamma,\omega) = \bigcup_{i\in N} C_i(\gamma,\omega)$$

A voting configuration $S \subseteq N$ is said to be a minimal critical winning configuration (MCWC), if

$$\sum_{i\in S} \omega_i \geq \gamma \quad and \quad \sum_{k\in T} \omega_k < \gamma \quad for\ any\ T \subsetneq S$$

Let us denote by $M(\gamma, \omega)$ the set of all MCWC in the committee $[\gamma, \omega]$. By $M_{is}(\gamma, \omega)$ we shall denote the set of all minimal critical winning configurations S of the size s such that $i \in S$, and by $M_i(\gamma, \omega)$ the set of all minimal winning configurations containing member i.

Lemma 1
A member $i \in N$ of a committee $[\gamma, \omega]$ is **dummy** if and only if for any s = 1, 2, ..., n

$$card\, C_{is}(\gamma,\omega) = 0$$

Proof: trivial

Remark
Since for all s

$$M_{is}(\gamma,\omega) \subseteq C_{is}(\gamma,\omega)$$

the member $i \in N$ is dummy if and only if for any s = 1, 2, ..., n

$$card\, M_{is}(\gamma,\omega) = 0$$

Lemma 2
Let $[\gamma; \omega]$ is a committee and $[\gamma; \sigma\omega]$ its **permutation**, then for any s = 1, 2, ..., n

$$card\, C_{is}(\gamma,\omega) = card\, C_{\sigma(i)s}(\gamma,\sigma\omega)$$

Proof: trivial

Remark

The same statement is true for the sets $M_{is}(\gamma, \omega)$; for any $s = 1, 2, ..., n$

$$\operatorname{card} M_{is}(\gamma,\omega) = \operatorname{card} M_{\sigma(i)s}(\gamma,\sigma\omega)$$

Lemma 3

If the members $k, r \in N$ $(k \neq r)$ of a committee $[\gamma, \omega]$ are **symmetric**, then for any $s = 1, 2, ..., n$

$$\operatorname{card} C_{ks}(\gamma,\omega) = \operatorname{card} C_{rs}(\gamma,\omega)$$

Proof:

Let k, r are symmetric and let $S \in C_{ks}(\gamma, \omega)$. We want to prove that either $S \in C_{rs}(\gamma, \omega)$, or $\{S \setminus \{k\} \cup \{r\}\} \in C_{rs}(\gamma, \omega)$, so to any configuration S from $C_{ks}(\gamma, \omega)$ there corresponds at least one configuration in $C_{rs}(\gamma, \omega)$. There are two possibilities:

a) If $r \in S$, then (from marginality of k with respect to S)

$$\sum_{i \in S \setminus \{\{r\} \cup \{k\}\}} \omega_i + \omega_r + \omega_k \geq \gamma, \quad \sum_{i \in S \setminus \{\{r\} \cup \{k\}\}} \omega_i + \omega_r < \gamma$$

and from symmetry of k and r

$$\sum_{S \setminus \{\{k\} \cup \{r\}\}} \omega_i + \omega_k < \gamma$$

That implies marginality of r with respect to S and $S \in C_{rs}(\gamma, \omega)$.

b) If $r \notin S$, then from marginality of k with respect to S

$$\sum_{i \in S \setminus \{k\}} \omega_i + \omega_k \geq \gamma, \quad \sum_{i \in S \setminus \{k\}} \omega_i < \gamma$$

and from symmetry of k and r

$$\sum_{i \in S \setminus \{k\}} \omega_i + \omega_k \geq \gamma \quad \Rightarrow \quad \sum_{i \in S \setminus \{k\}} \omega_i + \omega_r \geq \gamma$$

and we know that

$$\sum_{i \in S \setminus \{k\}} \omega_i < \gamma$$

hence r is marginal with respect to $\{S \setminus \{k\}\} \cup \{r\}$ and $\{S \setminus \{k\}\} \cup \{r\} \in C_{rs}(\gamma, \omega)$.

Resuming cases a) and b), to any $S \in C_{ks}(\gamma, \omega)$ there corresponds one configuration in $C_{rs}(\gamma, \omega)$ and

$$\operatorname{card} C_{ks}(\gamma,\omega) \leq \operatorname{card} C_{rs}(\gamma,\omega)$$

By the same way we can show that

$$\operatorname{card} C_{rs}(\gamma,\omega) \leq \operatorname{card} C_{ks}(\gamma,\omega)$$

what implies

$$card\,C_{ks}(\gamma,\omega) = card\,C_{rs}(\gamma,\omega)$$

QED.

Remark:

Using the same arguments we can show that for k, r symmetric and for any s

$$card\,M_{ks}(\gamma,\omega) = card\,M_{rs}(\gamma,\omega)$$

Conjecture to be proved: if for all s

$$card\,C_{ks}(\gamma,\omega) = card\,C_{rs}(\gamma,\omega)$$

then k, r are symmetric.

Lemma 4

Let $[\gamma, \omega]$ be a committee such that $\omega_k > \omega_r$, then for any $s = 1, 2, ..., n$

$$card\,C_{ks}(\gamma,\omega) \geq card\,C_{rs}(\gamma,\omega)$$

Proof:

For any $s = 1, 2, ..., n$ and any voting configuration $S \in C_{rs}(\gamma, \omega)$ we can consider two cases: a) $k \in S$, b) $k \notin S$. By definition of marginality

$$\omega_r + \sum_{i \in S \setminus \{r\}} \omega_i \geq \gamma, \qquad \sum_{i \in S \setminus \{r\}} \omega_i < \gamma$$

a) If $k \in S$, then from marginality of r with respect to S and assumption of the lemma it follows that

$$\sum_{i \in S} \omega_i = \omega_k + \omega_r + \sum_{i \in S \setminus \{\{r\} \cup \{k\}\}} \omega_i \geq \gamma$$

and

$$\sum_{i \in S \setminus \{k\}} \omega_i = \omega_r + \sum_{i \in S \setminus \{\{r\} \cup \{k\}\}} \omega_i < \gamma$$

what implies that $S \in C_{ks}(\gamma, \omega)$.

b) If $k \notin S$, then marginality of r with respect to S and $\omega_k > \omega_r$ implies

$$\omega_k + \sum_{i \in S \setminus \{r\}} \omega_i \geq \omega_r + \sum_{i \in S \setminus \{r\}} \omega_i \geq \gamma, \qquad \sum_{i \in S \setminus \{r\}} \omega_i < \gamma$$

what implies that $\{S \setminus \{r\}\} \cup \{k\} \in C_{ks}(\gamma, \omega)$.

Resuming cases a) and b) we receive

$$S \in C_{rs}(\gamma,\omega) \implies \begin{cases} S \in C_{ks}(\gamma,\omega) & \text{if } k \in S \\ \{S \setminus \{r\}\} \cup \{k\} \in C_{ks}(\gamma,\omega) & \text{if } k \notin S \end{cases}$$

hence

$$card[C_{ks}(\gamma,\omega)] \geq card[C_{rs}(\gamma,\omega)]$$

QED.

Remark:

Let $[\gamma, \omega]$ be a committee such that $\omega_k > \omega_r$, then

$$card\,C_k(\gamma,\omega) = \sum_{s=1}^{n} card\,C_{ks}(\gamma,\omega) \geq \sum_{s=1}^{n} card\,C_{rs}(\gamma,\omega) = card\,C_r(\gamma,\omega)$$

Lemma 5

Let $[\gamma, \alpha]$ and $[\gamma, \beta]$ be two committees of the size n such that for one $k \in N$

$$\alpha_k > \beta_k$$

and for all $i \neq k$

$$\alpha_i \leq \beta_i$$

then for any $s = 1, 2, ..., n$

$$card\,C_{ks}(\gamma,\boldsymbol{\alpha}) \geq card\,C_{ks}(\gamma,\boldsymbol{\beta})$$

Proof:

Let $S \in C_{ks}(\gamma, \beta)$, then by definition of marginality of k with respect to S

$$\beta_k + \sum_{i \in S \setminus \{k\}} \beta_i \geq \gamma, \qquad \sum_{i \in S \setminus \{k\}} \beta_i < \gamma$$

From definition of a committee

$$\alpha_k = 1 - \sum_{i \in N \setminus \{k\}} \alpha_i$$

$$\beta_k = 1 - \sum_{i \in N \setminus \{k\}} \beta_i$$

and from assumption of the lemma it follows that

$$\alpha_k - \beta_k = \sum_{i \in N \setminus \{k\}} (\beta_i - \alpha_i) \geq \sum_{i \in S \setminus \{k\}} (\beta_i - \alpha_i) \geq 0$$

Then

$$\alpha_k + \sum_{i \in S \setminus \{k\}} \alpha_i \geq \beta_k + \sum_{i \in S \setminus \{k\}} \beta_i \geq \gamma$$

$$\sum_{i \in S \setminus \{k\}} \alpha_i \leq \sum_{i \in S \setminus \{k\}} \beta_i < \gamma$$

(marginality of k with respect to S in the committee $[\gamma, \beta]$ implies marginality of k with respect to S in the committee $[\gamma, \alpha]$). Thus

$$S \in C_{ks}(\gamma, \beta) \Rightarrow S \in C_{ks}(\gamma, \alpha)$$

and

$$C_{ks}(\gamma, \beta) \subseteq C_{ks}(\gamma, \alpha)$$

and we have

$$\operatorname{card} C_{ks}(\gamma, \alpha) \geq \operatorname{card} C_{ks}(\gamma, \beta)$$

QED.

Remark:

Let $[\gamma, \alpha]$ and $[\gamma, \beta]$ be two committees of the size n such that

$$\alpha_k > \beta_k$$

$$\alpha_i \leq \beta_i \quad for\ i \neq k$$

then

$$C_k(\gamma, \beta) = \bigcup_{s=1}^{n} C_{ks}(\gamma, \beta) \subseteq \bigcup_{s=1}^{n} C_{ks}(\gamma, \alpha) = C_k(\gamma, \alpha)$$

hence

$$card C_k(\gamma, \alpha) \geq card C_k(\gamma, \beta)$$

4 Properties of Power Indices

The three most widely known power indices were proposed by Shapley and Shubik (1954), Banzhaf and Coleman (1965, 1971), and Holler and Packel (1978, 1983). All of them measure the power of each member of a committee as a weighted average of the number of his marginalities in CWC or MCWC.

The Shapley-Shubik (SS) power index assigns to each member of a committee the share of power proportional to the number of permutations of members in which he is pivotal. A permutation is an ordered list of all members. A member is in a pivotal position in a permutation if in the process of forming this permutation by equiprobable additions of single members he provides the critical weight to convert the losing configuration of the preceding members to the winning one. It is assumed that all winning configurations are possible and all permutations are equally likely. Thus SS power index assigns to the i-th member of a committee with quota γ and weight allocation ω the value of his share of power

$$\pi_i^{SS}(\gamma,\omega) = \sum_{S\subseteq N} \frac{(\text{card}\,S-1)!(\text{card}\,N-\text{card}\,S)!}{(\text{card}\,N)!} \qquad (1)$$

where the sum is extended to all winning configurations S for which the i-th member is marginal.

The Banzhaf-Coleman (BC) power index assigns to each member of a committee the share of power proportional to the number of critical winning configurations for which the member is marginal. It is assumed that all critical winning configurations are possible and equally likely. Banzhaf suggested the following measure of power distribution in a committee:

$$\pi_i^{BC}(\gamma,\omega) = \frac{\text{card}\,C_i(\gamma,\omega)}{\displaystyle\sum_{k\in N} \text{card}\,C_k(\gamma,\omega)} \qquad (2)$$

(so called normalized BC-power index).

The *Holler-Packel* (HP) power index assigns to each member of a committee the share of power proportional to the number of minimal critical winning configurations he is a member of. It is assumed that all winning configurations are possible but only minimal critical winning configurations are being formed to exclude free-riding of the members that cannot influence the bargaining process. "Public good" interpretation of the power of MCWC (the power of each member is identical with the power of the MCWC as a whole, power is indivisible) is used to justify HP index. Holler [1978] suggested and Holler and Packel [1983] axiomatized the following measure of power distribution in a committee:

$$\pi_i^{HP}(\gamma,\omega) = \frac{\text{card}\,M_i(\gamma,\omega)}{\displaystyle\sum_{k\in N} \text{card}\,M_k(\gamma,\omega)} \qquad (3)$$

Example 4

Let [γ, ω] = [0.6; 0.5, 0.3, 0.2]. The set of CWC and MCWC member 1 is marginal with respect to:

$$C_1(\gamma,\omega) = (\{1,2\},\{1,3\},\{1,2,3\})$$

$$M_1(\gamma,\omega) = (\{1,2\},\{1,3\})$$

The set of CWC and MCWC member 2 is marginal with respect to:

$$C_2(\gamma,\omega) = M_2(\gamma,\omega) = (\{1,2\})$$

The set of CWC and MWC member 3 is marginal with respect to:

$$C_3(\gamma,\omega) = M_3(\gamma,\omega) = (\{1,3\})$$

SS-index (weighted sum of number of pivotal position in one member, two member and three member configurations):

$$\pi_1^{SS}(\gamma,\omega) = 0(\frac{1}{3}) + 2(\frac{1}{6}) + 1(\frac{1}{3}) = \frac{2}{3}$$

$$\pi_2^{SS}(\gamma,\omega) = 0(\frac{1}{3}) + 1(\frac{1}{6}) + 0(\frac{1}{3}) = \frac{1}{6}$$

$$\pi_3^{SS}(\gamma,\omega) = 0(\frac{1}{3}) + 1(\frac{1}{6}) + 0(\frac{1}{3}) = \frac{1}{6}$$

BC-index (ratio of the number of CWC a particular member is marginal with respect to the total number of all CWC of all members):

$$\pi_1^{BC}(\gamma,\omega) = \frac{3}{5} \quad \pi_2^{BC}(\gamma,\omega) = \frac{1}{5} \quad \pi_3^{BC}(\gamma,\omega) = \frac{1}{5}$$

HP-index (ratio of the number of MCWC containing a particular member to the total number of all MCWC of all members):

$$\pi_1^{HP}(\gamma,\omega) = \frac{2}{4} \quad \pi_2^{HP}(\gamma,\omega) = \frac{1}{4} \quad \pi_3^{HP}(\gamma,\omega) = \frac{1}{4}$$

Which index is right? This is an issue for a rather extensive discussion that can lead to refinement of the original model. Each of the indices apparently answers a slightly different question and the problem is to formulate explicitly the relevant question as a part of the model of a committee. In this paper we are not going to follow this direction. The purpose of the paper is to confront the three indices with the five axioms of power introduced in section 2.

Theorem 1
Shapley-Shubik power index satisfies axioms D, A, S, LM and GM.

Proof
It is obvious that formula (1) for SS-index can be rewritten in the following way:

$$\pi_i^{SS}(\gamma,\omega) = \sum_{s=1}^{n} \frac{(s-1)!(n-s)!}{n!} \operatorname{card} C_{is}(\gamma,\omega)$$

Then from lemma 1 it follows that axiom D is satisfied, from lemma 2 axiom S, from lemma 3 axiom A, from lemma 4 axiom LM and from lemma 5 axiom GM.

Remark
Axioms D and A were explicitly used in axiomatic development of SS-power index (see Shapley and Shubik [1954]).

Theorem 2
Banzhaf-Coleman power index satisfies axioms D, A, S and LM.

Proof

It is obvious that formula (2) for BC power index can be rewritten in the following way:

$$\pi_i^{BC}(\gamma,\omega) = \frac{\sum\limits_{s=1}^{n} \operatorname{card} C_{is}(\gamma,\omega)}{\sum\limits_{k \in N} \sum\limits_{s=1}^{n} \operatorname{card} C_{ks}(\gamma,\omega)}$$

Then from lemma 1 follows axiom D, from lemma 2 axiom A, from lemma 3 axiom S, and from lemma 4 axiom LM.

Remark

The following example shows that Banzhaf-Coleman power index is not globally monotonic:

Let us consider two committees of the same size n = 5

$$[\gamma,\alpha] = [9/13;6/13,4/13,1/13,1/13,1/13]$$

$$[\gamma,\beta] = [9/13;5/13,5/13,1/13,1/13,1/13]$$

The weight of the first member in the committee $[\gamma,\ \alpha]$ is greater than his weight in the committee $[\gamma,\ \beta]$ while the weights of all other members in $[\gamma,\ \alpha]$ are less or equal than in $[\gamma,\ \beta]$. It can be easily verified that

$$\pi^{BC}(\gamma,\alpha) = \left(\frac{9}{19},\frac{7}{19},\frac{1}{19},\frac{1}{19},\frac{1}{19}\right)$$

$$\pi^{BC}(\gamma,\beta) = \left(\frac{1}{2},\frac{1}{2},0,0,0\right)$$

Theorem 3

Holler-Packel power index satisfies axioms D, A and S.

Proof:

It is obvious that formula (3) for HP power index can be rewritten in the following way:

$$\pi_i^{HP}(\gamma,\omega) = \frac{\sum\limits_{s=1}^{n} \operatorname{card} M_{is}(\gamma,\omega)}{\sum\limits_{k \in N} \sum\limits_{s=1}^{n} \operatorname{card} M_{ks}(\gamma,\omega)}$$

Then from remark to lemma 1 follows axiom D, from remark to lemma 2 axiom A, and from remark to lemma 3 axiom S.

Remark

a) Axioms D and A were explicitly used in axiomatic development of HP index (Holler and Packel [1983]).

b) Holler and Packel [1983] provided the following example showing that HP index is not locally monotonic:

Let us consider a committee of the size n = 5

$$[\gamma,\omega] = [0.51;0.35,0.20,0.15,0.15,0.15]$$

The weight of the second member is greater than the weight of the third, fourth and fifth member. It can be easily verified that

$$\pi^{HP}(\gamma,\omega) = \left(\frac{4}{15},\frac{2}{15},\frac{3}{15},\frac{3}{15},\frac{3}{15}\right)$$

so having the greater weight the second member has smaller value of evaluation of power than third, fourth and fifth members.

c) Using the same example as in remark to theorem 2 we can show that HP index is not globally monotonic.

Reconsidering two committees

$$[\gamma,\alpha] = [9/13;6/13,4/13,1/13,1/13,1/13]$$

$$[\gamma,\beta] = [9/13;5/13,5/13,1/13,1/13,1/13]$$

with the weight of the first member in the committee $[\gamma, \alpha]$ greater than his weight in the committee $[\gamma, \beta]$ while the weights of all other members in $[\gamma, \alpha]$ are less or equal than in $[\gamma, \beta]$, it can be easily verified that

$$\pi^{HP}(\gamma,\alpha) = \left(\frac{2}{6},\frac{1}{6},\frac{1}{6},\frac{1}{6},\frac{1}{6}\right)$$

$$\pi^{HP}(\gamma,\beta) = \left(\frac{1}{2},\frac{1}{2},0,0,0\right)$$

Theorem 4

If a power index satisfies axioms S and GM, then it also satisfies LM.

Proof:

Let π satisfies symmetry and global monotonicity axioms. Assume at the same time that π does not satisfy local monotonicity. Then, there exists a vector of weights

$$(\sigma_1,...,\sigma_r,...,\sigma_s,...,\sigma_n)$$

and a quota γ such that $\sigma_r > \sigma_s$ and

$$\pi_r(\gamma,\sigma) < \pi_s(\gamma,\sigma)$$

Let us set

$$\lambda_i = \sigma_i \qquad \text{if } i \neq r,s$$

$$\lambda_r = \sigma_r - \frac{\sigma_r - \sigma_s}{2}$$

$$\lambda_s = \sigma_s + \frac{\sigma_r - \sigma_s}{2}$$

Then, by global monotonicity

$$\pi_s(\gamma, \lambda) \geq \pi_s(\gamma, \sigma)$$

$$\pi_r(\gamma, \lambda) \leq \pi_r(\gamma, \sigma)$$

and, by symmetry

$$\lambda_r = \lambda_s \quad \Rightarrow \quad \pi_r(\gamma, \lambda) = \pi_s(\gamma, \lambda)$$

Then, from our assumption that local monotonicity is not satisfied, we obtain

$$\pi_r(\gamma, \lambda) = \pi_s(\gamma, \lambda) \geq \pi_s(\gamma, \sigma) > \pi_r(\gamma, \sigma)$$

and

$$\pi_s(\gamma, \lambda) = \pi_r(\gamma, \lambda) \leq \pi_r(\gamma, \sigma) < \pi_s(\gamma, \sigma)$$

which contradicts global monotonicity assumption of the lemma. Hence, global monotonicity and symmetry implies local monotonicity. QED.

References

Allingham, M. G. (1975): *Economic Power and Values of Games*. Zeitschrift für Nationalökonomie, **35**, 293-299.

Banzhaf, J. F. (1965): *Weighted Voting doesn't Work: a Mathematical Analysis*. Rutgers Law Review, 19, 1965, 317-343.

Coleman, J.S. (1971): *Control of Collectivities and the Power of a Collectivity to Act*. Social Choice (B. Liberman ed.), New York, 277-287.

Gambarelli, G. (1992): *Political and Financial Applications of the Power Indices*. Decision Process in Economics. G. Ricci (ed.), Springer Verlag, Berlin-Heidelberg, 84-106.

Gambarelli, G. (1990): *A New Approach for Evaluating the Shapley Value*. Optimization **21**, 445-452.

Holler, M.J. (1978): *A Priori Party Power and Government Formation*. Munich Social Science Review, **4**, 25-41.

Holler, M.J. and E.W. Packel (1983): *Power, Luck and the Right Index*. Journal of Economics, **43**, 21-29.

Holler, M.J. and Xiaoguang Li (1995): *From Public Good Index to Public Value. An Axiomatic Approach and Generalization*. Control and Cybernetics, **24**, 257-270.

Holubiec, J.W. and J.W. Mercik (1994): *Inside Voting Procedures*. Accedo Verlag, Munich.

Roth, A.E. ed. (1988): *The Shapley Value (Essays in honor of Lloyd S. Shapley)*. Cambridge University Press, Cambridge.

Shapley, L. S. and M. Shubik (1954): *A Method for Evaluating the Distribution of Power in a Committee System*. American Political Science Review, 48, 787-792.

Turnovec, F. (1994): *Voting, Power and Voting Power*. CERGE & EI Discussion Paper No. 45, Prague.

Turnovec, F. (1995): *The Political System and Economic Transition*. In: The Czech Republic and Economic Transition in Eastern Europe. Academic Press, New York, 47-103.

Turnovec, F. (1996): *Weights and Votes in European Union: Extension and Institutional Reform*. Prague Economic Papers, No. 2, 161-174.

Widgrén, M. (1994): *Voting Power in the EC and the Consequences of Two Different Enlargements*. European Economic Review, **38**, 1153-1170.

Widgrén, M. (1995): *Probabilistic Voting Power in the EU Council: the Cases of Trade Policy and Social Regulation*. Scandinavian Journal of Economics, **97**, 345-356.

Fairness and Equity via Concepts of Multi-Criteria Decision Analysis

F.A. Lootsma[1], R. Ramanathan[2], and H. Schuijt[1]

[1] Delft University of Technology, Faculty ITS, Mekelweg 4, 2628 CD Delft, The Netherlands
[2] IGIDR, Gen. Arun Kumar Vaidya Marg, Goregaon (East), Bombay 400 065, India.

Abstract. We briefly discuss principles of fairness and equity in order to incorporate them in a mathematical method for the allocation of benefits or costs (the output) in a distribution problem, on the basis of the effort, the strength or the needs (the input) of the respective parties. Usually, input and output are multi-dimensional, and proportionality seems to be the leading principle. Therefore we employ several algorithmic ideas of Multi-Criteria Decision Analysis in order to support the solution of distribution problems, in particular the ideas underlying the Multiplicative AHP which was designed to process ratio information. We extend the method in order to cover the principles of progressivity, priority, and parity as well. Two examples, (a) the establishment of the member state contributions to the European Union, and (b) the allocation of seats in the European Parliament to the member states, show that the proposed method produces contributions and allocations with a higher degree of fairness and equity than the actual solutions.

Key words. Fairness, equity, proportionality, progressivity, priority, parity, distribution criteria, desired ratios, logarithmic regression, geometric means.

1. Introduction

"All men agree that what is just in distribution must be according to merit in some sense, but they do not specify the same sort of merit" (Aristotle's Ethic, in the translation by J. Warrington, 1963). This statement briefly summarizes the two issues to be discussed in the present paper. First, the leading principle in fairness and equity is proportionality, which means that the benefits and the costs (the output) to be allocated to the parties in a distribution problem must be proportional to the effort, the strength, and/or the needs (the input) of the respective parties. Second, since input and output are usually measured under several criteria so that they are multi-dimensional, we have to weigh the distribution criteria in order to establish the aggregate quantities that must be proportional.

In this paper we present a general method to support a fair allocation of benefits and costs. As we will see (section 2), there are several principles of fairness. Proportionality is the leading one, not only in recent times but also in the Antiquity, see Aristotle (op. cit.) and the biblical parables of St. Matthew 25, 14 - 30 and St. Luke 19, 11 - 27. Many decisions, however, are based on the principles of progressivity, parity, or priority. According to the principle of progressivity, the output is increasingly assigned to the weaker or the stronger parties. Parity is the egalitarian principle whereby the output is divided into equal shares. The principle of priority awards the output in its entirety to one of the parties only. The choice of a particular principle is controlled by the assignment of specific values to certain parameters in the method. Since there is no clear border line between the above principles we design the method in such a way that the transitions from one principle to another one are as smooth as possible.

Since we are confronted with the subjective evaluation of multi-dimensional entities we turn to Multi-Criteria Decision Analysis (MCDA) for the calculation of fair allocations (section 3). By the principle of proportionality we are led to the algorithmic steps of the Multiplicative AHP because this method is particularly designed for the elicitation and the processing of ratio information. Thus, we use logarithmic regression in order to analyze desired-ratio matrices, and we apply geometric-mean calculations in order to work with a variety of distribution criteria. We will demonstrate that the method can easily be used under the other principles of fairness as well (section 4). All we have to do is to employ powers of the desired ratios. With exponents greater (smaller) than 1 we model the principle of progressivity (moderation), and in the limiting cases, when the exponents tend to infinity (zero), we proceed to the priority (parity) principle.

To illustrate matters, we first analyze the contributions of the member states to the European Union under the distribution criteria of population and Gross Domestic Product (section 5). With equal criterion weights we can reasonably enhance the fairness of the actual contributions. When we also take into account the national areas we obtain a really new, politically unexplored, set of contributions. This example highlights the typical features of our method. Similarly, we observe that the actual allocation of seats in the European Parliament deviates significantly from the allocation which is proportional to the size of the population. It can reasonably be approximated, however, by an allocation which is moderately proportional to the size of the population (section 6). Imposing certain additional constraints we can also prevent the total domination of the weaker countries by the stronger ones. We conclude this paper with a brief summary of open research questions (section 7).

2. Fairness and equity

Proportionality of input and output is by far the most widely discussed principle of fairness and equity. There are obvious utilitarian reasons for this. People are unwilling to make relatively high inputs unless they can look forward to relatively high outputs. Moreover, a person who can more effectively use a given scarce resource as a means of production should have a greater claim to its use. It is not unusual, however, that people moderate or amplify a proportional distribution. Young (1994) proposes four principles to determine a claimant's share: proportionality, progressivity, parity, and priority. The rationale for the progressivity principle, usually found in taxation schemes, is that those who are better off should pay at a higher rate because they can absorb the loss more easily. Under the egalitarian parity principle benefits and costs are allocated equally, even if the parties are unequal. The priority principle is an "all or nothing" principle which assigns absolute precedence to one party in the allocation of benefits and costs. It is usually applied to distribute indivisible goods.

Fishburn and Sarin (1994) discuss the issues of fairness and equity in the broad perspective of social choice. Fairness is based upon the preferences of individuals and groups, and upon the ways in which they perceive themselves in relation to others. Distributions that are fairest are those in which there is little or no envy among parties. Equity is based upon external ethical criteria, not on the specific preferences of individuals or groups. It is usually interpreted as some sort of equality, meaning that people are morally equal and should be treated with equal concern and respect. A distinction can also be made between fairness of outcome and procedural fairness (Linnerooth-Bayer et. al., 1994). In the first case the emphasis is on the results of the distribution process (do the parties agree with the proposed shares?), in the second case on the distribution process itself and on the role of the respective parties in it (did they receive a fair treatment, did they have a fair opportunity to explain their viewpoints?).

To conclude this section, we note that the literature on fairness and equity and on distributive justice is not only extensive but also confusing. Many highly similar concepts appear under different names. It is not always clear how they could be made operational (that is in fact the objective of the present paper). For more information we refer the reader to Deutsch (1975, 1985), Kasperson (1983), and Messick and Cook (1983).

3. A mathematical method for fair allocations

We consider a distribution problem with m criteria and n parties, first under the principle of proportionality. Let us take the symbol r_{ijk} to represent the desired ratio of the contributions c_j and c_k to be made by the respective parties under criterion i. Let us further introduce the symbol $R_i = \{r_{ijk}\}$ to stand for the matrix of the desired ratios under the i-th criterion. This matrix is positive and reciprocal, but not necessarily consistent, just like a pairwise-comparison matrix in the Analytic Hierarchy Process (AHP) of Saaty (1980). It may happen that $r_{ijk} \times r_{ikl} \neq r_{ijl}$ (for an example, see section 5). Let w_i stand for the weight assigned to the i-th distribution criterion. Following the mode of operation in the Multiplicative AHP of the first author (1993), we take the ratio c_j/c_k of any pair of contributions to approximate the desired ratios $r_{1jk},...,r_{mjk}$ simultaneously, in the sense that we solve the contributions from the logarithmic-regression problem of minimizing

$$\sum_{i=1}^{m}\sum_{j<k} w_i \left\{ \ln r_{ijk} - \ln c_j + \ln c_k \right\}^2. \tag{1}$$

Actually, we carry out the unconstrained minimization by solving the associated linear system of normal equations with the variables $u_j = \ln c_j, j = 1,..., n$. Obviously, the u_j have an additive degree of freedom. The c_j will accordingly have a multiplicative degree of freedom. A particular solution to the regression problem is given by

$$c_j = \prod_{i=1}^{m} \left(\sqrt[n]{\prod_{l=1}^{n} r_{ijl}} \right)^{w_i}, \tag{2}$$

which can be obtained if we calculate first the geometric row means of the matrices R_i and thereafter the geometric means of the row means. These operations may be interchanged without altering the final results. If the desired-ratio matrices happen to be consistent, the ratio of any pair of contributions is uniquely given by

$$\frac{c_j}{c_k} = \prod_{i=1}^{m}\left(\sqrt[n]{\prod_{l=1}^{n}\frac{r_{ijl}}{r_{ikl}}}\right)^{w_i} = \prod_{i=1}^{m}\left(r_{ijk}\right)^{w_i} \tag{3}$$

Let us illustrate the above results via the allocation of fair contributions to the European Union, to be paid annually by the member states. If the respective contributions must be proportional to the size of the population and the Gross Domestic Product, we have two diverging requirements that can only approximately be satisfied. Suppose that equal weights are assigned to these distribution criteria. We take the ratio c_j/c_k of any pair of contributions to approximate the desired ratios

$$r_{1jk} = \frac{Pop_j}{Pop_k} \quad \text{and} \quad r_{2jk} = \frac{GDP_j}{GDP_k}$$

simultaneously by the solution of the above logarithmic regression problem. On the basis of formula (3) the ratio c_j/c_k can now be written as

$$\frac{c_j}{c_k} = \sqrt{\frac{Pop_j}{Pop_k} \times \frac{GDP_j}{GDP_k}}. \tag{4}$$

If we also want to the use the national area as a yardstick to set the contributions, we introduce the desired ratios

$$r_{3jk} = \frac{Area_j}{Area_k}.$$

By formula (3) the ratio of any pair of contributions is now given by

$$\frac{c_j}{c_k} = \sqrt[3]{\frac{Pop_j}{Pop_k} \times \frac{GDP_j}{GDP_k} \times \frac{Area_j}{Area_k}}, \tag{5}$$

at least if equal weights are assigned to the distribution criteria. The choice of the criterion weights in general is still under investigation. It is unclear, for instance, how large the weights should be in order to represent various gradations of relative importance of the distribution criteria. A detailed discussion of fair contributions to the European Union may be found in section 4.

A refinement of the model is to replace the r_{ijk} by powers $(r_{ijk})^{q_i}$. The positive exponent q_i introduces a moderation (amplification) of the desired ratios under the i-th distribution criterion if $q_i < 1$ ($q_i > 1$). For very small (very large) values of q_i there is a transition from the principle of proportionality to the principle of parity (priority). An application of the idea is presented in section 5, where we concern ourselves with the fair allocation of seats in the European Parliament.

4. Fair contributions to the European Union

The most recent data concerning the European Union may be found in the Eurostat Yearbook (1995). They reflect the situation until 1993. Table 1 shows the size of the population and the Gross Domestic Product of the respective member countries in that year, as well as the national area.

The resources of the European Union (63.75 billion ECU) consisted of customs revenues (16.8%), levies on agricultural imports and sugar storage (2.9%), a VAT-based levy (52.2%), a GDP-based contribution (25.2%), and non-attributable income (2.6%). The total contribution of each of the member countries in 1993 is exhibited in Table 2, which also shows how the contributions are related to the national economies. The contributions expressed as a percentage of the GDP vary between 1.07 (United Kingdom) and 2.21 (Portugal), the contributions in ECU per capita between 92 (Portugal) and 420 (Luxemburg). Since the ratio of the smallest to the largest contribution per capita is roughly 1:5, fairness is far to seek.

The Tables 3 and 4 show that contributions proportional to the size of the population or the GDP only (one single distribution criterion) are also unfair. When only the size of the population is used, there is indeed a uniform contribution of 184 ECU per capita, but the contributions expressed as a percentage of the GDP vary between 1.00 (Luxemburg) and 4.41 (Portugal). Similarly, with the GDP as the unique yardstick, the contributions are uniformly set to 1.55% of the GDP, but the contributions in ECU per capita vary between 64 (Portugal) and 300 (Luxemburg). So, the ratio of the smallest to the largest contribution per capita is roughly 1:5, although the GDP is generally considered to be a proper yardstick for a country's ability to pay (see also Beckermann (1980)).

A considerable improvement is obtained when we allocate the contributions on the basis of the size of the population and the GDP simultaneously, with equal weights so that formula (4) applies. Table 5 shows that the contributions expressed as a percentage of the GDP now vary between 1.25 (Luxemburg) and 2.63 (Portugal), whereas the contributions in ECU per capita vary between 109 (Portugal) and 250 (Luxemburg). Under both distribution criteria the ratio of the smallest to the largest contribution is reduced to 1:2.

The third distribution criterion that may come up in the discussions on fairness and equity is the national area. A large area has many possible advantages for a country: a large amount of arable land and fresh water to support agriculture, large mountainous regions to support tourism and water winning, large spaces to enhance the quality of life, and/or large mineral deposits or fossil-fuel supplies. Table 6 shows the possible contributions to the European Union when the size of the population, the GDP, and the national area are used with equal weights to distribute the total burden. Formula (5) is clearly applicable. The contributions now vary between 1.08% (The Netherlands) and 3.61% (Portugal) of the GDP, and between 146 ECU (The Netherlands) and 264 ECU (Ireland) per capita.

Table 1. *Population, Gross Domestic Product, and National Area of the* 12 *Member States of the European Union in* 1993. *Data from the Eurostat Yearbook* 1995, *pages* 72, 168, *and* 196.

	Population 1993 in millions	GDP 1993 in 10^9 ECU	National Area in 1000 sq km
Belgium	10.07	125	31
Denmark	5.18	85	43
Germany	80.98	1105	357
Greece	10.35	59	132
Spain	39.05	277	505
France	57.53	810	549
Ireland	3.56	37	70
Italy	56.96	658	301
Luxemburg	0.40	8	3
The Netherlands	15.24	205	41
Portugal	9.87	41	92
United Kingdom	58.10	709	244
European Union	347.29	4119	2368

Table 2 *Actual Contributions of the Member States to the European Union. Data from the Eurostat Yearbook* 1995, *page* 402.

	Contribution, 10^9 ECU	Percentage of GDP	ECU per capita
Belgium	2.39	1.91	237
Denmark	1.21	1.42	233
Germany	19.03	1.72	235
Greece	1.01	1.72	98
Spain	5.19	1.88	133
France	11.56	1.43	201
Ireland	0.57	1.53	159
Italy	10.08	1.53	177
Luxemburg	0.17	2.10	420
The Netherlands	4.02	1.96	264
Portugal	0.91	2.21	92
United Kingdom	7.61	1.07	131
European Union	63.75	1.55	184

Table 3 *Possible Contributions of the Member States to the European Union according to the Size of the Population.*

	Contribution, 10^9 ECU	Percentage of GDP	ECU per capita
Belgium	1.84	1.47	184
Denmark	0.95	1.11	184
Germany	14.87	1.34	184
Greece	1.90	3.22	184
Spain	7.16	2.58	184
France	10.56	1.30	184
Ireland	0.66	1.78	184
Italy	10.45	1.59	184
Luxemburg	0.08	1.00	184
The Netherlands	2.80	1.37	184
Portugal	1.81	4.41	184
United Kingdom	10.67	1.50	184
European Union	63.75	1.55	184

Table 4 *Possible Contributions of the Member States to the European Union according to the Gross Domestic Product.*

	Contribution. 10^9 ECU	Percentage of GDP	ECU per capita
Belgium	1.93	1.55	192
Denmark	1.31	1.55	253
Germany	17.10	1.55	211
Greece	0.91	1.55	88
Spain	4.28	1.55	110
France	12.53	1.55	218
Ireland	0.57	1.55	160
Italy	10.18	1.55	179
Luxemburg	0.12	1.55	300
The Netherlands	3.17	1.55	208
Portugal	0.63	1.55	64
United Kingdom	10.97	1.55	189
European Union	63.75	1.55	184

Table 5 *Possible Contributions of the Member States to the European Union according to the Geometric Mean of the Population and the Gross Domestic Product.*

	Contribution. 10^9 ECU	Percentage of GDP	ECU per capita
Belgium	1.91	1.53	190
Denmark	1.13	1.32	218
Germany	16.07	1.45	198
Greece	1.32	2.24	128
Spain	5.58	2.01	143
France	11.60	1.43	202
Ireland	0.62	1.68	174
Italy	10.40	1.58	183
Luxemburg	0.10	1.25	250
The Netherlands	3.01	1.47	198
Portugal	1.08	2.63	109
United Kingdom	10.91	1.54	188
European Union	63.75	1.55	184

Table 6 *Possible Contributions of the Member States to the European Union according to the Geometric Mean of the Population, the Gross Domestic Product, and the National Area.*

	Contribution. 10^9 ECU	Percentage of GDP	ECU per capita
Belgium	1.50	1.20	149
Denmark	1.18	1.39	228
Germany	14.07	1.27	174
Greece	1.92	3.25	186
Spain	7.80	2.82	200
France	13.06	1.61	227
Ireland	0.94	2.54	264
Italy	9.94	1.51	175
Luxemburg	0.10	1.25	250
The Netherlands	2.22	1.08	146
Portugal	1.48	3.61	150
United Kingdom	9.56	1.35	165
European Union	63.75	1.55	184

The ratio of the smallest to the largest percentage of the GDP is higher than 1:2 now, but this may have a good reason: the population densities in the European Union vary widely, between 51 inhabitants per km² (Ireland) and 370 (The Netherlands). The third distribution criterion is clearly an incentive for the member states to exploit their natural resources more effectively.

Note. In the present paper we ignored the attempts of various European countries to use the principle of "juste retour" in order to regain their contributions to the Union as much as possible.

5. Fair seat allocations in the European Parliament

With some moderations the principle of proportionality can also be used to explain the seat allocation in the European Parliament as a function of the size of the population of the member states. Because we have only one distribution criterion here we can simplify formula (1). We carry out the study via unconstrained and constrained minimization of the sum of squares

$$\sum_{j<k} \left\{ \ln r_{jk} - \ln s_j + \ln s_k \right\}^2, \tag{6}$$

where s_j and s_k denote the size of the delegations of the member states j and k. Let us first introduce new symbols to represent the actual population ratios

$$\alpha_{jk} = \frac{Pop_j}{Pop_k}. \tag{7}$$

Table 7 shows the respective seat allocations obtained by unconstrained minimization of the sum of squares (6) when the desired ratio of any pair of national delegations is taken to be

$$r_{jk} = \alpha_{jk}^q, \quad q = 1, 0.9, 0.8, 0.7.$$

The non-integer solution of the minimization process must be adjusted to an integer seat allocation via an apportionment rule (Webster's rule, for instance, see Young, 1994). With $q = 1$, we have pure proportionality on the basis of the size of the population. Smaller values of q lead to more equality between the delegations. Remarkably enough, the seat allocation for $q = 0.7$ practically coincides with the actual seat allocation in the European Parliament. At the extreme ends of the spectre, however, we find some discrepancies. There are two countries with an over-representation, Ireland and Luxemburg, and one with an under-representation, Germany. An issue to be put on the political agenda?

Let us now use the seat-allocation problem in order to illustrate the introduction of certain constraints in a distribution problem. Barzilai and Lootsma (1997), considering the power relations in groups, proposed to limit the ratio of any two delegations to the range between $^1/16$ and 16 in order to prevent the total domination of the stronger parties over the weaker ones. Column 6 in Table 7 shows that these constraints are actually satisfied in the European Parliament despite the discrepancies between Luxemburg and Germany. We have tried to achieve this via unconstrained and via constrained minimization of the sum of squares (6) as well.

Table 7 *Number of seats in the European Parliament, proportional to the size of the population of the member states (column 2), moderately proportional to the size of the population (columns 3 - 5), as well as the actual allocation of seats in 1993 (column 6). Data from the Eurostat Yearbook 1995.*

	Pop	$Pop^{0.9}$	$Pop^{0.8}$	$Pop^{0.7}$	Actual
Belgium	16	19	22	25	25
Denmark	8	10	13	16	16
Germany	132	124	116	108	99
Greece	17	20	22	26	25
Spain	64	65	65	65	64
France	94	91	88	85	87
Ireland	6	7	10	12	15
Italy	93	91	88	84	87
Luxemburg	1	1	2	3	6
The Netherlands	25	28	30	33	31
Portugal	16	19	22	25	25
United Kingdom	95	92	89	85	87
Eur. Parliament	567	567	567	567	567

Table 8 *Number of seats in the European Parliament, proportional to the size of the population (column 2), calculated with truncated population ratios (column 3) and with constrained population ratios (column 4), as well as the actual allocation of seats in 1993 (column 5). Data from the Eurostat Yearbook 1995.*

	Pop	Pop. truncated	Pop. constrained	Actual
Belgium	16	19	21	25
Denmark	8	10	11	16
Germany	132	122	96	99
Greece	17	19	21	25
Spain	64	65	69	64
France	94	92	95	87
Ireland	6	7	8	15
Italy	93	92	94	87
Luxemburg	1	2	6	6
The Netherlands	25	27	30	31
Portugal	16	19	20	25
United Kingdom	95	93	96	87
Eur. Parliament	567	567	567	567

In the first approach we take the desired ratios r_{jk} such that

$$r_{jk} = \begin{cases} \frac{1}{16}, & \alpha_{jk} < \frac{1}{16}, \\ 16, & \alpha_{jk} > 16, \\ \alpha_{jk}, & \frac{1}{16} \leq \alpha_{jk} \leq 16. \end{cases}$$

This matrix is inconsistent. Unconstrained minimization of the sum of squares (6) provides a non-integer solution so that one needs again an apportionment rule to obtain an integer seat allocation. The result is shown in column 3 of Table 8. Some ratios are still beyond the desired range, however, see the size of the delegations of Luxemburg (2 seats) and Germany (122 seats). In the second approach we minimize the sum of squares (6) with $r_{jk} = \alpha_{jk}$ under the constraints

$$- \ln 16 \leq \ln s_j - \ln s_k \leq \ln 16, \tag{8}$$

for any j and k. This is a convex quadratic-programming problem when we take the logarithms of the delegations to stand for the variables. In the non-integer optimal solution Luxemburg appears to have a delegation of 5.5 seats whereas the large countries France, Germany, Italy, and the UK have 88 seats each. An apportionment rule will not change the size of the delegations of the large countries, but if the delegation of Luxemburg is reduced to 5 some constraints in (8) will be violated, and if the delegation of Luxemburg is rounded off to 6 the constraints (8) will be inactive. Column 4 of Table 8 shows the result when the delegation of Luxemburg has been set to 6 before the start of the constrained minimization process. The ratio of the delegations of Luxemburg (6 seats) and Germany (96 seats) is now precisely within the required range.

6. Epilogue

We conclude this paper with a brief sketch of some open research questions and some possible applications of our proposal to model the principles of proportionality, progressivity, parity, and priority via desired-ratio matrices and logarithmic regression.

Our method is clearly applicable only as soon as a major decision has been made: what are the relevant distribution criteria? This question is beyond the scope of the method. It is one of our objectives, however, to model the relative importance of the criteria (see Lootsma (1996)) and to study the parameters controlling the convergence towards the principle of parity (moderation) or priority (amplification).

Just like in MCDA, we plan to extend the method so that it can be used in group-decision making. The method should not only identify the opinions of the individual members of the group. It should also come up with a possible compromise solution which takes into account the relative power of the members.

It sometimes happens that additional constraints are imposed on the possible allocations (Ramanathan and Lootsma (1996)). Each party should have a minimum contribution, for instance, and certain coalitions of parties should not be permitted to dominate the scene. This may lead to a problem formulation where we are confronted with constrained minimization of the logarithmic-regression function (1). In general, as soon as we have a problem which is formulated in terms of contributions and also in terms of their logarithms, we may be running up against a non-convex problem with local, non-global solutions.

A possible application is the allocation of hazardous waste to certain regions with widely varying levels of urbanization, industrialization, affluence, natural beauty, unemployment, and rural development. The potential of the method is still to be explored here. It can also be applied on any world issue where consensus has to be obtained after negotiations by all countries involved in a distribution problem. This is a typical problem that is regularly encountered in the global negotiations on actions to deal with the environment and with climatic change. We could possibly identify the important criteria and the preferences of the individual countries, and we could analyze the power structure of the negotiating parties, in order to arrive at a fair compromise solution.

Acknowledgement

It is a pleasure to thank Dr. J. Linnerooth-Bayer and Dr. M. Makowski, both at the International Institute of Applied Systems Analysis, Laxenburg, Austria, for the inspiring ideas and discussions. In fact, when we planned to study budget allocation via MCDA, they suggested us to model the wider issues of fairness and equity. We also thank Drs M. Kreuk and Drs. F. Phillipson (Faculty of Mathematics and Informatics, Delft University of Technology) for the careful analysis of the seat allocation problem.

References

1. Aristotle, *"Ethics"*, in the translation by J. Warrington. Dent and Sons, London, 1963.

2. Barzilai, J., and Lootsma, F.A., "Power Relations and Group Aggregation in the Multiplicative AHP and SMART". *Journal of Multi-Criteria Decision Analysis* 6, 155 - 165, 1997.

3. Beckerman, W., *"An Introduction to National Income Analysis"*. Weidenfeld and Nicolson, London, 1980.

4. Deutsch, M., "Equity, Equality, and Need: What Determines which Value will be Used as the Basis of Distributive Justice?" *Journal of Social Issues* 31, 137 - 149, 1975.

5. Deutsch, M., *"Distributive Justice"*. Yale University Press, New Haven, 1985.

6. *"Eurostat Yearbook* 1995, *a Statistical Eye on Europe* 1983 - 1993". Office for the Publications of the European Communities, Luxemburg, 1995.

7. Fishburn, P.C., and Sarin, R.K., "Fairness and Social Risk I: Unaggregated Analysis". *Management Science* 40, 1174 - 1188, 1994.

8. Kasperson, R.E. (ed.), *"Equity Issues in Radioactive Waste Management"*. Oelschlager, Gunn, and Hain, Cambridge, 1983.

9. Linnerooth-Bayer, J., Davy, B., Faast, A., and Fitzgerald, K., "Hazardous Waste Cleanup and Facility Siting in Central Europe: the Austrian Case". Technical Report GZ 308.903/3-43/92, IIASA, Laxenburg, Austria, 1994.

10. Lootsma, F.A., "Scale Sensitivity in the Multiplicative AHP and SMART". *Journal of Multi-Criteria Decision Analysis* 2, 87 - 110, 1993.

11. Lootsma, F.A., "A Model for the Relative Importance of the Criteria in the Multiplicative AHP and SMART". *European Journal of Operational Research* 94, 467 - 476, 1996.

12. Messick, D.M., and Cook, K.S. (eds.), *"Equity Theory, Psychological and Sociological Perspectives"*. Praeger, New York, 1983.

13. Ramanathan, R., and Lootsma, F.A., "Fairness Issues in Group Decision Making, and a Model using the Multiplicative AHP". Report 96-05, Faculty of Mathematics and Informatics, Delft University of Technology, Delft, The Netherlands, 1996.

14. Saaty, T.L., *"The Analytic Hierarchy Process: Planning, Priority Setting, and Resource Allocation"*. McGraw-Hill, New York, 1980.

15. Young, H.P., *"Equity in Theory and Practice"*. Princeton University Press, New Jersey, 1994.

Part 5
Systems Thinking and Philosophical Issues

Structuring and Weighting Criteria in Multi Criteria Decision Making (MCDM)

Cathal M. Brugha

Department of Management Information Systems, Michael Smurfit Graduate School of Business, University College Dublin, Blackrock, Dublin, Ireland.

Abstract: The implications of qualitative distinctions between multiple criteria are considered. Some contributions to theory about the Analytical Hierarchy Process (AHP) are challenged. Experiments on alternative criteria structures are reported. These suggest that confusing structures are bad, but good structures are better than none. Guidelines on how to develop a structure are given for a well known case of the purchase of a house. It is suggested that differences between decision alternatives should provide a first phase basis for discovering criteria. A criteria tree should be structured 'top down' as a second phase by clustering criteria on the basis of qualitative difference. On any level the differences between criteria should follow relatively simple patterns. The rules used suggest the relevance of work on the structure of qualitative decision-making which is determined by Nomology, the science of the laws of the mind. Implications are considered for weighting trade-offs between homogeneous clusters of criteria. This should be done as a later 'bottom up' phase. The AHP scoring system is challenged. Some tests of alternative scoring methods are reported.

Keywords. Analytic Hierarchy Process, Decision theory, Nomology, Qualitative structuring

1. Introduction

This article develops from work by the author on the structure of qualitative decision-making (Brugha 1998a, 1998b and 1998c) which is determined by Nomology (Hamilton, 1877), the science of the laws of the mind. It provides a basis for modelling the way people might differentiate multiple criteria. Also relevant to this paper is psychophysics (Gescheider, 1985) which has provided an empirical basis for the comparison of objects by means of relative measurement, and is the basis of the Analytical Hierarchy Process (AHP) (Saaty, 1996).

The AHP has been the subject of much controversy, particularly to do with the question of it apparently being the cause of rank reversal in some circumstances. One problem that has concerned this author is the use by the AHP of the Right Eigenvector method to synthesise multiple scores of relatively measured objects.

Crawford and Williams (1985), Barzilai et al (1987, 1992, 1994), Holder (1990) and Lootsma (1993, 1996) have considered this at length. Lootsma's proposed variation, Geometric AHP, is now widely accepted and so this issue will not be considered here. Two other issues which have concerned this author will be considered in this article. The first is what this author sees as the inadequate treatment of qualitative difference between criteria particularly in some applications of the AHP. There seems to be some fundamental confusion about what is meant by having multiple criteria. In the next section we illustrate some of these confusions. We then review some criteria structuring experiments which were carried out in University College Dublin. We then propose some guidelines for structuring criteria using a well-known case for illustration.

The other issue considered here is the calculation of the weights of relative importance of criteria and, in particular, the 1 to 9 semantic differential scoring system used in the AHP (Saaty, 1980, 1990a). A semantic differential is used to get a consensus on agreement about the importance of one item compared to another, or in the AHP case agreement on how different two things are from each other. This author has been concerned about this strength-of-agreement figure being used to score the actual weight of that relative importance. Some suggestions about the calculation of weights are given. Also some tests of alternative weighting systems are reported, leading to some further guidelines.

2.   Criteria Structuring

The AHP grew out of a need to accommodate *qualitative* differences between criteria. Some difficulties with AHP have focused on the *quantitative* issue of rank reversal. Saaty (1994), for example, introduced a new form of normalisation, the 'Ideal' form, to avoid rank reversal being caused by the entry of irrelevant alternatives or numerous copies of one alternative. He also used an argument about the introduction of copies of a hat, for example, to justify the change in rank of existing alternatives caused by the AHP model (Saaty, 1994). For instance, if A is preferred to B and B is preferred to C, possibly the introduction of multiple copies of C could lead to B being preferred to A, hence rank reversal.

There seems to have been a confusion here about the interaction that occurs between alternatives and criteria. The above problem arose only indirectly because of the multiple copies. Clearly the introduction into the mind of the decision maker of the possibility of copies revealed a previously hidden criterion based on the issue of uniqueness. We would suggest that, if the other criteria have been modelled correctly, such a new criterion should be easy to fit into the existing structure, and the weighting of that issue should be possible to do as an extension of the existing system of comparisons of criteria.

In general, criteria should be included in a goal modelling process on the basis of their relevance to choices between the existing alternatives. This can mean that new criteria can be included or old ones dropped as new alternatives are considered or excluded. The process is essentially about the interaction between objectives and alternatives, i.e. between criteria and attributes. The proper structuring of criteria facilitates their change as the necessity arises.

2.1 Car Choice Example

Schoner and Wedley (1989) have proposed 'Referenced AHP' and 'B-G modified AHP' to deal with quantitative differences in attributes such as in the case of the purchase of a car. The attributes which are relevant to the decision to choose which car to buy were related quantitatively. From this author's point of view there was only one criterion, cost. Consequently the AHP should not have been used. The car was expected to have a life of five years. Hence five years maintenance was factored in with the price. The cars were expected to do 10,000 miles each year at an estimated cost of $1.50 per gallon. This was also factored in. Each car had a different price, maintenance cost and fuel usage in gallons per mile. For any car it is easy to calculate its estimated cost over the five years. Thus, for any two cars it is easy to calculate their relative costs directly without reference to the other cars. Applying the AHP in the usual way leads to rank reversal when a new car alternative is introduced. 'Referenced AHP' reinstates the correct relative relationship between alternatives. Before the new car was introduced the following were the alternative cars and their attributes:

Car	Price ($)	Maintenance ($/yr.)	Fuel (gal./mi.)
1	14,000	2,000	0.05
2	5,000	4,000	0.03
3	6,000	4,000	0.05
Total	25,000	10,000	0.13

The scale factors are $q_1 = 1$ for price, $q_2 = 5$ for maintenance and $q_3 = 10,000 \times 5 \times 1.5 = 75,000$ for fuel costs over the five years. The ratio of cost of car 1 to car 2 is

$$R_{1,2} = \frac{14,000 + 5*2,000 + 75,000*.05}{5,000 + 5*4,000 + 75,000*.03} = \frac{27,750}{27,250} = 1.018$$

The vectors of local priorities were produced using the AHP procedure, but clearly can be viewed from the data. The three cars score as follows on price, maintenance and fuel:

$$\begin{bmatrix} 14/25 \\ 5/25 \\ 6/25 \end{bmatrix} \quad \begin{bmatrix} 1/5 \\ 2/5 \\ 2/5 \end{bmatrix} \quad \begin{bmatrix} 5/13 \\ 3/13 \\ 5/13 \end{bmatrix}$$

The vectors of the weights of importance of the three attributes are calculated similarly using the AHP and can be seen to be derived from the total costs from all three attributes over all three cars: \$25,000 plus 5 × \$10,000 plus \$75,000 × .13 equals \$84,750. Proportionately these are 0.295, 0.590 and 0.115. In AHP these weights should remain the same. So, when a new car is included and these weights are used a conflict can lead to a rank reversal between cars 1 and 2. 'Referenced AHP' prevents this conflict happening by restoring the true weights.

The strength of an AHP type procedure is that it can be used to synthesise scores on qualitatively distinct criteria which might be measured most easily in terms relative to one another. Not only was there no need to apply the somewhat complicated AHP procedure to measure a relationship that was already known quantitatively, it was not appropriate to apply it in this situation because all three attributes were based on cost. The implication that the AHP procedure is usable in every multi-attribute situation is more than just wrong, it distracts from its contribution as a method for synthesising criteria whose multiplicity arises in the mind of the decision maker.

'Referenced AHP' produces scaling factors that convert quantitatively related attribute scores from AHP tables back into a common score. Because the AHP should not be used to synthesise qualitatively similar attributes we believe there is no need to use Schoner and Wedley's (1989) proposed 'Referenced AHP' or 'B-G modified AHP' which they adapted from Belton and Gear (1983, 1985).

2.2   Farmer's Field Example

Schenkerman (1994) shows similar difficulties when the AHP is applied to a field where the goal is the perimeter and the attributes are the distances on the rectangular sides North-South (N) and East-West (E).

	Unnormalised				Normalised		
Field	N	E	Perimeter	Priority	N	E	Synthesised
A	100	800	1800	0.450	$\frac{1}{5}$	$\frac{8}{15}$	0.367
B	200	100	600	0.150	$\frac{2}{5}$	$\frac{1}{15}$	0.233
C	200	600	1600	0.400	$\frac{2}{5}$	$\frac{6}{15}$	0.400
Total	500	1500	4000	1.000	$\frac{5}{5}$	$\frac{15}{15}$	1.000

Prioritisation on the basis of the perimeter using the AHP puts field A first. Synthesisation using AHP normalisation puts field C first. Schenkerman shows that the normalisation method causes a difficulty with absolute measurement (such as this case), and that Referenced AHP, B-G modified AHP, the linking-pin method and the supermatrix method simply undo the eigenvector normalisation.

In this case, the N and E attributes are brought back to the same dimension by re-introducing weights of 5 and 15. The resolution to this problem is to recognise that lengths and breadths of fields are not qualitatively distinct criteria for any decision maker's goal. Consequently the conflict implied by the illustration does not apply.

2.3   Experiments With Alternative Trees

As part of their course assignments some alternative trees were tested by several groups of final year students of the Bachelor of Commerce Degree in University College Dublin. The decisions modelled mainly had to do with the choice they had to make about what to do the following year. Part of their task was to compare *Naive AHP* with *Structured AHP*, and then to compare both with SMART. *Naive AHP* corresponds to the usual approach as in Saaty's (1990b) house example (below), which means pairwise comparisons of each criterion with every other criteria. *Structured AHP* meant clustering the criteria into a hierarchy based on qualitative similarities.

One group carried out the process with final year students of engineering on the issue "What to do next year?" after graduating. In the first phase they identified the following alternative activities the graduating engineers would consider: get a job (Job), do a post-graduate course in Ireland (P/G Here), do a post-graduate course abroad (P/G Abroad), or take a year off (Bum). They also identified their main objectives as make money (Cash), develop work experience (Work Exp), get better qualifications (B/Q), an issue to do with pressure and free time (Pres F/T), and one to do with personal development (Per Dev). They then applied Naive AHP to those five issues with one group of eight students, and Structured AHP to another group of eight students, both randomly chosen and tested individually. The latter was structured with Cash, Work Exp and B/Q clustered together and called Economic (Figure 1). This author suggested that comparisons on any level should be between things of a corresponding level of generality. Consequently PRES F/T was called Personal Feelings and Per Dev was called Personal Growth.

The AHP package Expert Choice was used to synthesise the results. The most striking outcome was that Pres F/T had a global weight of 0.128 using Naive AHP for one group and 0.474 using Structured AHP under the heading of Personal Feelings for the other group. It was felt that there may have been some other sub-criteria under Personal Feelings which had been left out. Hence a possible benefit of the structured approach would be that general categories such as Personal Feelings might subsume other important but unarticulated feelings. The overall inconsistency score was 0.19 for the naive approach and 0.07 for the structured approach. The better score for the latter may have been partly due to having fewer comparisons, but was more likely due to the structuring.

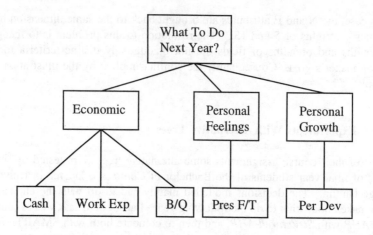

Figure 1. Graduating Engineers' Choices

The following is a summary of the comments from all the groups. The *Naive AHP* caused a problem when there were many criteria. This led to an excessive number of questions, a drop off in interest towards the end of the questioning process and higher inconsistencies. The *Structured AHP* led to higher consistency scores. The grouping of criteria helped respondents to become aware of criteria or attributes that they might have forgotten. The questions were more specific and more easily understood. However, if the structuring was poorly done there was a danger of confusion. It required a greater understanding of the issues in order to do the structuring. Also, the added levels might lead to distortions in the global weightings.

2.3.1   Comparison of Three Criteria Tree Structures

In a subsequent running of the course the above test was extended to a comparison of three alternative structures. Several class groups were given the following task specification. In the first phase of the study they had to find a homogeneous group of similar respondents who had a real choice to make, interview them, determine their three most preferred alternatives, and their four most important criteria. The rest of this section describes one of the groups which based its project on the choice a final year college student faces in the year following their departure from college. The first phase determined the three alternatives to be Employment, Travel and Postgraduate Study. The criteria were Income, Challenge, Education and Experience. Three alternative trees were then constructed. The first tree was comprised of the four criteria in *Naive AHP* form. Each of the respondents was asked to group the four criteria into two pairs based on which were closest. Some formed clusters of Income / Education and Challenge / Experience, others had clusters of Income / Challenge and Education / Experience. This provided the second tree. Each was then asked was there any criterion in one of the pairs which seemed closer to the combined other pair. This provided the basis of the third tree, as in Figure 2.

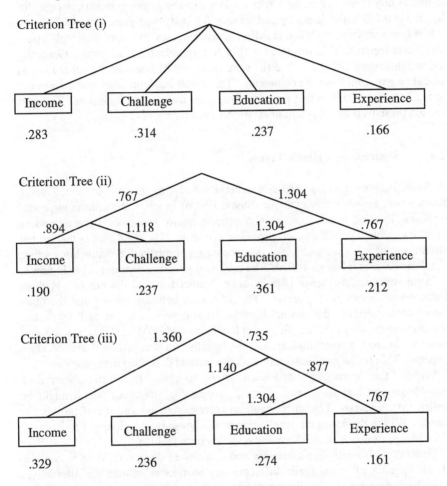

Figure 2. Alternative Criteria Trees

The respondents were asked to rate the different trees, to comment on the process, particularly the ease of understanding it, and to evaluate the exercise. Of the nine respondents in this case only one liked the first tree, two liked the second, and six liked the third. Most of this group isolated Income as the most different as in Criterion Tree (iii). These results fit some theoretical ideas developed by this author from his work in applying nomology to MCDM, which are discussed in the next section.

The numbers in Figure 2 come from one of the respondents in this group. Geometric AHP was used to synthesise the data. The below the line figures are normalised to total one so as to allow comparison. The figures on the branches of the trees multiply to one. Weights are distributed downwards so that they multiply to one at any level. Thus, for Criterion Tree (ii) the global geometric weight for Income is 0.783 which is got by multiplying 0.894 by the square root of 0.767.

It is apparent from the test that different criteria trees have a considerable effect on criteria weights. This was true for all nine respondents in this group. Generally the weights using Criterion Tree (ii) were more out of line than the other two, as would be expected from the comments. This result confirms what was discovered in the first test, which is that a poorly structured tree causes confusion. The next section presents some suggestions on proper structuring.

2.4   Structuring Criteria Trees

A formalisation of some ideas on criteria structuring is now presented using Saaty's well known house example (Saaty, 1990b) in which the criteria were size of house, location to bus lines, neighbourhood, age of house, yard space, modern facilities, general condition, and financing available. The main idea is that proper structuring of criteria should take into account any qualitative distinctions which the decision-maker has identified. The proposed structure is presented in Figure 3.

In this case the *financing* issue is quite distinct from all the others. Without finance you cannot get the house. The difference between finance and the other issues corresponds to the conflict between what is *needed* to get each house and the *preferences* for each one. Some applications of MCDM treat this as a separate issue to the rest of the problem and use benefit-cost analysis as a second stage process. This becomes unnecessary with the method described here.

Within the remaining issues, location to bus lines (*transportation*), *neighbourhood* and *yard space* form a comparable subgroup; which might be called *surroundings*. The other subgroup corresponds to more specifically *house* issues. Within the *house* cluster the size of the house is very different to issues to do with age of house, general condition and modern facilities.

Generally this qualitative clustering and ordering of the criteria is done in a 'top down' manner after the main differentiating attributes between the alternatives have been discovered. By distinguishing out the criteria which are very different, such as financing in this case, one can help to clarify the decision maker's thinking particularly. This helps later with trade-offs. A frequent comment from the students was that having to do this forced them to clarify their thoughts.

Presenting the issues on any level on a qualitatively logical order may also help. Thus *financing* comes before *new home*. Likewise within *surroundings* one should begin with the more technical *transportation*, then *neighbourhood* which has more to do with other people and lastly *yard space* which is more a description of the situation of the house. The latter two could be interchanged if *yard space* meant a play area for children and *neighbourhood* corresponded to ambience.

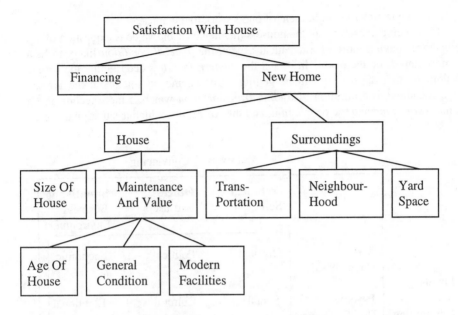

Figure 3. Hierarchy for Satisfaction with House

2.5 Relevance of Nomology

Nomology (Brugha 1998a, 1998b and 1998c) offers a basis for understanding the structures which underlie qualitative difference. It has three main dimensions: Adjusting, Convincing and Committing. The kinds of problems considered here, course choice and house purchase, involve a combination of convincing oneself about various issues and committing oneself to making a choice. Committing has three levels: need, preference and value. Convincing has three levels: technical or self, other people and situational. These are presented in Figure 4 along with a revision of Maslow's Hierarchy of Needs to show the generic nature of this structure. Proper structuring of a multi criteria problem should take account of the nomological rules based on which the criteria clusters are formed. In this case it provides the basis for differentiating the various levels.

Consider the structuring that was done for the house choice. The hierarchy in Figure 3 shows that the biggest qualitative distinction is at the top of the hierarchy with the issue of financing determining whether or not one can afford the house or not. The major trade-off is about commitment and the *financing* the buyers have against the *new home* they would prefer. At the second level there is a trade-off to be made between the house itself and the situational aspect of the surroundings. At the third level down there is the issue about the *size of house* one will have versus the work one has to do in terms of *maintenance* and adding *value* to it.

This is a question of commitment of one's work to the house.

Convincing oneself about the house's maintenance and value is a question of *age of house* having more of a technical aspect, the *general condition* likely to have more impact on the people in it, and the *modern facilities* determining the type of situation that the people will be getting. Convincing oneself about the house's surroundings is a question of trading *transportation* which is more technical, the *neighbourhood* and the people in it, and the *yard space* which is situational.

Levels of Convincing

	Technical Self	**Others** End-users	**Situational** Business / Environment
Somatic Have / Need	Physical	Political	Economic
Psychic Do / Preference	Social	Cultural	Emotional
Pneumatic Are / Value	Artistic	Religious	Mystical

Levels of Committing

Figure 4. Structure of Development Activities

3. Criteria Weighting

Criteria arise from the differences between the alternatives. We would propose that the third phase of the process should start with the comparative evaluation of the alternatives with respect to the lowest level criteria. Intuitively, it makes sense to not mix evaluations which are qualitatively quite different and distinct from each other, for instance, in the case of the house above, to compare *general condition* with *neighbourhood*, or *yard space* with *financing available*. We would propose that the identification of criterion weights within each category and on each level that is relevant to a choice process should be carried out at the one time, working from the bottom up, if possible. Then the more macro weighting process could be done between the synthesised sets. Occasionally this could lead to only two items being compared at one level. For example, in the house case there would be only one comparison between *house* and *surroundings*, and then one between *financing* and *new home*. This has the benefit of requiring the decision-maker to answer fewer questions. It might create anxiety that the result would be over-dependent on single judgements such as on the *financing* versus *new home* question. In our opinion this problem can be dealt with most easily by using sensitivity analysis. As it was, in the actual application (Saaty 1990b, p.17) there was some surprise about the emergence of the least desirable house with respect to *financing* as most preferred. It is possible that our suggested approach might have helped in this case.

The benefit of working from the bottom up is that, through working with the alternatives and the criteria, the decision-maker learns more about what they mean and so scores them accurately. We would also suggest that it makes more sense to have decision-makers work firstly with criteria which are qualitatively close and finish up with those which are very different from each other. These latter are the more difficult judgements, and the trade-offs most crucial to the decision. Decision makers should be able to see the synthesised scores of the alternatives for each criteria cluster as these scores are synthesised. Working interactively can allow for 'outranking' or elimination of alternatives which scored very poorly on some criterion. It also facilitates sensitivity and benefit cost analyses.

A possible alternative scoring approach is to compare the importance of the group of *surroundings* issues with one another and then extend the comparison to the individual house issues. This might be difficult because of mixing different qualitative changes from criteria to criteria. It also makes more demands on the patience of the decision-maker because of the increased number of questions.

A variation on this would be to order the criteria qualitatively within any level and carry out comparisons between levels using the criteria that are qualitatively nearest to each other. For instance, higher criteria within an one category could be compared with lower criteria within a higher category, hopefully producing a seamless join. Within *surroundings transportation* would be lower than *neighbourhood* or *yard space* and so closer to some of the *house* criteria. Within *maintenance and value modern facilities* is a higher criterion than *age* or *general condition*. Thus one would expect it would be relatively easy for a decision-maker to compare the relative importance of the *modern facilities* and *transportation* criteria. In fact the first is about the convenience within the home and the second is about convenience of travel to and from the home, two qualitatively similar issues. Likewise *size of house* is very close to *age of house*. From Figure 3 one can see that a 'qualitative frontier' has been created. If one was uncomfortable about developing clusters of criteria one could compare criteria which were nearby on the frontier.

Structuring criteria by working from the bottom up the tree in clusters, the first and recommended method, has operational advantages. The number of comparisons are few because the clusters are usually small. This facilitates information gathering and reduces inconsistency. The challenge it presents the decision advisor particularly is to analyse the structure of the decision maker's criteria before gathering the data. The increase in the complexity of the analysis offsets the operational advantages. The benefits from qualitatively structuring the objectives come more from the validity of the process, how it is truer to the processes of the decision maker's mind.

Bottom-up evaluation of alternatives may convert the weighting process into two phases. It may be necessary to convert quantitative scores on attributes of alternatives into scores which match some criterion. If one of the houses is very much bigger than the others it might be appropriate to use a utility function to take account of diminishing returns from house size. Where the yard space of one of the houses was very inconvenient or wasteful of space one might convert its space into terms which were expressed in terms of the yard space of another house. The many modelling possibilities for scoring alternatives on particular attributes should not influence how one should synthesise the relative importance of multiple criteria, that is where the distinctions incorporate a qualitative aspect.

In the second set of student tests the criteria were compared using AHP type relative measurement, while the attributes of the alternatives with regard to each lowest level criterion were scored using a SMART type utility scale. If adopted generally, this approach would have the effect of introducing compatibility between AHP and multi-attribute utility theory (MAUT), as mentioned by Dyer (1990).

3.1   Naming Clusters

In the first test (Figure 1) and in the house purchase example (Figure 3) the criteria clusters were given new names. The group running the second test preferred not to do this because of the fear that it would introduce new issues into the decision. It is most important that the respondents understanding of the issues be used when naming clusters. Respondents generally seem to be more comfortable if clusters are presented in the language they have been using in the process. The greatest difficulties arose when comparing clusters whose meanings were not homogeneous and consequently did not facilitate the trade-offs. For example with Criterion Tree (ii) in Figure 2, comparing Income / Challenge against Education / Experience was seen as difficult. On the other hand comparing Income against Challenge / Education / Experience was seen as easier.

3.2   Experiments With Alternative Forms of Weighting

Some of the tests carried out in University College Dublin involved comparing the AHP with SMART. AHP's 1 to 9 scale caused difficulties leading respondents to re-consider some of their first answers. SMART's visible scale was liked.

An second experiment was carried out whereby respondents were asked to rank their criteria from the least preferred to the most preferred. Then they were asked to fill in an AHP type table with numbers greater than one corresponding to how much more they preferred one to another, either directly or in percentage terms. Respondents found this difficult.

This experiment was refined by giving respondents scale bars to look at with the lower one given a "length" of 100; they were required to fill in the value, over 100, for the one being compared with it. This was welcomed. This author had the suspicion that it might work well for relative scores in the order of 1 to 1.5 times the lower value, but might break down for higher values.

The current belief is that scoring the relative importance of criteria from the highest downwards might more easily deal with large variations in relative scores.

4.   Conclusion

The research described here focused on personally relevant multi criteria decisions. By emphasising respect for the decision processes within the mind of decision makers some developments have been made and tested with regard to the procedures used for constructing criteria trees, for synthesising the weights of criteria, and for choosing preferred alternatives. Some small tests mainly amongst college students indicate a good response and enthusiasm for the improvements from both respondents and students who carried out the tests as part of their course assignments. Criteria trees that are structured to reflect the various small and large qualitative distinctions in the minds of the decision makers help decision makers to clarify their thoughts. They also make scoring trade-offs easier and reduce inconsistency. The use of visible bands or bars to help score trade-offs was also seen as an improvement on using numbers alone, including the AHP 1 to 9 scale.

The suggestions proposed here can be summarised as follows. Criteria and alternatives should be seen as intertwined. Differences between the alternatives generate the relevant criteria. Consequently including or excluding an alternative may have a consequence for the criteria. The initial list of criteria generated by the alternatives should be structured on a 'top down' basis from the most qualitatively distinct to the least. Alternatives should be scored on the lower level criteria. Then trade-off scores should be calculated on the criteria moving 'bottom up' towards the overall goal. Decision makers should be made aware of the entire process as it happens. In this way they can revise their judgements where necessary. Also, occasionally it may be possible to 'outrank' or eliminate an alternative which scores poorly on some clustered criterion. At the end of the process, if the major issue is a trade-off of benefits against costs, the procedure takes on the appearance of a benefit-cost analysis.

It is believed that these conclusions apply where the multiplicity of criteria in a decision is based on qualitative difference, particularly to models such as the AHP whose foundations are based in psychophysics. The research also indicates the relevance of Nomology for guiding the qualitative structuring of a criteria tree.

References

Barzilai, J.; Cook, W. D. and Golany, B. (1987) *Consistent Weights for Judgements Matrices of the Relative Importance of Alternatives,* Operations Research Letters, **6** (3), 131-134.

Barzilai, J.; Cook, W. D. and Golany, B. (1992) "The Analytic Hierarchy Process: Structure of the Problem and its Solutions" in *Systems and Management Science by Extremal Methods*, Phillips, F. Y. and Rousseau, J. J. (eds.), Kluwer Academic Publishers, 361-371.

Barzilai, J. and Golany, B. (1994), *AHP Rank Reversal, Normalisation and Aggregation Rules,* INFOR, **32** (2), 5 -64.

Belton, V. and Gear, A. E. (1983), On the Shortcoming of Saaty's method of Analytic Hierarchies, Omega, **11**, 228-230.

Belton, V. and Gear, A. E. (1985) *The legitimacy of Rank Reversal - A comment,* Omega, **13**, 143-144.

Brugha, C. (1998a), The structure of qualitative decision making, *European Journal of Operational Research*, **104** (1), pp 46-62.

Brugha, C. (1998b), The structure of adjustment decision making, *European Journal of Operational Research*, **104** (1), pp 63-76.

Brugha, C. (1998c), The structure of development decision making, *European Journal of Operational Research*, **104** (1), pp 77-92.

Crawford, G. and Williams, C. (1985) *A Note on the Analysis of Subjective Judgement Matrices,* Journal of Mathematical Psychology, **29**, 387-405

Dyer, J.S. (1990), *Remarks on the Analytic Hierarchy Process,* Management Science, **36** (March), 249-258.

Gescheider, G.A. (1985) *Psychophysics Method, Theory, and Application*, Lawrence Erlbaum Associates, Publishers, New Jersey.

Hamilton, W. (1877), *Lectures on Metaphysics*, Vols. 1 and 2, 6th Ed., in *Lectures on Metaphysics and Logic,* London: William Blackwood and Sons.

Holder, R. D. (1990), *Some comments on the Analytical Hierarchy Process*, J. Opl. Res. Soc. **41** (11), 1073-1076.

Lootsma, F. A. (1993), *Scale sensitivity in the Multiplicative AHP and SMART,* Journal of Multi-Criteria Decision Analysis, **2**, 87-110.

Lootsma, F.A. (1996), "A model of the relative importance of the criteria in the Multiplicative AHP and SMART", *European Journal of Operational Research,* **94**, 467-476.

Saaty, T.L. (1980), *The Analytic Hierarchy Process,* McGraw-Hill, New York.

Saaty, T.L. (1990a), *Multicriteria Decision-Making: the Analytic Hierarchy Process,* The Analytic Hierarchy Process Series Vol. 1, RWS Publications.

Saaty, T.L. (1990b), *How to make a decision: the Analytic Hierarchy Process,* European Journal of Operational Research, **48**, 9-26.

Saaty, T.L. (1994), Highlights and critical points in the theory and application of the Analytic Hierarchy Process, European Journal of Operational Research, **74**, 426-447.

Saaty, T.L. (1996), "Ratio Scales are Fundamental in Decision Making", *ISAHP 1996 Proceedings*, Vancouver, Canada, July 12-15, 146-156.

Schenkerman, Stan (1994), *Avoiding rank reversal in AHP decision-support models,* European Journal of Operational Research, **74**, 407-419.

Schoner, B. and Wedley, W.C. (1989), *Ambiguous criteria weights in AHP: consequences and solutions,* Decision Sciences, **20**, 462-475.

The Analytic Hierarchy Process and Systems Thinking

D.Petkov[1] and O. Mihova-Petkova[2]

[1] Department of Computer Science and Information Systems,
   University of Natal, Pietermaritzburg 3209, South Africa
[2] Department of Information Systems, University of Durban Westville
   Private bag X54000, Durban 4000, South Africa

Abstract. The purpose of this research is to determine the similarities between the Analytic Hierarchy Process (AHP) and various strands of thought in the systems movement, and where AHP stands in the System of Systems Methodologies, proposed by Jackson and Keys. It is claimed that there can be found a common foundation for the complementarist use of AHP with other well known systems methodologies.

Keywords. Systems Thinking, AHP, Multicriteria Decision Making

1. Introduction

Traditionally the Analytic Hierarchy Process (AHP) has been viewed mainly as a multicriteria decision making (MCDM) theory (Saaty, 1990). There have been some attempts by Saaty and Kearns to show the links between systems ideas and the AHP, especially with regard to the issue of problem complexity (Saaty and Kearns, 1991). After outlining the prerequisites for an effective systems methodology they come to the conclusion that AHP should be a promising and powerful tool in the study of systems. It must be noted that their analysis is mainly with reference to a more traditional systems approach known in the past 15 years as hard systems thinking. Meanwhile, in the last decade we have observed a wider understanding of the systems field, which is seen as one that incorporates a number of approaches. Each of them fits in a particular way in the System of Systems Methodologies, a two dimensional grid of problem contexts used to classify different Systems Approaches, suggested by Jackson and Keys (Jackson, 1991:31). Thus the question arises, where does AHP stand with regard to them?

There have been essential developments in AHP since the first edition in 1985 of the Saaty and Kearns book and most of those are still rarely quoted either by AHP critics or by its supporters. Regarding the general literature on AHP, there have been numerous publications, some of which raise controversial issues and debates, the review of which falls outside the scope of this paper. In particular, a major development has been the elaboration of the theory of AHP for systems with feedback. Very little is done to highlight the relevance of systems thinking and cybernetics to this type of AHP modelling. Recently Marzen proposed the theory of sustainable systems as a jigsaw puzzle of significance of contemporary scientific concepts. She allocates an important role to AHP which, according to her, incorporates the systemic-evolutionary and biocybernetic/taoistic philosophy of sustainable systems (Marzen,1994:86). Nevertheless, a recent review of the diversity of systems thinking (Lane and Jackson, 1995) makes no mention of AHP, and it is difficult to pinpoint any references to it in the mainstream systems literature so far. This is probably partially due to the fact that the main and widely quoted source books on AHP such as Saaty (1990) and Saaty (1994a), in contrast to Saaty and Kearns (1991), do not discuss the relationship between AHP and systems ideas in general.

Soft Operational Research or Problem Structuring Methods (PSM) also seek a demarcation from the Analytic Hierarchy Process, according to Rosenhead, because of their transparency of method, restricted mathematisation and their focus on supporting judgement rather than representing it (Rosenhead, 1989). As Jackson (1994) points out, while soft OR compensates for one of the major flaws in hard systems thinking by accepting subjectivity, it does not address the others, and our subsequent analysis will show that the above demarcation between AHP and PSM can be viewed as strengthening the case for AHP from the point of view of systems thinking since in practice it is important both to support and to measure human judgement. One example of utilisation of concepts from both fields is the work of Verkasalo who has applied some of Checkland's ideas on soft systems methodology (SSM) to the chain of AHP modelling processes as a human activity system (Verkasalo, 1994). It can be noted that his brief analysis ignores a major feature of SSM and AHP: their support for the participation of the stakeholders in decision making as a learning process.

The purpose of this study is to determine the similarities between AHP and various strands of thought in the systems movement, and where AHP stands in the System of Systems Methodologies. Since AHP is a decision making approach, it would be fair to compare it to a framework for decision making that is widely recognised In the next section, some comparisons will be made with Simon's model of decision making. The following one contains an analysis of problem structuring in AHP in order to attempt to reveal further its systemic nature. The latter are described briefly in Lane and Jackson (1995), and for that reason they will not be discussed in this paper. Then follows an evaluation of two dimensions of the problems handled by AHP, complexity and divergence of values or interests which will result in determining its place within the Jackson and Keys' System of Systems Methodologies.

2. AHP as a systemic methodology for decision making

2.1. AHP and Simon's model of the process of decision making

A widely recognised model of the decision making process is the one proposed by H. Simon. It consists of four stages: Intelligence, Design, Choice and Implementation of solution. Successful implementation results in solving the original problem, while failure leads to a return to the modelling process (Turban, 1991). Saaty gives another definition of the steps in the decision making process which seem to be strongly influenced by the characteristics of AHP (Saaty,1994b):

(1) Structure a problem with a model that shows the problem's key elements and their relationships.
(2) Elicit judgements that reflect knowledge, feelings or emotions.
(3) Represent those judgements with meaningful numbers.
(4) Use these numbers to calculate the priorities of the elements.
(5) Synthesise these results to determine an overall outcome.
(6) Analyse sensitivity to changes in judgement.

The above definition has as a drawback, when compared to Simon's model of decision making, the fact that it assumes that the problem has been identified somehow and a hierarchy can be designed for it. Though Saaty indicates how to structure a hierarchy (eg see Saaty,1990 and 1994a) this is still the least formalised aspect of AHP and which has been the subject of very little research.

According to Dyer and Forman, the Analytic Hierarchy Process focuses on the choice phase of decision making (Dyer and Forman, 1992:100). It seems that such a statement makes unnecessarily narrow the scope of AHP since it ignores a number of considerations on problem structuring which are inherent to AHP. There have been some attempts to use AHP in a broader context. As long ago as 1982 Arbel and Tong applied AHP for the generation of options in decision analysis problems (Arbel and Tong, 1982). Their procedures are relevant both for the Intelligence and the Design phases of Simon's model. It is not the purpose of this paper to claim that AHP is the single best existing technique for these two phases. It may be rather expected that a complementarist approach involving some of the Problem Structuring Techniques (see Rosenhead, 1989) like SODA or SSM in combination with AHP might account better for these two phases, though there are no reports on that in the literature. However there is one reference showing such a complementarist application of AHP and another well known systems approach, the Viable Systems Model (VSM), in which various aspects of organisational design, developed through VSM, are prioritised using AHP (Steenge et al., 1990).

AHP supports the Implementation phase of Simon's model of decision making, not just through the priorities that are derived in the Choice Phase, but also through the possibility for "what-if" type of analysis of the influence on the final decision by a number of small changes in the assumptions guiding the decision. These provide further insight into the problem under consideration, and can be viewed also as part of the identification phase. It is especially important also to underline the support provided by AHP for group decision making processes (see Dyer and Forman, 1992 as well as Forman and Peniwati, 1996). The support offered by AHP for group decision making makes it an extremely powerful tool for consolidating opinions in complex situations that usually involve a number of stakeholders. It can be noted that most of the Problem Structuring Methods aim mainly at development of a consensus within the group, but that may lead to the suppression of original individual opinions. Research in the field of Group Decision Support Systems has shown that anonymity in the expression of opinions is often essential for the sincerity of judgements (Nunamaker et al., 1991). The possibility for aggregation of opinions within AHP provides a mechanism for preserving that anonymity. On the other hand, through proper clustering of the value tree, it is possible to preserve the initial judgements if necessary in order to reflect real power relationships. The above analysis shows that AHP is applicable to all the stages of the decision making process concerning either individuals or groups and this is significant for revealing its systemic nature.

Since the least formalised part of AHP is related to how we structure problems with it, the following subsection will examine this issue more closely.

2.2 Problem structuring in AHP

The first issue here is to what class of problems the Analytic Hierarchy Process is applicable. According to a survey on the status of Multiple Criteria Decision Making, the problems to which this group of techniques is suitable can be classified as "messy" (see Stewart, 1992:569). This leads to decision making goals being increasingly imprecise. According to the same source, MCDM attempts to represent imprecise goals in terms of a number of individual (relatively precise, but generally conflicting) criteria. It is interesting to note that "messes" are also claimed to be the area of applicability of a range of systems approaches including Soft Systems Methodology but these do not necessarily follow the philosophy of MCDM as underlined above. We will not discuss here the notion of a "mess" since it is widely publicised in the works of Ackoff and Checkland (see Jackson, 1991).

Problems are structured in AHP either in the form of a hierarchy, or as a network in the case of existing interdependencies between the elements of one level, or feedback from lower levels in the hierarchy. The hierarchy has long been seen as a suitable representation for handling complexity (Simon, 1962), and is widely used in general systems theory, cybernetics and hard systems thinking. According to Saaty (1994b), the basic principle to follow in creating this structure is always to see if one can answer the following question: Can I compare the elements on a lower level using some or all of the elements on the next higher level as criteria? Hierarchies are only special cases of more general network models that can capture the interdependencies between elements within a level, or feedback from a lower level to a higher one. The number of links in a system that can be represented and measured in a network AHP model is far greater than what is possible in Systems Dynamics or soft systems thinking models. Being a multicriteria decision making theory, AHP is suitable for evaluating issues revealing cultural and political differences within the context of a given problem (see Saaty and Alexander, 1990). It is interesting to note that the recent changes in Soft Systems Methodology also take account of culture and power in organisational problem solving (Checkland and Scholes, 1990). However, like most systems approaches, AHP does not cover all the aspects of the organisation, like processes and organisational design, though it can be applied to prioritise the factors affecting these two areas.

AHP models include both qualitative and quantitative data, reflecting in this way the subjectivity of the decision maker as an important element in complex problems, in which not all factors can be precisely measured in quantitative terms. There is a certain analogy between the criticism made by Saaty on the applicability of other Operations Research models to issues like strategic arms control or scenario planning (Saaty, 1990), and the analysis of the weaknesses of Hard Systems Thinking provided by Checkland and Scholes (1990), Jackson (1991), Flood and Jackson (1991).

As stated above AHP allows the incorporation of subjective data in a decision problem. It has to be stressed that this is a controlled subjectivity through the Consistency Ratio (Saaty, 1990). The latter provides an effective feedback mechanism for the quality of decisions based on the pairwise comparisons of decision makers. AHP allows also the incorporation of uncertainty in the decision making process through interval judgements (Saaty,1990), and recently through emerging fuzzy logic extensions of AHP. The above features can be seen as contributing to the systemic nature of AHP too.

It has been shown recently that some ideas of Beer's VSM may be operationalised through a modification in the procedure of the Analytic Hierarchy Process, leading to the possibility for propagation of managerial policies in a decentralised decision making environment (Petkov, 1994:1192). According to Jackson's interpretation of VSM, the parts of a system must be granted autonomy so that they can absorb some of the massive environmental variety that would otherwise overwhelm higher management levels. The only degree of constraint permitted is that necessary for overall systemic cohesion and viability (Jackson, 1991:119). It can be easily observed that higher levels in a hierarchy usually reflect policy factors that are of consideration predominantly by top level management. On the other hand, lower levels in an AHP model comprise features reflecting more specific knowledge about the problem. In a decentralised decision making environment such operational level knowledge is typical for the decisions that are taken by individual organisational units. In reality, whenever decisions are taken in a decentralised manner, there is usually a distortion of the policy considerations of the top management. Petkov (1994) reports of a decision support system developed using AHP, where judgements about the higher levels in a hierarchy are supplied by top management and "fixed" before the model is made available to the decentralised units, thus solving the problem of propagation of management policies in such an environment. This means that if somebody directly affected by a decision to be made, like the actual user of expensive equipment, applies such a model for its selection, she/he will be only able to make comparisons at the lower levels of the hierarchy according to her/his particular needs. After these are supplied to the model, the global or overall priorities of the lower level elements in the hierarchy are affected by the "fixed" comparisons of top management contained in the higher level judgements. The propagation of top level policies is achieved through the AHP principle of hierarchical composition. This seems to be an important aspect of problem structuring using AHP in decentralised organisations. It acts as the vehicle ensuring systemic cohesion and viability in terms of the Viable Systems Model. It can be concluded that AHP has a number of features that can be classified as contributing to its systemic nature and therefore it is appropriate to ask the question where it could possibly be classified within the System of Systems Methodologies.

3. The place of AHP in the System of Systems Methodologies

The System of Systems Methodologies is an ideal-type grid of problem contexts that was used by Jackson and Keys in 1984 to classify systems methodologies according to their assumptions about problem situations, and later extended by Jackson (Jackson, 1991:27). The vertical axis in Figure 1 (based on Jackson, 1991 and Jackson, 1995) reflects the complexity of problems according to such factors as number of elements, rate and character of the interactions between the elements, attributes of the elements, nature of the subsystems, and the environment. The horizontal axis reflects the increasing divergence of values among those interested in or affected by a problem situation. People are considered to be in a unitary relationship if they share values and interests; in a pluralist relationship if their values and interests diverge but they have enough in common to make it worthwhile for them to remain members of the coalition that constitutes the organisation, and in conflictual or coercive relationship if their interests diverge irreconcilably (Jackson, 1994).

According to Figure 1, traditional hard systems methodologies as Operations Research are considered to be simple and unitary. The reason is that they assume it is relatively easy to establish clear objectives for the system. Then they represent the problem with a quantitative model which is possible only for simple systems.

Soft Systems Approaches are viewed by Jackson (1991) as suitable for situations where the values or interests of the stakeholders are pluralist, while the complexity of the problem can range. When a problem is relatively simple then the Strategic Assumption and Surfacing Method by Mason and Mitroff is recommended, while for more complex situations Ackoff's Interactive Planning and Checkland's Soft Systems Methodology are suitable (Jackson,1991). There are only two approaches that seem to be suitable for conflictual situations, namely the metamethodology Total Systems Intervention by Flood and Jackson and Critical Systems Heuristics by W. Ulrich. It seems that new developments can be expected in that area, labelled by Jackson as Emancipatory Systems Thinking (see Figure 1). The System of Systems Methodologies demonstrates the interrelationship between different systems approaches, and the relationship these have to ideal-type problem contexts.

As mentioned already, AHP allows the incorporation of both quantitative and qualitative variables in a particular problem. That enables its usage in much wider fields, including socio-technical and management problems. The analysis in the previous section, and in particular the incorporation of qualitative variables, multiple conflicting criteria and of subjectivity, provides the prerequisites to consider AHP as capable of handling on its own simple problem contexts. A synthesis of the systems approach and AHP for multiple subsequent decision making tasks (somewhat similar to the ideas in Verkasalo (1994)) is suggested in Liao (1998). It however is based on the traditional functionalist understanding of AHP and therefore lacks the power to deal with complex "messy" problems. Regarding such problem contexts AHP can still be applicable provided it is used in combination with an appropriate systems approach facilitating the problem structuring stage in AHP. One of the first examples in that regard is the application of SAST and AHP in Banville et al. (1998). Our paper attempts to generalise their idea to the concept of complementarist use of MCDA with any of the problem structuring methods, described in Rosenhead (1989).

```
                INCREASING DIVERGENCE OF VALUES/INTERESTS------->
I
N              UNITARYPLURALIST        CONFLICTUAL
C
R
E  S
A  I
S  M           OPERATIONS
I  P            RESEARCH      SOFT           EMANCIPATORY
N  L
G  E

C                            SYSTEMS        SYSTEMS
O
M  C
P  O           DESIGN OF
L  M           ADAPTIVE      THINKING       THINKING
E  P           SYSTEMS
X  L
I  E
T  X
Y
```

Figure 1. The grid of problem contexts as a foundation for
The System of Systems Methodologies

The theory and practice of Group Decision Making using AHP may shed light to the question about the possible divergence of interests between the stakeholders. For the case when there are no radical differences in the opinions of the decision makers, a geometric mean of their judgements can be considered an aggregate of those judgements within the group. According to Saaty (1994a), this approach is suitable when a small group of individuals work closely together and the group becomes a new "individual" and behaves like one. Consequently there is no synthesis for each individual and the Pareto principle is irrelevant in this case (Forman and Peniwati, 1996:386). Ramanathan and Ganesh (1994) have explored the case when the group consists of individuals with different value systems. They propose to use the arithmetic mean for aggregating individual priorities. When a large number of geographically scattered individuals provide the judgements we need a statistical procedure to deal with variation among several people to surface a single value for the weights of the alternatives. Saaty (1994a) shows how this is implemented as well as procedures that can be applied both for constructive or retributive conflicts.

On the basis of the above one can make the conclusion that AHP is suitable for unitary, pluralist and conflictual problem situations, following the horizontal axis in the diagram showing the System of Systems Methodologies. The many examples of AHP applications in Saaty(1990) and Saaty(1994a) indicate that it may be suitable for both simple and medium complex problems. All this shows that AHP may cover a very broad range of problem areas as specified in the System of Systems Methodologies, though in its more traditional applications it concerns unitary or pluralist simple and medium complex problems.

Conclusion

The Analytic Hierarchy Process has been analysed in order to find evidence for its systemic nature. It seems that in many instances this has proved to be the case and AHP can be recognised as relevant to the systems field, and its place can be identified in the System of Systems Methodologies. Certain conclusions from this research might be generalised to the entire field of MCDM. Future research on these issues may concentrate on which organisational metaphors are supported by AHP, to which sociological paradigm it relates, whether it serves Habermas's technical, practical or emancipatory interests, and where it stands in the modernism versus post-modernism debate along the analysis of other systems methodologies provided in Jackson (1991).

On the other hand it is not appropriate to claim that AHP can be used as a single systems methodology, mainly for the fact that problem structuring within it is not formalised (if this task can be formalised at all - the same is valid also for the Multi Attribute Utility Theory as is evident from Keeney,1994) and there are better techniques for the Intelligence stage in Simon's model of decision making. That should not be viewed as a rejection of AHP as a systems science tool. On the contrary, it is possible to claim that on the basis of Critical Systems Thinking one can find a common foundation for the complementarist use of AHP or other MCDM techniques with soft systems approaches belonging to the field of problem structuring methods. As Jackson (1995) writes, Critical Systems Thinking is characterised by critical awareness of the strengths and weaknesses of available systems techniques, social awareness, which considers the organisational and societal climate that determines the popularity or otherwise of particular systems approaches at particular times, and complementarism between various systems methodologies, both at the level of methodology and at theoretical level. This paper attempts to raise critical awareness about the strengths and weaknesses of AHP as a possible strand of Systems Thinking as well as to highlight some issues in support of its possible complementarist use with other approaches. The need for such pluralist application of different methods or parts of them is highlighted also in a recent metatheory called Multimethodology (Mingers and Gill, 1997). It can be concluded that there is considerable scope for new and fruitful combined applications of AHP with different strands of systems thinking, enriching both Systems Theory and MCDM.

References

Ackoff R. and J. Gharajedaghi (1995): Reflections on Systems and their Models, *Systems Research*, 13 (1), 13-24.

Arbel A. and R Tong, (1982): On the Generation of Alternatives in Decision Analysis Problems, *Journal of the Operational Research Society*, 33, 377-387.

Banville C., M Landry, J-M Martel and C. Boulaire (1998): A Stakeholder Approach to MCDA, *Systems Research and Behavioural Science,* 15 (1), 15-32.

Checkland P. and J.Scholes,(1991): *Soft Systems Methodology in Action,* J. Wiley & Sons.

Dyer R.F. and E.H.Forman (1992): Group Decision Support with the Analytic Hierarchy Process, *Decision Support Systems*, 8, No.l, 99-123.

Flood R.L. and M.Jackson (1991): *Creative Problem Solving.Total Systems Intervention*, Wiley

Forman E. and K. Peniwati (1996): Aggregating Individual Judgements and Priorities with AHP, *Proceedings of the Fourth ISAHP*, Burnaby B.C., 383-391.

Jackson M. (1991): *Systems Methodologies for the Management Sciences*, Plenum.

Jackson M. (1994): Critical Systems Thinking: Beyond the Fragments, *Systems Dynamics Review*, Vol. 10, No.2-3, 213-229.

Jackson M. (1995): Beyond the Fads: Systems Thinking for Managers, *Systems Research*, Vol. 12, No.1, 25-43.

Keeney, R.(1994): Using values in Operations Research, TIMS/ORSA Joint National Meeting, Boston.

Lane D. and M. Jackson (1995): Only Connect! An Annotated Bibliography Reflecting the Breadth and Diversity of Systems Thinking, *Systems Research*, 12,No.3, 217-228.

Liao Z. (1998): A Systematic Integration Model to Support Complex Decision Making in a Dynamic Environment, *Systems Research and Behavioural Science,* 15 (1), 33-46.

Marzen V. (1994): The Enlargement of the Analytic Hierarchy Process in Respect To Global Planning and Sustainable Development, *Proceedings of 3rd Int. Symposium on AHP*, George Washington University. Washington DC, 75-88.

Mingers J. and A. Gill (1997): *Multimethodology,* J. Wiley & Sons.

Nunamaker, J.F., A.R Dennis, J.S.Valacich and D.R Vogel, (1991):Information

Technology for Negotiating Groups:Generating Options for Mutual Gain, *Management Science*, Vol 37, No.l0, 325-1346.

Petkov D. (1994): One Way to Propagate Management Policies in a Decentralised Decision Making Environments, in R Trappl(ed), *Cybernetics and Systems'94*, World Scientific Publ. Singapore, 1191-1198.

Ramanathan R and L.S.Ganesh, (1994): Group Preference Aggregation Methods Employed in AHP: An Evaluation and Intrinsic Process for Deriving Members' Weightages, *European Journal of Operational Research*, 79, 249-266.

Rosenhead, J. ed. (1989): *Rational Analysis for a Problematic World*, Wiley.Saaty T. and J. Alexander (1989): *Conflict Resolution: The Analytic Hierarchy Process*, Praeger Publ. Co..

Saaty T. (1990): *Multicriteria Decision Making - The Analytic Hierarchy Process*, 2nd ed. RWS Publications, Pittsburgh.

Saaty T. and K.P.Kearns (1991): *Analytical Planning. The organisation of systems*, 2nd ed.,RWS Publications, Pittsburgh.

Saaty T.(1994a): *Fundamentals of Decision Making and Priority Theory with the Analytic Hierarchy Process*, RWS Publications, Pittsburgh

Saaty T. (1994b): How to Make a Decision: The Analytic Hierarchy Process, *Interfaces*, Vol 24,No. 6, 19-46.

Simon H. (1962):The Architecture of Complexity, *Proc. Am. Philos. Society*, 106, p.467-486.

Steenge A.E., A. Bulten and F.G. Peters (1990): The decentralization of a sales support department in a medium-large company, *European Journal of Operational Research*, Vol 48, 120-127.

Stewart, T. (1992): A Critical Survey of the Status of Multiple Criteria Decision Making Theory and Practice, *OMEGA*, 20 (5-6), 569-586.

Turban T. (1991): *Decision Support and Expert Systems-Management Support Systems*, 2nd ed.,New York, Macmillan.

Verkasalo M.(l994): Repetitive Use of the AHP Hierarchy, in *Proceedings of the 3rd Int Symposium on AHP*, G. Washington University, Washington DC, 89 - 102.

The Analytic Hierarchy Process and Systems Thinking: A Response to Petkov & Mihova-Petkova

Denis Edgar-Nevill and Michael Jackson

Lincoln School of Management
Lincoln University campus
Brayford Pool
Lincoln LN6 7TS
United Kingdom

Abstract. Petkov and Mihova-Petkova in their paper "The Analytic Hierarchy Process and Systems Thinking" make a well argued case for the addition of AHP to the methods and tools in Systems Thinking. The claim is made that AHP is capable of dealing with complexity and pluralism. In this paper the authors consider the argument put forward and raise concerns about the effectiveness of AHP in this regard. These centre around the particular characteristics of complexity in social systems and the ability of AHP to allow each subjective viewpoint to be fully represented. It is argued that AHP can only reveal it's full potential when used alongside other systems approaches which do pay more attention to these matters. On the basis of this argument the case for pluralism in Systems Thinking is made with AHP an important element in this pluralism.

Keywords AHP, Systems Thinking, Complexity, Pluralism

1 Introduction

Petkov and Mihova-Petkova propose a role for the Analytic Hierarchy Process (AHP) as a useful tool for Systems Thinking (Petkov & Mihova-Petkova 1996). They claim that AHP can be particularly useful in situations where complex systems are being considered in a pluralist regime (reflecting the divergent interests of many subjective viewpoints commonly found in social systems). The question which must be considered is if AHP should be best used as a generic tool in understanding complex systems, modelling whole situations, or rather it is one of many useful techniques available in the armoury of Systems Thinking tools. Does AHP emerge as a unitary or pluralist technique useful for simple or complex system investigations? Where in the Jackson Grid (Jackson 1994) is AHP usefully classified?

2 The Analytic Hierarchy Process

Building a AHP model is essentially a reductionist approach to understanding a complex situation. Many of the arguments raised in this paper centre around the excessive nature of this reductionist approach when applied to the understanding of social systems. As Petkov and Mihova-Petkova agree, AHP has some parallels with hard systems description methodologies which create hierarchical descriptions of systems; such as Object-Oriented Design (Booch 1991). It develops representations as a set of levels organised as a hierarchy or as a network/plex structure with interdependencies and linking between levels representing elements of a problem. Petkov and Mihova-Petkova argue that the freedom to represent so many relationships & links in this kind of model elevates AHP above other methods. The ability to store quantitative information at each level of representation makes AHP a useful tool in Multiple Criteria Decision-Making (Saaty and Alexander 1990); (Saaty 1992). Once an AHP model is populated with quantitative and qualitative data, well-ordered processes may be used to combine quantities in ways to account for degrees of subjectivity (e.g. Consistency Ratio (Saaty 1992)) and different degrees of uncertainty (eg making use of fuzzy logic techniques (Kosko 1994)).

The key claims Petkov and Mihova-Petkova make centre on the capability of AHP in dealing with system complexity and pluralism.

2.1 Complexity & Pluralism in Social Systems

Understanding what is meant by the "complexity" is important to our discussions in this paper. There is no one definition which is entirely satisfactory to set a mark below which a system is "simple" and above which a system is "complex". These are terms which denotes movement towards different ends of a grey-scale. This is why the ideas of complexity, pluralism and social systems are used together to identify the types situation being studied.

When viewing social systems which involve complex interactions and stake-holdings which are difficult to identify, attempts to rationalise and build views have considerable hurdles to overcome.

An important idea to the systems thinker must be the acceptance of the limitations of any single rationalisation and viewpoint. A "solution" (a proposed future state of a system) to one stake-holder may be less attractive to others with alternative perceptions about the organisation and different personal goals. Making use of Soft Systems Methodology requires a number of views to be identified, discussed and reconciled in formulating solutions. Solutions are developed by stakeholders as important agents of change in the process (Checkland and Scholes 1991).

As human beings, we have a desire to simplify and order complex situations. Simple representative forms, such as hierarchies, are attractive because they can help to order and control complexity (Simon 1962). The mistake made by many developers is to severely underestimate the complexity of social systems. We can be faced with situations where using crude generalisations or pre-judgements can mean vital facets of a problem, central to successful resolutions, are unseen, ignored or misrepresented. There are numerous examples of large information systems developments which have failed because developers have not identified or understood the goals of one or more groups of stakeholders. Simple techniques may satisfy our realistic desires to understand complex systems but do not guarantee a successful outcome will result if we apply them to solve a problem. Overextending a simple model may mislead us into making false assumptions and conclusions.

3 The Limitations of AHP

The authors suggest three areas of concern with Petkov and Mihova-Petkova's claim that AHP can help resolve problems of system complexity & subjectivity.

3.1 Missing Subjective Views of Complex Systems

Consider an attempt to understand a crime scene using Systems Thinking ideas. Some viewpoints are easily identified (e.g. the victim(s) of the crime, the police). Some will be only guessed at (e.g. the general public, the law courts). Some viewpoints may never be studied at all. The criminal(s) committing the crime may never be identified. Important witnesses to the crime may not come forward (figure 1).

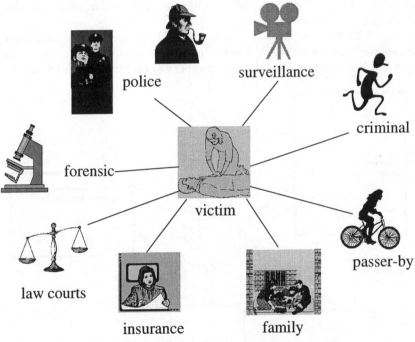

Figure 1 - Understanding a Crime Scene

Some viewpoints come into existence and become important at different stages/times. If the scope of the description is broadened we soon run into a social system with potentially hundreds or thousands of viewpoints.

AHP may help in allowing some of these perspectives to be represented and ordered but we must accept that much of the information will not be available. The picture of the crime is incomplete. Any AHP hierarchy will represent an incomplete picture. The incompleteness here is not a question of the uniform crudeness of any modelling regime but a reflection that complete parts of the model will usually be missing; calling into question the whole basis of using the model to understand a situation.

The hierarchies constructed within this partial view of a complex social interaction have then to be reconciled.

3.2 Reconciling Subjective Views of Complex Systems

Building models of any situation is difficult. Many researchers in Systems Thinking would even doubt the ability to ever "model" social systems satisfactorily. The key to success in this endeavour is the recognition of the limitations of any modelling technique used and structure created.

When building models of complex situations we are attempting to view an N dimensional system (where N is very large) with a modelling process which has M dimensions (where M is very small). To illustrate this we can refer to a simple modelling example from mathematics.

Consider taking conic sections. A 3 dimensional object is being viewed using 2 dimensional slices. With only a minor adjustment to perspective we can develop views which are very different from one another and hard to reconcile (figure 2).

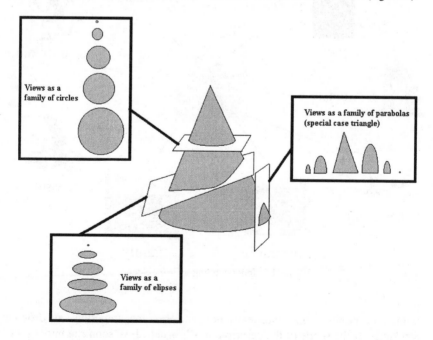

Figure 2- Modelling a 3D Object with 2D Slices

In this case we have only lost one dimension in our view of the system described. The analogy here using AHP is in taking a family of slices each related to the other much as the levels in a AHP hierarchy are related. Even in this very simple case we can imagine an infinite number of different families of slices (AHP hierarchies) resulting from a systematic slicing process based on the same simple system.

In a complex social system we are loosing many more dimensions between reality and the crude model we construct. Consider our model of a 3D cone modelled by 1D line segments of different orientations. Making sense out of these views becomes almost impossible. The same must be true in social systems as the number of missing dimensions increases (N is much greater than M).

The process of reconciling different subjective views to obtain a consensus AHP model must be considered. If this cannot be achieved all we have succeeded in doing is creating a number of distinct views - separate distinct AHP hierarchies. Conflict resolution and forming consensus requires a clear understanding of different subjective views within a common frame of reference. In AHP the numbers and qualitative information can be recorded within levels. The third concern raised with AHP relates to the true meaning of the information which might be recorded. Can a AHP hierarchy is to act as a repository providing a common frame of reference?

3.3 Reconciling Subjective Assessments

Without a clear agreement between viewpoints (an almost inevitable consequence of subjectivity) gaining sensible conformity on numeric values ascribed to elements within levels of a model is difficult. Each of the numbers recorded is a subjective judgement on an attitude scale. The development of coherent attitude scales requires large sample populations to be analysed if anything other than broad generalisations are to be drawn. In most cases large sampling is not available. A wealth of research has been done on the development and use of subjective assessment measurement (Edwards 1957). This work highlights the possibilities for measurement processes to introduce ambiguity and error.

For example, suppose two people agree on a subjective judgement of "5" on a scale 1 to 6 for an element describing the "Usability" of a part of the system being studied. The first person might be considering the usability from the experience they have of using the subsystem to perform a task. The second person might base their judgement on the range of different situations the subsystem can be used in. The "agreed" value recorded is thus an amalgam of different viewpoints, definitions and meanings of the number "5". We have no real reason for combining these alternative viewpoints simply because the same numeric value has been recorded. A resolution of any potential divergence of opinion cannot be implied. Experience also shows that when recording attitude assessments factors such as the time of day, the questions asked and the perspective of the person collecting the information can all vary the outcome. There may be many reasons why the same value is recorded but meaning different things to those agreeing the value. There may be many reasons why different values are recorded but meaning the same thing by those not agreeing on a value. Similar problems arise for the reverse case where two people give very different scores for the same situation; biased by their own subjective view of the system being assessed or their own subjective view of the assessment scale.

Having constructed an AHP model we must consider what happens when we apply our well-ordered processes to analyse it. The further one progresses down through the levels of an AHP hierarchy the more likely that the numbers in any scale mean different things to different people - each ambiguity of meaning being amplified in the simple arithmetic processes applied to consider data. Collective decision-making algorithms will be based upon ambiguous results with large error-bands. Spurious accuracy giving false confidence in results and decision-making is an inevitable consequence of processes of this kind. This seems to fail Saaty's basic principle in creating structures of this kind (Saaty 1994) (Can I compare the elements on a lower level using some or all of the elements on the higher level as criteria?) as quoted by Petkov and Mihova-Petkova.

So how is one model, one AHP hierarchy, to be agreed by those involved in the system? By adopting a pluralist approach one might see the way forward. Making use of a range of different processes (including AHP), with the limitations and generalisations of each paradigm/technique being recognised, a clearer insight into the system being studied will result when their collective views are compared. This lies at the heart of the System of Systems Methodologies (Jackson 1991).

4 The Role of AHP in Systems Thinking
As we have seen, if AHP is used in isolation to develop a complete system model we have to face the difficulties of:

- incomplete representation with missing views
- fragmented multiple hierarchies
- combined views based on unwarranted assumptions
- numbers with large error bands of limited use in analysis

AHP would be useful to structure the information/judgements from individual perspectives. As a vehicle for ordering a single person or groups views, AHP may provide a useful framework for controlling the complexity of the model of information we build. AHP therefore has a valuable role to play as one of a number of investigative tools in a pluralist systems thinking process.

However, it relies upon simple unitary assumptions which is a limitation in describing complex systems. As a method of reconciling different views and resolving conflicts there seem to be definite objections to overcome. The authors would suggest the development of AHP has some way to go to meet the role proposed by Petkov and Mihova-Petkova (figure 3).

We must recognise that systems involving social interaction are complex and any system models we build are, at best, pale reflections of them.

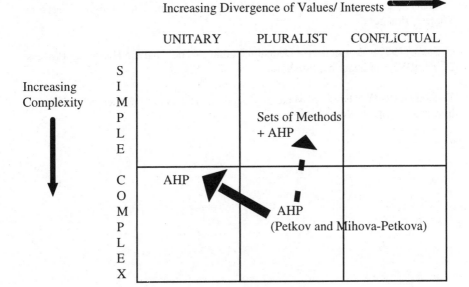

Figure 3 - AHP and Jackson's Grid

References

G. Booch (1991) : Object Oriented Design with Applications, The Benjamine/Cummings Publishing Company Inc.

P. Checkland and J. Scholes (1991) : Soft Systems Methodology in Action, Wiley

H. Edwards (1957) : Techniques of Attitude Scale Construction, Appleton-Century-Cofts Inc, New York

M. Jackson (1991) : Systems Methodologies for the Management Sciences, Plenum Press, New York

M. Jackson (1994) : Critical Systems thinking: Beyond the Fragments, Systems Dynamics review, 10, No 2-3, 213-229

B. Kosko (1994) : Fuzzy Thinking - The New Science of Fuzzy Logic, Harper Collins Publishers

D. Petkov and O. Mihova-Petkova (1996) : The Analytic Hierarchy Process & Systems Thinking, 13th Conference on MCDM, Cape Town, South Africa, January 1997

T. Saaty and J. Alexander (1989) : Conflict Resolution: The Analytic Hierarchy Process, Praeger

T. Saaty (1992) : Multicriteria Decision Making - The Analytic Hierarchy Process, 2nd ed. RWS Publications, Pittsburgh

T. Saaty (1994) : How to Make a Decision: The Analytic Hierarchy Process, Interfaces, Vol 24, No 6, 19-46

Interpretative Structural Modelling: Decision Making and the Method of Science

Tom Ryan[1]

[1] School of Engineering Management, Univesity of Cape Town, Private Bag, Rondebsch 7700, Republic of South Africa.

"Such is the method of science. Its fundamental hypothesis, restated in more familiar language is this: There are real things whose characters are entirely independent of our opinions about them; these reals affect our senses according to regular laws, and though our sensations are as different as our relations to the object, yet, by taking advantage of the laws of perception we can obtain by reasoning how things really and truly are; and any man, if he has sufficient experience and he reasons enough about it, will lead to one true conclusion."   Charles Saunders Peirce (Almeder 1980)

Abstract. This paper briefly explores how the method of science may be of use to management. It concludes that the abductive phase in the method of science, as described by Charles Saunders Peirce, offers opportunities for improving the effectiveness of decision making and planning. It goes on to describe a group consensus reaching methodology Interpretative Structural Modelling and argues that that it can be used a useful basis for management abduction.

Keywords. Decision making, Method of science, Abduction, Group work.

1 Introduction

Few people involved in the practice of management would dispute any proposition implying that the task of managing our institutions has become significantly more complex over the past few decades. Our ability to deal with such complexity has not developed at the same rate. Dwindling resources and increasing demands threaten the very foundations of our institutions. One of the sources of this complexity is the increase in competing and contradictory objectives in decision making. Traditional decision science can only cope with a single objective. This problem has led to the development of Multiple Criteria Decision Making (MCDM). Much of this development has revolved around the later stages in decision making, once the problem has been formulated and available alternatives have been identified. The early stages have largely been neglected. These early stages are important as they set the context and the direction for the later stages. Russell Ackoff has pointed that it is better to arrive at the wrong solution to the right problem than to get the right solution to the wrong problem, since a process of rigorous scientific inquiry will eventually led us to the right solution.

In the business arena increasing globalisation and intensive competition have added a further dimension to this complexity. Harry Levinson has captured the nature of this problem beautifully in the following words:

> *"A major issue that is getting practically no attention in the management literature is the reality that in many cases the chief executive officer does not have the conceptual capacity to grasp the degree of complexity that he or she must now confront. In short, they do not know what they are really up against and what is happening to them and their organisations, let alone know what to do about it. They simply cannot absorb the range of information they should and organise it from its multiple sources and focus it on the organisation's problems in a way that would both become vision and strategy."*.
>
> (quoted in Keidel 1995)

John Warfield refers to this as the problem of "underconceptualisation" and has spent much of his life on the task of developing ways of dealing with this problem (Warfield 1990). This paper suggests that one of the reasons for this is that management has turned its back on the *Method of Science (MoS)* and explores a way in which we can recapture *MoS* and integrate it into management practice.

This paper is divided into three parts:

- In Section 2 the first part discusses *MoS* and how it may be integrated into management practice. It draws on the work of Revans (1984) and CS Peirce (Misak 1991, Reilly 1970, Almeder 1980, Goudge 1948).
- In Section 3 the second part introduces Interpretative Structural Modelling (ISM) which offers a methodology for introducing *MoS* into management practice. John Warfield developed the process. (Warfield 1990, 1994).
- The last part covered in Section 4.0, which discusses how ISM can be used in *MoS* and integrated into management practice.

2 Management and the Method of Science

2.1 A model of Scientific Inquiry

Charles Saunders Peirce (CSP) was the founder of the Pragmatist school of philosophy popularised by William James and John Dewey. He is recognised as one of the most important American philosophers and has produced one of the most complete and coherent systems of philosophy. We find his system to be particularly relevant and useful to the practice of management because it offers a way of coping with both "hard" and "soft" problems. The term "hard" refers to problems where there is meaningful agreement on the nature of the problem and the decision criteria while the term "soft" refers to problems where there is little agreement. Unfortunately CSP never wrote up his philosophy in a single definitive document. Scholars have been at work for the last five decades in the task of interpreting and integrating his seemingly fragmented notes and papers into a coherent whole.

One of the most enduring management ideas that has emerged over the past decade in the idea of the' learning organisation' popularised by the work of Peter Senge (Senge 1990). These ideas are largely based on the work done by Chris Argyris and his colleagues over the past *25* years (Argyris 1993) under the general heading of Action Science. CSP pre-empted this work with his Belief – Inquiry-Model (BIM) towards the end of the last century (Misak 1991, Brent 1993, and Reilly 1970). He claimed that we only know the world to the extent of the stable beliefs we hold about it. A stable belief is a habit of expectation leading to a certain behaviour. A belief about something will lead us to expect certain things under certain conditions and this will lead to a certain behaviour under those conditions. Our beliefs are therefore the basis of our behaviour and the actions we take.

Our beliefs will cause us to expect certain consequences resulting from our actions. If these do not occur we are surprised and our beliefs are destabilised and we are cast into doubt. Doubt leaves us perplexed and unable to act, which in turn causes a dissonance within us. This dissonance moves us to undertake an inquiry in order to establish new or modified beliefs. Doubt raises questions which inquiry attempts to answer by establishing stable beliefs. The process of inquiry may be "hard", "soft" or a mix of the two. CSP calls this the process of fixing beliefs. He goes to great pains to point out these beliefs are fallible and are likely to fail us in the future and that we should be ready to cast them aside when they no longer conform to experience. A key part of CSP's pragmatist philosophy is the crucial role of experience in learning and inquiry.

CSP suggest that the many different processes for fixing belief which can be classified into four categories:

1   *The Method of Tenacity:* which covers the many processes we use to "hang on" on to the beliefs we already hold at all costs because they are convenient, expedient or familiar.
2   *The Method of Authority:* which covers the processes that lead to beliefs based on what is well spoken for by authoritative advocates seen as 'experts' or as people 'in the know'.
3   *The 'A Priori' Method:* which covers processes based on what is seen as plausible and reasonable in terms of general principles.
4   *The Method of Science (MoS):* covers those methods that align beliefs with experience.

CSP argues that in the long term *MoS* is the more effective of these methods for establishing stable beliefs. His arguments for this include (Rescher *1995,* Reilly1970):

1   It has the better track record in supplying us with more reliable information. An analysis of any inquiry process that that has been applied continuously over time and that has yielded good results over time will show that it conforms to *MoS*. A good example is the Shewhart/Deming inquiry cycle called the *P*lanning-*D*oing-*C*hecking-*A*djusting cycle.. It is better known as the *PDCA* cycle, which forms the basis of successful Japanese manufacturing management.

2   It is rigorous in that it relies on some external frame that is largely independent of the thought of the inquirer and is therefore not subject to the whims of the inquirer.

3   It is self-correcting in the long term because it responds to experience and will eventually eliminate errors that it has incurred. This is often referred to "fact management" in the management literature.

4   CSP convincingly argues that if any method of inquiry can lead us to the truth then the method of science can and therefore it is the more reliable alternative.

CSP highlights the following important characteristics of *MoS* (Goudge 1948):

1   Scientific inquiry is a co-operative social process pursued by a community of inquirers in which contribution build on prior contributions in the process of settling opinion.

2   The process is persistent and strives to be disinterested in pursuit of a shared truth.

3   Its data is obtained through some form of observation.

4   It deals with the data through rational and logic thought.

5   It conclusions must be verified by experience.

6   Its conclusions are fallible in that they are provisional and are susceptible to further refinement or correction

2.1 The Method of Science

Reg Revans' interpretation *of MoS* is perhaps the better known one (Revans 1982). He recognises the *MoS* as the foundation of human achievement and he describes it as consisting of the following five steps:

1   Observation of the external world

2   Formulation of theories based on these observations

3   Design and conduct experiments to test these theories

4   Comparison of the experimental results with those predicted by the theories

5   Rejection, modification or confirmation of the theories in accordance with the results of these comparisons

The process consists of two distinct phases: (1) the development of a theory or a hypothesis that explains the observation and (2) the verification of the hypothesis. Karl Popper sees the second phase as a process of refutation since it is not possible to prove a theory (Miller 1983). Any proof can only cover a limited sample of the potentially infinite number of instances of the theory. A theory is therefore only tentative and it is a good theory if it can withstand refutation over time. We must however accept that it will eventually be toppled. This is consistent with CSP's. concept of fallibility.

CSP argued that there are three radically different kinds of inferences involved and that these make up the stages of a scientific inquiry (Reilly 1970):

1 *Abduction*: the process of generating an explanatory hypothesis that plausibly explains a surprising phenomenon.
2 *Deduction*: the process that predicts the practical consequences of the hypothesis being true under certain conditions.
3 *Induction*: the process of realising the conditions specified in step 2 and testing the hypothesis

CSP insisted that any real inquiry starts with an unanswered question or doubt that flows from a destabilised belief. He developed a comprehensive philosophical system to support his process of inquiry. He saw deduction as explicative inference, which explicates its premises. It does not produce new knowledge, it merely rearranges existing knowledge. Abduction and induction are ampliative inferences. They amplify their premises and produce new knowledge, particularly abduction.

2.3 Management and the Method of Science

Reg Revans is physicist turned management thinker. He left the Cavendish Laboratory at Cambridge University where he worked in the company of 11 Nobel prize winners to become training manager at the National Coal Board in the UK. He has spent the rest of his life finding ways in which management can learn from the success of *MoS*. He suggested that good management practice is nothing more than the practical expression of MoS. It is simply "a statement of the intelligent fulfilment of human need" and involves (Revans 1982, page 97):

1 Finding out the needs of the situation
2 Deciding what to do to meet these needs
3 Doing it in what seems to be the best way
4 Testing to see how well it was done, and
5 Reviewing the whole situation and if necessary modifying the decision about what needs to be done in the future.

In CSP's terms management can be seen as:

1 Abduction: the process of understanding and explaining an unsatisfactory result.
2 Deduction: planning an intervention, based on this understanding, to improve the situation.
3 Induction: implementing and monitoring the plan, taking corrective action where necessary, and in the process establishing good management practice.

Clearly the success of a management intervention depends to a large degree on the understanding of the initial problem situation i.e. the abductive phase. This in an area of management that is poorly understood and therefore poorly practised, particularly in the case of soft problems. The next part of this paper introduces Interpretative Structural Modelling (ISM) and then argues that it can be effectively used as an abductive process for management.

3 Interpretative Structural Modelling

3.1 CSP's Realist Philosophy

CSP's realist philosophy, as reflected in the quote at the beginning of this paper, deals with the "hard"/"soft" debate by acknowledging an external objective world independent of our thought, but at the same time recognising that people experience it and interpret it differently. The different interpretations reflect different realities. The implication is that anyone who has an interest in a particular situation has a legitimate but limited perspective of the situation and will behave in accordance with it. This has far reaching implications for managing complexity and diversity.

Organisations and institutions that keep society operating are having to deal with increasing complexity and diversity. It is becoming common practice to develop and employ teams to deal with the complex issues that face the management of these institutions. In countries like South Africa these teams have people with widely differing backgrounds, life experiences, cultures and value systems. The challenge is to mobilise the combined intelligence and experience of such teams in the interests of the institutions that they serve. The past two decades has seen the emergence of a wide range of consensus seeking processes which have met with varying degrees of success. ISM is one such process which we have use with success. It is based on the *MoS* and Systems Thinking. Before introducing ISM the ideas of systems thinking and complexity are briefly introduced.

3.2 Systems Thinking

It is beyond the scope of this paper to develop the ideas of systems thinking in any detail. I will briefly introduce some of the core ideas that are relevant to the topic of this paper. This introduction will largely be based on the ideas of Russell Ackoff (Ackoff 1994).

A central idea in systems thinking is that any effect or behaviour is emergent and is the product of a set of interacting co-producers. The behaviour is not the sum of the outputs of the individual producers, it is the product of their interaction. The producers are seen as the interacting parts of a system and the behaviour is the output of the system. Each part can influence the output and the way in which it influences it depends on its interactions with the other parts. The parts may or may not exhibit a measure of intelligence. An intelligent part:

1   Has purposes of its own
2   Is able to interpret its situation
3   Can make choices of its own based on its purposes and interpretations.

This leads to three categories of systems;

1   Mechanistic systems in which neither the systems or their parts are intelligent.

2   Organismic systems in which the systems have intelligence but their parts do not.
3   Social systems in which both the systems and their parts have intelligence

The ideas of hierarchy; boundaries and environment; and communication and regulation play an important part in systems thinking. A systems based inquiry into the any effect or behaviour under consideration will seek to answer the following questions:

1   What is the is the behaviour of interest and why does it require attention?
2   How is the behaviour co-produced?
3   Why does the system behave in the way it does?

3.3 Complexity

There are number of way of viewing and describing complexity. One way is to do it on three levels:

1   Static complexity which is a measure of the number of parts and their range of behaviours that make up the system.
2   Dynamic complexity which is the measure of the level of the dynamic interaction between the parts.
3   Psychological complexity, which is a measure of the number of interpretations and interests, exhibited by those observing the behaviour.

The situation is further complicated by the limited span of human recall. Miller (Miller 1956) has suggested that we humans can only bring to mind about seven items at a time from our short-term memory. Warfield (Warfield 1994) has shown that when this includes the relationships between interacting parts we can only bring to mind, unaided, three interacting parts. This severely limits our ability to deal with complexity.

3.4 Interpretative Structural Modelling (ISM)

John Warfield (Warfield 1970) developed a process called Interpretative Structural Modelling (ISM) in an attempt to deal with the problem of managing complexity. It is a group-based process for exploring the different parts and their relationships that co-produce the complex issue under consideration. As the name implies the process produces a structured model of the groups understanding of the complex issue. It essentially consists of two phases:

1   A creative process (such the Nominal Group Technique) in which the group generates a list of the possible parts of a complex issue, and

2   A computer aided process for structuring the parts into a model of the underlying system that produces the complex issue. It is essentially a process for organising the group's thinking and understanding of the issue. The structuring is done using a pair-wise comparison process

Warfield's major contribution lies in the second process. He developed a mathematical basis for rigorously and coherently synthesising a groups' fragmented knowledge and experience into a whole. Two parts of the issue under consideration are compared at a time with respect to their relationship to the issue. This involves three elements, which fall within the limits of the span of our immediate recall. Six categories of relationships are recognised and are taken to be transitive: definitive, comparative, causal, temporal, spatial and mathematical relations.

The computer poses questions to the group about two parts at a time regarding their relationships to the issue. Set theory and matrix methods are used to infer other relationships assuming transitivity. A structured model using digraphs is produced once the comparisons are done. Based on the different types of relationships it is possible to generate one of the following types of structure:

1   Intent structures which show the relationships between intentions, objectives. goals purposes, aims etc. A typical question would be: *'In the context of the issue would part A help to achieve part B?'*.

2   Priority structures which structures the parts in order of priority. A typical question would be *'In the context of the issue is part A more important than part B?'*.

3   Attribute enhancement structures which show the relationships between a set of factors, problems or opportunities. A typical question would be. *'In the context of the issue would part A contribute to part B?'*.

4   Process structures which sequence the parts of activities e.g. *".... must Activity A precede activity B?'*.

5   Mathematically dependant structures which structure relationships between quantifiable parts.

4 Using ISM in Management Abduction

We have found ISM very useful approach to developing a shared understanding of a complex issue or problematic situation as a group abductive process. Consider the situation in which a manager is required to deal with a complex issue which is an unsatisfactory result of previous management action or policy. The details of the process may vary from one situation to another but in general most processes include the following steps:

1   Describe the issue or problem. Determine why it is thought to be problematical and why is it important that something be done.

2   Assemble a group people, numbering six to twelve people, who have an interest in the issue. It is useful to include representatives of the following stakeholder groups: those who have some authority in the situation, those who are involved in the situation and have detailed knowledge of it, those who are affected by the issue, and those who are particularly concerned about the situation. It is also important to have a facilitator who has a working

knowledge of ISM and has access to the necessary resources like the software and hardware.

3   Phrase the issue as question that if appropriately answered will provide a basis for dealing with the issue, e.g. *'in what way do we need to change our current management practices to maintain our competitiveness?'*.

4   Each member of the group silently prepares a list of responses to this question.

5   On a round-robin basis each member has an opportunity to present one of their responses which is copied on a flipchart in sight of the whole group. The process continues until all the responses have been recorded.

6   Each response is then discussed by the group, clarified and edited. The final list must contain a set of clear, unambiguous and coherent problem statements perceived as relevant to the initial trigger question. These are seen as potential co-producers of the complex issues under consideration.

7   Each member privately votes what that person sees as the five most important problems.

8   A new list containing all the problems that have at least one vote is entered into the computer.

9   A contextual description of the issue, a relationship appropriate to the issue and trigger question is entered into the computer. This relationship is used to structure the questions the computer poses to the group. This relationship will determine the type of structure that is finally produced.

10   The computer then poses the first question to the group e.g. "In the context of the issue does problem A contribute to problem B". The group is required to answer yes or no.

11   A new question containing a new problem is posed and step 10 is repeated. The software then determines what additional relationships between the problem statements can be inferred from these answers. It then decides on the next question to ask the group. This process is repeated until all possible relationships have been established.

12   The software then graphically models the logical relationship pattern developed by the group using digraphs.

This is essentially a group abductive process that produces a model reflecting the group's understanding of the situation surrounding the issue. This then becomes the basis for the design of possible management interventions to deal with the issue. We have also used ISM successfully in the deductive and inductive phases of inquiry.

5 Discussion

Management has always been about change, but change is not what it used to be. In his book "The Age of Unreason" Charles Handy (Handy 1989) argues that change has accelerated to such an extent that it no longer makes patterns, it has become discontinuous. This means the management policies and practices that we have developed over time are unlikely to be the most appropriate to deal with the challenges we face today. What we used to take for granted can no longer be held as 'true'. If our institutions are to survive and remain viable we will have to develop new and more appropriate management policies and practices. We will have to move our focus from the 'back end' to the 'front end' of planning and decision making. This means that we need to introduce abduction along side deduction and induction in our planning and decision-making processes.

Every time we get our planning and decision-making is wrong it costs society both economically and socially. We need to develop abductive management processes that improve the frequency of getting it right first time. The Japanese have demonstrated this to some degree. ISM offers us a process that can do this in a rigorous and scientific way. It does so by helping to deal with three challenging issues that characterise organisational life today:

1   The knowledge and skills needed to deal with the problems that plague organisations can no longer be found in one person, we need teams to do the job. ISM helps to integrate the fragmented knowledge and experience of team members into a coherent whole.
2   Successful implementation of a management intervention requires the co-- ordinated action of many diverse stakeholders that make regular individual discretionary decisions. ISM helps to develop a shared understanding of the intervention which gives coherence to these decisions.
3   ISM offers a comprehensive framework for developing a rich and appropriate conceptualisation of the issues that decision-makers have to deal with.

A wide range of ISM practitioners has documented their experiences that corroborate these claims. References to these experiences can be sourced from the following documents:
Annotated Bibliography on Complexity Research, Annotated Bibliography on interpretative Modelling and Related Work, and Annotated Bibliography on Generic systems Design and Interactive Management. Information on these documents can be got from: Institute for Advanced Study in the Integrative Sciences, George Mason University, Fairfax, Virginia, USA.

References

Ackoff, Russell L (1994): *The Democratic Corporation,* Oxford University Press.
Argyris, Chris (1993): *On Organisational Learning,* Blackwell Publishers.
Almeder, Robert (1980): *The Philosophy of Charles Peirce: a Critical Introduction,* Basil Blackwell Press, Oxford.
Brent, Joseph (1993): *Charles Saunders Peirce: a Life,* Indiana University Press.
Goudge, Thomas A. (1949): *The Thought of CS Peirce,* Dover Publication, New York
Handy, Charles (1989): *The Age of Unreason,* Business Books Ltd.
Keidel, Robert W. (1995) *Seeing Organisation Patterns: a new theory and language of organisational design.* Barrett-Koehler Publishers Inc.
Miller, David (1983) ed.: *A Pocket Popper.* Fontana Press.
Miller, G.A. *(1956): The Magical Number Seven, Plus or Minus Two: Some Limits on our Capacity for Processing Information.* Psychological Review 63 81-89.

Misak, Cl. (1991): *Truth and the End of Inquiry: A Peircian account of Truth,* Clarendon Press, Oxford

Rescher, Nicholas *(1995): Peirce on the Validation of Science,* in Ketner, Kenneth Lain (Ed): *Peirce and Contemporary Thought,* Fordham University Press.

Reilly, Francis E. (1970): *Charles Peirce 's Theory of Scientific Method,* Fordham University Press, New York

Revans, Reg (1982): *The Origins and Growth of Action Learning,* Chartwell Bratt, Bickley, Kent.

Senge, Peter (1990): *The Fifth Discipline,* Doubleday New York.

Warfield, John N. (1976): *Societal Systems: Planning, Policy and Complexity,* Wiley, New York

Warfield, John N. *(1990): Underconceptualisation,* Institute for the Advanced Study of the Integrative Sciences, George Mason University.

Warfield, John N. (1994); *A Science of Generic Design: Managing complexity through Systems Design,* Iowa State University Press.

A New Conflict-Based Redundancy Measure

Per J. Agrell

Department of Operations Research, University of Copenhagen, Universitetsparken 5, DK-2100 Copenhagen Ø, DENMARK. On leave from the Department of Production Economics, Linköping Institute of Technology, S-581 83 Linköping, SWEDEN.

Abstract: The concept of criterion redundancy in MCDM has been sparsely addressed by previous research in the field, albeit being of inevitable interest in any practical application. This paper presents a new conflict-based framework for redundancy that is argued better suited for interactive approaches than the definition proposed by Gal and Leberling (1977). Redundancy in the continuous case is defined similarly to the definition of mutual preferential independence in multi attribute utility theory, a connection that brings additional insights to the conceptual problem of criterion selection. Finally a simple test for the linear case is described, intended for initial screening of models.

Keywords. Redundancy, Multi-Criteria Decision Making

1. Introduction

Despite its deceptive triviality, the question of which criteria to select in a multi-criterion model provokes a number of interesting issues related to behavioral and mathematical modeling of preferences. Where textbooks in the field devote a substantial proportion to the technicalities involved in solving an already predefined mathematical program, the question how to formulate and test criteria is often left to the analyst's discretion. Nevertheless, practitioners and students deserve a more thorough and systematic answer to such a fundamental concern.

This paper does not purport to address the general problem of criterion selection, but advances the discussion of one important aspect, criterion redundancy. Intuitively, non-redundancy among criteria is a necessary but not sufficient requirement for a representative model. As opposed to the conceptually more difficult concern of relevance, redundancy has potentially quantitative properties which may be used in the construction of a test. As will be shown further on, the proposed framework is consistent with the related issue of preferential independence in multi-attribute utility theory. First, a brief summary of some of the few previous works in the area.

The seminal paper by Gal and Leberling (1977) defined and investigated a linear redundancy relation very close to the conventional single objective linear programming convention. Amador and Romero (1989) investigated published applications of Lexicographic Goal Programming for redundancy and found it abundant. In all but one of twenty reviewed cases one or more goals were redundant. The existence of goals with primarily ornamental significance is not satisfactory in MCDM, since the decision-maker is falsely made to believe that the model incorporates his stated preferences. van Herwijnen *et al* (1994) discuss the interdependency between criteria and conclude that in many cases criteria are strongly correlated. Their evaluation, mainly aimed at a multi-attribute utility framework, outlines different methods for sensitivity analysis in the case of stochastic interdependence. Romero (1994) recently reopened the discussion with a new survey, also aimed at Lexicographic Goal Programming.

1.1 Motivation of Research

As opposed to the case of Information Theory, where redundancy can be seen as added security in communication, the effects of superfluous model components in MCDM are highly negative. Redundant criteria, especially in Lexicographic Goal Programming, may easily deceive the decision-maker with the impression that more is considered in a specific model than the actually case. Formulations of models by inexperienced practitioners may also lead to redundant formulations, which are not only misleading but also computationally cumbersome. In effect, the decision-maker may end up without decision support, although a reformulation would achieve the desired result with less complexity. Thus, the issue of redundancy deserves to be seriously addressed and incorporated in texts on multi-criteria modelling and analysis.

1.2 Linear Multi-Criteria Optimization

Assume a general continuous MCDM minimisation formulation as [MCDM], where a *decision vector* $\mathbf{x}$ is to be chosen from a compact and non-empty *decision space* X, $X \subseteq \Re^n$. The *criterion vector* $\mathbf{y}$ is composed of continuously differentiable *criterion functions*, $y_i: X \to \Re$, $i = 1, ..., p$.

PROBLEM [MCDM]

$$\min \mathbf{y}$$

s.t.

$$\mathbf{x} \in X.$$

A priori it is known by standard results that if the noninferior set exists it is an image of an efficient edge, point or ray in the decision space $X \subset \Re^n$. As a basis for our linear discussion we choose a standard problem, [MOLP].

PROBLEM [MOLP]

$$\min \mathbf{y} = \mathbf{Cx}$$

s.t.

$$\mathbf{Ax} \geq \mathbf{b},$$

$$\mathbf{x} \geq 0.$$

where $\mathbf{A} = \begin{bmatrix} \mathbf{a}_1^T \\ \vdots \\ \mathbf{a}_m^T \end{bmatrix} \in \Re^{m \times n}$, $\mathbf{b} \in \Re^m$ and $\mathbf{C} = \begin{bmatrix} \mathbf{c}_1^T \\ \vdots \\ \mathbf{c}_p^T \end{bmatrix} \in \Re^{p \times n}$, $\mathbf{c}_j \in \Re^n$, $j = 1, ..., p$,

with $y_j(\mathbf{x}) = \mathbf{c}_j \mathbf{x}$. At least one element in each row of $\mathbf{A}$ and $\mathbf{C}$ is non-zero. X_E is assumed as a non-empty, connected, bounded, convex set, and subsequently Y_E is non-empty, connected, bounded, and convex.

The problem *ideal* (utopia point) $\mathbf{y}$, cf. Steuer (1976) is of interest in the context, i.e., the lexicographic minimum attained by repeated scalar minimisation.

Def. 1:   *Ideal point,*

$$\mathbf{y}^* = \min \left\{ \mathbf{z} \in \Re^p \middle| \forall \mathbf{y} \in Y_E : \mathbf{z} \geq \mathbf{y} \right\}$$

3. Redundancy Revisited

In single-objective Linear Programming the concept of redundancy for constraints is readily defined. A constraint $g_j(\mathbf{x}) \leq 0$ is called (strongly) redundant iff $g_j(X) \leq 0$ and X denotes the feasible decision space with respect to all other constraints than j. A (strongly) redundant constraint can thus be removed without affecting the feasible space X. The concern here is clearly of computational nature.

The ultimate justification for multiple criteria models is some inherent conflict in the decision. The conflict is evident by the inability to simultaneously optimise all relevant criteria, or technically, the infeasible ideal point. The dimensional problem in criterion space prohibits the use of an equivalent definition to decision space redundancy. The addition of a criterion per definition changes the criterion space by adding one dimension. However, the addition is irrelevant with respect to the conflict as long as it does not concern the ranking of points in X. Thus, we easily arrive at a theoretical (negative) definition as a basis for further (operational or constructive) definitions.

Three immediate remarks are of general interest. First, as in decision space, redundancy is detectable only by pairwise comparison and no prior judgement is possible with regard to which criteria to eliminate. Definitions focusing at the entire space Y are not constructive and must be followed by pairwise comparison to draw any conclusion. Second, the analogy with the decision space is not exhaustive. A constraint added to a program, which does not affect the optimal solution, is clearly unnecessary by any means. An added criterion that correlates highly with some other criteria may be of semantical, if not mathematical, interest. If such semantical interest motivates a redundant criterion, the model benefits from excluding such criterion from the actual (computational) model, but retaining it in the interaction with the decision-maker. Third, the decision whether to retain or eliminate a particular criterion in a redundant pair is to be judged by other properties such as relevance, measurability, causal implications, etc. From a mathematical viewpoint, the choice is arbitrary once redundancy has been detected.

3.1 Gal-Leberling Redundancy

The definition suggested by Gal and Leberling for the linear case is based on the notion of a minimal spanning cone. Colloquially, a criterion is redundant if its inclusion in the model does not affect the set of efficient solutions, Y_E. Mathematically, we introduce the relation $\mathfrak{L}$ for the constructive definition.

Def: $\mathfrak{L}$ -Redundancy, (Gal and Leberling, 1977)

Criterion i is absolutely redundant if it is linearly dependent. Or,

$$\mathfrak{L}i \Leftrightarrow \exists \lambda \in \mathfrak{R}_+^{p-1} : c_i^T = \begin{vmatrix} c_1 & \dots & c_{i-1} & c_{i+1} \dots & c_p \end{vmatrix} \lambda.$$

Elementary algebra yields that the criterion matrix C may have at most n linearly independent criterion vectors. Gal and Leberling show several interesting properties of $\mathfrak{L}$ and in addition present an algorithm for determining the minimal spanning cone and subsequently the redundant criteria. However, their proposed definition completely oversees an important property of multi criteria programs, the concept of conflict. Especially in interactive methods, a characteristic feature of the decision analysis is the successive exploration of inherent conflicts in the model. Trade-offs and reservation levels are frequent actions in such methods, as well as in the actual decision. To eliminate a criterion in high conflict with other criteria merely on the basis of a geometrical property does not seem advantageous from the viewpoint of the decision-maker. The framework suggested by Gal-Leberling is applicable primarily to methods for generation of the entire efficient set, which are only a subset of currently available methods.

3.2 $\mathfrak{S}$-Redundancy Relation

We argue that a redundancy concept must be rooted in the MCDM modelling tradition, with closer bounds to Decision Theory than to vector optimization. Agrell (1995, 1997) define a conflict based redundancy, $\mathfrak{S}$, where a criterion j is said to be $\mathfrak{S}$-*redundant* if it is not in conflict with any other criterion. In the absence of conflict, the ranking of all elements in X is only a matter of scaling. Thus, there is no trade-off possible between two redundant criteria and in any particular decision between any other criteria, the two redundant criteria would strive towards the same direction. Clearly, this notion corresponds to an undesirable property of a criterion in the context of interactive decision support.

Mathematically, the definition is stated as:

Def. 2:   *Redundancy Relation $\mathfrak{S}$*

$j\,\mathfrak{S}i \Leftrightarrow$ For all efficient decisions **x** criterion j is redundant with respect to criterion i,

$$j \mathfrak{S} i \Leftrightarrow \forall \mathbf{x}', \mathbf{x}'' \in X_E \exists \xi > 0 : y_j(\mathbf{x}') - y_j(\mathbf{x}'') = \xi \big(y_i(\mathbf{x}') - y_i(\mathbf{x}'') \big).$$

The pairwise relation is independent of the addition or elimination of other criteria to the problem. Agrell (1997) shows the relation to be reflexive, symmetric, and transitive.

The $\mathfrak{S}$ definition is negatively stated, i.e., knowledge of the entire efficient frontier is necessary in order to detect redundancy. However, a few straightforward results help to reduce the number of suspicious cases effectively:

Remark 1:   *If the lexicographic minimum y^*_j is unique then criterion j is not redundant,*
$$\forall i \neq j : y_j(\mathbf{x}^{*i}) \neq y_j(\mathbf{x}^{*j}) \Rightarrow \forall i : -j \mathfrak{S} i .$$

PROOF: Trivial, the ideal point *per se* falsifies the $\mathfrak{S}$-redundancy condition in Def. 2, by letting $\mathbf{x}' = \mathbf{x}^{*i}$ and $\mathbf{x}'' = \mathbf{x}^{*j}$. Q.E.D.

Thus, a quick look in the pay-off matrix would probably eliminate the most redundancy candidates in a typical MCDM model. For the remaining subset of criteria, a test is presented and illustrated using $\mathfrak{S}$.

4. An Extreme-Point Test of Redundancy in MOLP

The principle behind the test is surprisingly simple. In each efficient corner point the preferred edge for each criterion under scrutiny is recorded. Whenever variations in the choice are noted, the suspicion of redundancy is refuted. The test continues until either all efficient corner points are examined or the decision maker has enough evidence to revise the model anyway.

The efficient extreme points may be sequentially generated, e.g., by means of ADBASE (Steuer, 1974) or some other algorithm. For each point $\mathbf{x}'$ the gradients of the criteria are compared with the gradients of outgoing efficient edges. The active constraints in the non-degenerate point are distinguished by their associated negative dual variables.

Assume that in an efficient extreme point $\mathbf{x}'$, a number of constraints 1, 2, ... , m' are active. The edges most aligned with the criterion gradient ∇y correspond to efficient edges. The *most preferred edge l* for criterion j is characterised by (Agrell, 1997)

$$-\mathbf{a}_l^T \nabla y_j = \max_{b=1,\dots,m'} \left\{ -\mathbf{a}_b^T \nabla y_j \right\}, \tag{1}$$

where $\mathbf{a}_l$ denotes the lth row in $\mathbf{A}$. In this case, the direction is not relevant, thus for any point the edges may be renumbered 1, 2, ... , m' in some arbitrary order. The procedure is illustrated for a two-dimensional case in Fig. 1.

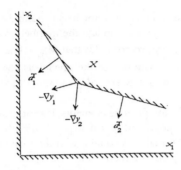

Figure 1: Extreme Point Test of Redundancy.

To prove that the procedure is adequate, we first need to establish a connection between the criterion gradients, c, the constraint gradients, a, and the desired directions, v. A two-dimensional case is depicted in Fig. 1, which is used in the formulation of Lemma 2. It is evident directly from Fig. 1 that the angle between c and v must always be acute to of interest, or else $c^T v > 0$, implying a worsening of the criterion value.

Lemma 1: $\qquad j \otimes i \wedge c_j^T v_{b'} = \min_b \{c_j^T v_b\} \Rightarrow c_i^T v_{b'} = \min_b \{c_i^T v_b\}$

Lemma 1 (Agrell, 1997) states a necessary condition for redundancy in terms of attainable directions along the adjacent edges. However, we now need a proxy for v expressed in the directly accessible vectors a and c. Thus, we formulate Lemma 2 (Agrell, 1997).

Lemma 2: $\qquad \forall b : \|a_b\| = \|v_b\| = 1 \wedge c^T v_l = \min_b \{c^T v_b\} \Leftrightarrow -a_l^T c = \max_b \{-a_b^T c\}$

Due to the construction of the proofs above, non-redundancy is shown by counter-proof, i.e., existence of divergent directions. Thus, the data storage requirements are limited to some provision against looping between already marked corners, x. In practice, the matrix $-A_l C^T$ is calculated once and corresponding elements are then compared for each corner point. Let us summarize the findings in an algorithmic manner, where an (optional) matrix M is used to record the number of coinciding preferred outgoing edges.

1. *Initialization* Find an initial efficient corner solution x^0. Also, establish the payoff-matrix and find the ideal points y_j^* for all criteria. Criteria with unique ideal points are excluded from the study as non-redundant. For all criteria j not eliminated, set $M_{ji} = 0$ for all i. Calculate $-A_l C^T$.

2. *Iteration* Sequentially explore corner points and mark the corner. Whenever different directions (as in Lemma 2) are recorded for two criteria i and j, set the

corresponding elements M_{ji} and M_{ij} to some negative number, since non-redundancy is proven. When directions coincide, update the number of correlated directions for both elements M_{ji} and M_{ij}. Symmetry yields that $M_{jk} < 0$ and $M_{ji} > 0$ implies $M_{ki} < 0$. In general, $\mathbf{M}$ is easy to update due to the previously stated properties of $\mathfrak{C}$. The diagonal elements M_{ii} can be used to record the number of visited corner points.

3. *Evaluation* If $\mathbf{M}$ contains any positive elements, except in the diagonal, continue by finding an adjacent corner point and repeat step 2. If no other efficient corner can be identified, or if the decision-maker is content with the total number of trials, stop and consider the corresponding criteria redundant.

4. *Stop-criterion* If all elements in $\mathbf{M}$ are negative, stop. Non-redundancy is proven, as of above.

4.1 Illustration of the Algorithm

EXAMPLE 1

$$\min y = \begin{bmatrix} 4 & -1 \\ 4 & 1 \\ -3 & -2 \end{bmatrix} x$$

s.t.

$$\begin{bmatrix} -2 & 1 \\ 1 & 1 \\ -3 & -2 \\ 2 & -1 \\ -3 & -1 \end{bmatrix} x \geq \begin{bmatrix} 500 \\ 1000 \\ -1400 \\ 500 \\ -1000 \end{bmatrix}$$

$$x \geq 0.$$

The illustrative problem has three criteria, y_1, y_2 and y_3, and five constraints. The decision space is depicted in Fig. 2. By lexicographic minimization the pay-off matrix (Table 1) is established. We immediately note that $x^{*1} = x^{*2}$, hence there is reason to suspect redundancy between criteria 1 and 2. However, a superficial look at the gradients in Fig. 3 would probably not have caused the inquiry for an inexperienced practitioner. The redundancy test is initiated with a normalisation of $\mathbf{A}$ and the calculation of $-\|\mathbf{A}\|\mathbf{C}^{\mathsf{T}}$.

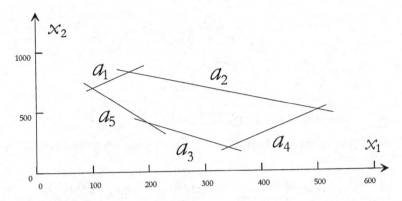

Figure 2: The Decision Space in Example 1.

$$
\|\mathbf{A}\| = \begin{bmatrix}
-2/\sqrt{5} & 1/\sqrt{5} \\
1/\sqrt{2} & 1/\sqrt{2} \\
-3/\sqrt{13} & -2/\sqrt{13} \\
2/\sqrt{5} & -1/\sqrt{5} \\
-3/\sqrt{10} & -1/\sqrt{10}
\end{bmatrix}
\tag{2}
$$

$$
-\|\mathbf{A}\|\mathbf{C}^{\mathrm{T}} = \begin{bmatrix}
4.02 & 3.13 & -1.79 \\
-2.12 & -3.54 & 3.54 \\
3.05 & 3.61 & -3.05 \\
-4.02 & -3.13 & 1.79 \\
3.48 & 4.11 & -3.48
\end{bmatrix}
\tag{3}
$$

Table 1. Pay-off matrix in Example 1.

	Lexicographic minima		
	$\mathbf{y}(\mathbf{x}^{*1})$	$\mathbf{y}(\mathbf{x}^{*2})$	$\mathbf{y}(\mathbf{x}^{*3})$
y_1	−300	−300	−166.67
y_2	1 100	1 100	1 500
y_3	1 700	1 700	−2 167

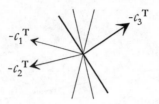

Figure 3: Criterion Gradients and Dominance Cone in Example 1.

Graphically, or by means of any MOLP (computer) algorithm the following corner-points are detected,

$$\mathbf{x}^1 = \begin{pmatrix} 342.86 \\ 185.71 \end{pmatrix}, \quad \mathbf{x}^2 = \begin{pmatrix} 200 \\ 400 \end{pmatrix}, \quad \mathbf{x}^3 = \begin{pmatrix} 100 \\ 700 \end{pmatrix}, \quad \mathbf{x}^4 = \begin{pmatrix} 166.67 \\ 833.33 \end{pmatrix}, \quad \mathbf{x}^5 = \begin{pmatrix} 500 \\ 500 \end{pmatrix}.$$

$X_E = X$, i.e., the entire decision space is efficient due to the high conflict between criteria 1, 2 and 3.

Iteration 1

Corner point $\mathbf{x}^1$ is studied. The Simplex-tableau (or corresponding) yields constraints 3 and 4 as active. The choices of most preferred direction for criterion j, $d_j(\mathbf{x})$, in the iteration are now (numerical values are rounded off):

$$d_1(\mathbf{x}^1) = 3 \Leftrightarrow -\mathbf{a}_3^T\mathbf{c}_1 = \max\left\{-\mathbf{a}_3^T\mathbf{c}_1 = 3.05, -\mathbf{a}_4^T\mathbf{c}_1 = -4.02\right\} \tag{4}$$

$$d_2(\mathbf{x}^1) = 3 \Leftrightarrow -\mathbf{a}_3^T\mathbf{c}_2 = \max\left\{-\mathbf{a}_3^T\mathbf{c}_2 = 3.61, -\mathbf{a}_4^T\mathbf{c}_2 = -3.13\right\} \tag{5}$$

$$d_3(\mathbf{x}^1) = 4 \Leftrightarrow -\mathbf{a}_4^T\mathbf{c}_3 = \max\left\{-\mathbf{a}_3^T\mathbf{c}_3 = -3.05, -\mathbf{a}_4^T\mathbf{c}_3 = 1.79\right\} \tag{6}$$

Thus, the updated **M**-matrix becomes,

$$\mathbf{M} = \begin{bmatrix} 1 & 1 & -1 \\ 1 & 1 & -1 \\ -1 & -1 & 1 \end{bmatrix}. \tag{7}$$

At this stage, further evaluation of the non-redundant criterion 3 is unnecessary. Note that this exclusion is implied already from Table 1 and Remark 1. Formally, we conclude that **M** contains positive elements and the list of corner points is not exhausted. Hence, iteration 2 is initiated.

Iteration 2

The manner in which corner points are selected is arbitrary, but a simple and handy rule is to follow an efficient edge emanating from the previous corner point until the next efficient corner point. Here, x^2 is obtained and constraints 3 and 5 are active. The choices are edge 5 for both criteria, i.e., still no difference in ranking among points.

M is updated to,

$$\mathbf{M} = \begin{bmatrix} 2 & 2 & -1 \\ 2 & 2 & -1 \\ -1 & -1 & 1 \end{bmatrix}. \tag{8}$$

The redundancy hypothesis cannot be rejected and yet an iteration is necessary.

Iteration 3

The succeeding efficient corner is x^3 with active constraints 1 and 5. The choices in iteration 3 are:

$$d_1(\mathbf{x}^3) = 1 \Leftrightarrow -\mathbf{a}_1^T \mathbf{c}_1 = \max\{-\mathbf{a}_1^T \mathbf{c}_1 = 4.02, -\mathbf{a}_5^T \mathbf{c}_1 = 3.48\} \tag{9}$$

$$d_2(\mathbf{x}^3) = 5 \Leftrightarrow -\mathbf{a}_5^T \mathbf{c}_2 = \max\{-\mathbf{a}_1^T \mathbf{c}_2 = 3.13, -\mathbf{a}_5^T \mathbf{c}_2 = 4.11\} \tag{10}$$

The conflict is finally detected and non-redundancy shown. *Pro forma*, the $\mathbf{M}$-matrix is updated as

$$\mathbf{M} = \begin{bmatrix} 3 & -1 & -1 \\ -1 & 3 & -1 \\ -1 & -1 & 1 \end{bmatrix}. \tag{11}$$

Evaluation

All non-diagonal elements in $\mathbf{M}$ are negative, hence Example 1 contains no redundant criteria. The reader is left with the exercise to investigate the remaining points for conclusion.

5. Conclusion

Redundancy in multi criteria decision making, defined as the existence of criteria functions with identical ranking in decision space, is deleterious to modelling and analysis of MCDM problems. Especially in interactive methods, where the decision-maker is required to concurrently inspect and decide upon preferred directions and levels, the information load is considerable. Any relief in this stress is as welcome to the decision-maker as to the analyst (or algorithm), with less computational burden. Generally, redundancy is primarily detected by using the ideal-points. The paper has proposed a simple redundancy test for linear models, based on the mathematical background in Agrell (1995, 1997). A suggested use would be to include it as an option in MOLP computer software in the case of non-unique ideal points.

References

Agrell, P. J. (1995) *Interactive Multi-Criteria Decision Making in Production Economics*, Production-Economic Research in Linköping (PROFIL 15), Linköping, Sweden.

Agrell, P. J. (1997) On Redundancy in MCDM. *European Journal of Operational Research*, 98(3), 571-586..

Amador F. and Romero C. (1989) Redundancy in Lexicographic Goal Programming: An Empirical Approach. *European Journal of Operational Research*, 41, 347-354.

Gal T. and Leberling H. (1977) Redundant Objective Functions in Linear Vector Maximum Problems and their Determination. *European Jounal of Operational Research*, 1(3), 176-184.

Grubbström R. W. (1988) *Linear Programming by Vertex Analysis - Preliminary Ideas*. WP 150, Department of Production Economics, Linköping Institute of Technology.

van Herwijnen M., Rietveld P., Thevenet K. and Tol R. (1994) Sensitivity Analysis with Interdependent Criteria for Multicriteria Decision Making: The Case of Soil Pollution Treatment. *Journal of Multi-Criteria Analysis*, 1(1), 57-70.

Ignizio, J. P. (1976) *Goal Programming and Extensions*. Lexington, New York.

Romero C. (1994) Redundancy in Lexicographic Goal Programming. *European Journal of Operational Research*, 78(3), 441-442.

Steuer R. E. (1974) *ADBASE: An Adjacent Efficient Basis Algorithm for Solving the Vector-Maximum and Interval Weighted Sums Linear Programming Problems*. (in FORTRAN) College of Business and Economics, University of Kentucky.

Steuer R. E. (1976) A Five Phase Procedure for Implementing A Vector-Maximum Algorithm for Multiple Objective Linear Programming Problems. In *Multiple Criteria Decision Making*, Thiriez H. and Zionts S. (Eds.), Springer, Berlin.

Changkong V. and Haimes Y. (1983) *Multiobjective Decision Making: Theory and Methodology*. North-Holland, Amsterdam.

Part 6
MCDA in Environmental and Natural Resource Management

Comparing Contingent Valuation, Conjoint Analysis and Decision Panels: An Application to the Valuation of Reduced Damages from Air Pollution in Norway

Bente Halvorsen[1], Jon Strand[2], Kjartan Sælensminde[3] and Fred Wenstøp[4]

[1]Central Bureau of Statistics, Oslo, Norway, E-mail: btl@ssb.no

[2] Department of Economics, University of Oslo
Box 1095, Blindern, N-0317 Oslo, Norway, E-mail: jon.strand@econ.uio.no

[3]Institute of Transport Economics
Box 6100 Etterstad, N-0622 Oslo, Norway, E-mail: kjartan.salensminde@toi.no

[4]Norwegian School of Management
Box 580, N-1301 Sandvika, Norway, E-mail: fred.wenstop@bi.no

Abstract. We discuss three stated-preference approaches to eliciting willingness to pay (WTP) for environmental goods, namely open-ended contingent valuation (OE-CVM), conjoint analysis (CA) and multi-attribute utility theory applied to decision panels of experts (EDP). We point out that each approach has advantages and disadvantages relative to the others, and that the relation between WTP estimates from the approaches cannot be predicted on prior arguments alone. The three approaches are applied in three different surveys, where we seek valuation of specific damages due to air pollution in Norway. In all cases studied, OE-CVM clearly yields the lowest average WTP estimates, while those from CA and EDP are similar. Possible explanations are offered, based on the principal differences between the approaches and on particular features of the three surveys.

Key words. Contingent Valuation, Conjoint Analysis, Decision Panels, Environmental Goods, Stated Preference

1 Introduction

The need for assessing values of non-market goods has led to the development of two main groups of valuation approaches, those based on revealed preferences (RP), and those on stated preferences (SP). Among SP approaches, the Contingent Valuation Method (CVM) has been most frequently applied among environmental economists. Lately, however, this approach has been heavily criticized on various grounds (Desvouges 1992; Diamond and Hausman 1994; Hausman 1993; Kahneman and Knetsch 1992). This has brought about alternative SP valuation approaches. For instance, McFadden (1993) recommends Conjoint Analysis (CA), which has been popular within marketing and transport economics. Gregory et al. (1993) emphasize a third procedure based on Multi-Attribute Utility Theory

(MAUT). This approach has so far had most of its applications in connection with decision analysis, largely using so-called Expert Decision Panels (EDP).

Most studies to date, have compared approaches based on SP versus RP (Adamowicz, Louviere, and Williams 1994; Brookshire et al. 1981; Hausman 1993; Hensher 1994; Seller, Chavas, and Stoll 1985), or results from different variations of the CVM approach (Kealy and Turner 1993; Kristrøm 1993; Loomis 1990; Shechter 1991). A few studies have compared different SP approaches, mainly CVM versus CA, or other methods based on ranking (Desvouges 1983; Mackenzie 1993; Magat, Viscusi, and Huber 1988). A common finding in all these studies is that different valuation approaches often yield systematically different willingness to pay (WTP) estimates for identical or closely related goods. This has led to the need to analyze more carefully the reliability of the different valuation approaches. Still, in the literature there is to our knowledge no systematic comparison, theoretical or empirical, between the three SP approaches that we regard as most relevant in environmental, health and transportation economics contexts, namely CVM, CA, and MAUT (or its applications through EDP).

The purpose of the present paper is to provide a unified framework for such a comparison, and to tie this framework to an empirical application where the methods can be compared for a given good to be valued. We argue that all three approaches have their potential strengths and weaknesses, and that they cannot be ranked unambiguously in terms of reliability or appropriateness for valuation. Rather, they complement each other, and can all be applied to the same valuation issue, thereby potentially raising the overall reliability of the valuation estimates.

In the following we will focus on particular types of application of the CVM and MAUT approaches. We will consider only open-ended CVM valuation (OE-CVM), and MAUT only as applied to panels of experts (EDP). This does not mean that other varieties of the approaches are not relevant or prevalent in use. In particular, a large number of discrete-response CVM studies have been made to date, and many favor this variety over the OE-CVM variety studied here; see Hanemann and Kanninen (1996) for a discussion. On the other hand, very few MAUT studies for the valuation of environmental goods, applied to random population samples, have been done to date; we are only aware of Heiberg and Hem (1989). Our main purpose is to emphasize differences between the three main SP approaches. For instance, discrete-response CVM would resemble the CA approach more closely than OE-CVM, while MAUT as applied to random population samples would also be more closely related to CA than is EDP. Our focus also corresponds to the way in which the three studies that we compare have actually been carried out.

2 Description and comparison of the three approaches

2.1 Description

2.1.1 Contingent valuation (CVM)

The idea behind CVM is to elicit the maximum willingness-to-pay (WTP) or willingness-to-accept (WTA) related to the commodities one seeks to value. This is either done in an open-ended (OE-) or a dichotomous choice (DC-) WTP question, or by some kind of bidding game. In the OE-CVM, focussed on here, the WTP estimate is calculated directly, as this approach asks the respondents to simply state their maximal WTP or WTA. A detailed overview of this approach is found in Mitchel and Carson (1989).

2.1.2 Conjoint analysis (CA)

In CA studies, respondents are asked to rank or rate combinations of two or more attributes, describing the good to be valued. Some CA studies have a complicated structure due to relatively large numbers of attributes describing the goods to be compared. Often, a sequence of such choices is made for each respondent. This procedure makes it possible to construct the preferences over all attributes for each respondent, and obtain an estimate of the individual valuation for each attribute describing the good, provided that one attribute is a monetary cost. The WTP is calculated implicitly, applying the estimated preference structure.

Two different types of application of CA may be distinguished in our particular study, namely a) a choice between different combinations of goods that do not directly involve actual consumption decisions by the respondents; and b) a situation that closely simulates a real-life choice or consumption decision. Alternative b) is the more traditional way of applying CA, at least in the most familiar transportation economics context, while alternative a) is more similar to CVM, at least with respect to the objects for valuation.

Much of existing research using CA is related to transportation and marketing issues. Classic references are Louviere and Hensher (1982) and Louviere and Woodworth (1983), with a review of the method's historical development as applied to transportation in Hensher (1994). Some applications have also been made of the method for valuing environmental commodities (Adamowicz, Louviere, and Williams 1994; Magat, Viscusi, and Huber 1988; Roe, Boyle, and Teisl 1996), and the interest seems to be increasing.

2.1.3 Expert decision panels (EDP)

When using EDP, a group of agents would be selected and asked to act as decision-makers on behalf of the population, in a real or constructed decision problem, where different attributes are weighted through pairwise comparisons. The weights are used to tentatively rank different policies. Through feedback processes, the panels reassess the weights, finally arriving at a consistent set of rankings, and thus a "corrected" set of weights attributed to the different criteria. Finally, if costs are one attribute, the WTPs implied in the weights can be computed from the relative weights.

With EDP, panel members are not supposed to let their private pecuniary situation influence their valuations. They judge what the *society* should be willing to pay for environmental goods. An EDP may play three alternative roles, the role of a politician, a citizen, or that of a stakeholder. In the role of a politician, the EDP is requested to interpret and represent the wishes of the population as a whole. In the citizen role, the EDP represents themselves, expressing their own personal preferences and values, amended by the insights acquired through their professional work. In the stakeholder role, the EDP represents an organization with particular viewpoints on the decision issue, e.g. an environmentalist organization. EDPs may also be composed of representatives with a known "high" valuation of a particular good (and who would be likely to overstate their valuations), and others with a known "low" valuation (and who would be likely to understate). Such compositions may be advantageous in order to handle, and control for, possible misrepresentation of preferences among panel members.

The approach has its intellectual roots in management and decision sciences, and is used as a tool for making decisions in corporate and government planning (Keeney and Raiffa 1976). Recently, several applications have been made of the method in environmental and energy policy contexts (Janssen 1992; Karni, Feigin, and Breiner 1991; Stam, Kuula, and Cesar 1992; Wenstøp and Carlsen 1987).

2.2 Methodological problems with the different approaches

We will now discuss a number of principal issues that are relevant for comparing OE-CVM, CA and EDP. The grouping of the issues is somewhat arbitrary, as there are overlaps with respect to the topics dealt with under each of them. It serves as a starting point for sorting out differences between the approaches.

2.2.1 Strategic bias and other incentives to misrepresent preferences

The greatest early skepticism to CVM was based on potential problems with strategic behavior in CVM surveys. Such problems may arise when the individuals surveyed have incentives, out of motivation based on self-interest, to answer untruthfully. In principle such biases can go both ways, i.e., lead to either over- or undervaluation. Incentives depend on the mechanism for payment that the subjects believe would be actually used in implementing the project. If, for example, people believe that a proportional tax rate on all individuals is to be used, those with very high valuations may have incentives to overstate their true preferences and those with low valuations to understate. If people on the other hand believe that they will pay an amount that strongly depends on their statement, all may have incentives to understate. One possible problem with OE-CVM is that the subjects may find themselves in a sort of bargaining situation with the interviewer whereby the stated bid defines the starting point for the bargain. This may lead to understatement, in particular when the subjects expect it to be a positive relationship between stated WTP and actual subsequent payments (Loomis 1990; Magat, Viscusi, and Huber 1988). Strategic bias may be less of a problem with CA, as it is often more difficult to see through since the valuation procedure is more complex. This makes it less obvious to the subjects how to answer strategically. EDP may have greater problems with potential strategic answers, as experts may have interest in, and know how to, manipulate their answers to their advantage.

Sponsor and interviewer biases may be of similar magnitude with CVM and CA, but may be lower with EDP as the respondents here are more knowledgeable about the good in question, and thus less easily misled by intrinsically irrelevant design issues.

2.2.2 Ordering, embedding and scope effects

Much of recent criticism against CVM, put forth in Desvouges (1992), Kahneman and Knetch (1992), Hausman (1993) and Diamond and Hausman (1994), is related to problems with ordering, embedding and scope effects. Ordering effects occur when the sequence in which the valuation questions are asked, affects the final valuation of different goods, when a package of goods is valued in a sequence. Embedding effects occur when the valuation is insensitive to the inclusiveness of the package of goods for which valuation is sought. A particular version of the embedding problem, illustrated by Desvouges (1992), is when the valuation of a good is insensitive to the amount in which it is provided. This effect is what Carson and Mitchell (1995) denote "problems with scope".

Some of these problems apply to CA and EDP surveys as well. For CA, this is especially true with regard to embedding and ordering, while problems with scope are likely to be less serious than for CVM. This is because the respondents are presented with scenarios in which the magnitudes of the different attributes vary. However, as most scope tests in CVM are external tests on different sub-samples, whereas the tests in CA are internal for each respondent, problems with scope may be less severe in CVM than many studies indicate. Order effects are in principle no problem in CA when consisting of one choice only, as all attributes are weighted simultaneously and not in a sequence. However, if several choices are made, such effects may occur, but for different reasons than in the CVM procedure. In a series of choices, some respondents may underrate presumed popular alternatives to ensure that their most preferred alternative is chosen. With EDP, each attribute is weighted in a sequence of paired comparisons. This may lead to biases due to ordering. Such ordering problems may be less severe than those occurring with CVM, for several reasons: First, all attributes are described to the respondents or panels up front. This reduces ordering effects due to imperfect information (Halvorsen 1996). Second, EDP surveys often search for possible inconsistencies in the responses. Since more time is spent on each panel than is customary with respondents to CVM surveys, the preferences elicited by the EDP approach will probably be more accurate. Also the fact that expert panel members are chosen just because they are more familiar and experienced with the issues at stake than the overall population should tend to reduce this type of problem with EDP.

2.2.3 Scenario misspecification bias and implied value cues

Biases due to problems of scenario misspecification and implied value cues may occur in CVM surveys when respondents misinterpret the contingent scenario described by the interviewer (Mitchel and Carson 1989). These may be due to problems with properly describing the good to be valued, or amenity misspecification (symbolic, part-whole bias, metric bias) and/or context misspecification (misinterpretation of payment vehicle, property rights, method of provision etc.). Implied value cues represent another potential source of bias. This occurs when the respondents treat specifications in the survey as indications of the "correct" response. The order in which the goods are valued suggests their relative importance. An indicated cost or range of costs may indicate the range of "expected" answers. The description of the good may suggest its relationship with other goods, or the act of being interviewed about a good may suggest to the respondent that it is important.

Most of these types of problems are found in CA and EDP as well. This is especially true for amenity and context misspecification bias using CA. With EDP, such biases may be less important, as more time is normally spent on the preference elicitation process, and experts are likely to be more familiar beforehand with the types of tradeoffs that have to be made. Problems with implied value cues may be severe for CA, as the magnitude and range of the changes in all attributes are specified in the survey design. Such problems are also likely to be less important with EDP where more generous time restrictions allow for countermeasures, for instance through repeated trade-off questions with randomized cues.

2.2.4 Econometric and statistical problems

In OE-CVM studies where maximum WTP for a particular commodity is stated, we obtain a continuous WTP variable. In this case, econometric problems in the analysis of the results are relatively small compared to valuation approaches based on discrete choices. With CA only information about the ranking of different alternatives is obtained. The result is less efficient estimation methods that require larger samples to obtain valuation estimates with a given level of reliability. See Kanninen (1995) and Hanemann and Kanninen (1996) for discussions. EDP represents a particular case here, since the "sample" is not representative by construction or definition, and is also usually very small. Thus statistical tests do not apply when analyzing such data. Rather, the valuation provided by each decision panel can be viewed as the panel's own point estimate of the average valuation in the population (in the absence of biases caused by the DP mechanism).

Furthermore, additional assumptions about the functional form and distribution of the utility structure are necessary to obtain valuation measures from discrete responses. The assumptions will influence both the WTP estimates, and their estimated variance. Inappropriate assumptions will lead to large residuals and poor prediction power (Greene 1993) and potentially biased WTP estimates. As a result, the estimates are very sensitive to omitted variables that are correlated with the attributes (Hanemann and Kanninen 1996). This may result in apparent differences in the valuation, which would have disappeared if we had more information about the utility structure in the discrete choices. One way to detect these problems is to test and correct the models for heteroscedastic error terms. These problems may also make comparisons of WTP estimates between different valuation approaches difficult, as the magnitude of the problems differs between the surveys.

2.2.5 Other issues

There are a few additional points relevant for comparison of the three approaches. It is generally recognized that the less familiar the respondent is with the valuation procedure and the good to be valued, the more prone it is to error and systematic biases. These are problems with all valuation approaches based on SP, and they are of course related to the problems discussed above. Most likely, EDP suffers less from such problems than does CVM.

An EDP is in principle not representative of the underlying population in a statistical sense, while CA and CVM as a rule use representative samples. Representativeness may, however, sometimes be a problem for CA when applied to a particular choice situation, since the sub-population of individuals who are familiar with the situation may not be representative of the entire population. In OE-CVM surveys the WTP question is asked directly and thus not derived indirectly, from answers to other questions, as in CA and EDP. Compared with CA and EDP, OE-CVM normally also uses relatively simple questionnaire formats. This may have advantages in terms of better intuitive understanding of the issue to be valued. On the other hand, a longer and more elaborate survey may better educate respondents and give them time for reflection and calculation. While directness of the valuation question can be said to be an unambiguous advantage, great simplicity is not an obvious advantage. A particular problem in CVM and CA, is that a failure to remind respondents of the budget constraint may seriously bias the stated WTP in an upward direction. Including such reminders should thus be an important aspect of the study design.

So far, we have said nothing about the likely direction of possible biases with the different approaches. However, as noted above, strategic bias may lead to too high or too low values, depending on individuals' incentives to understate or overstate own values. In any given context one may have an idea of what these incentives are, and thus a certain control of the direction of the bias. The same can probably be said about bias due to lack of representativeness, and implied value cues, biasing final valuations in the direction of the initial cues provided, but then it must be known on which side of the target the cue is. Lack of familiarity and simplicity, on the other hand, is likely to lead to increased uncertainty without much guidance as to the direction of possible biases. Embedding and scenario misspecification problems have, by researchers critical to CVM, been argued to lead to overvaluation, at least in cases where the goods to be valued are not very inclusive. Since such problems are particularly great with CVM, the argument implies too high valuations, and higher valuations than under the two other approaches.

3 Description of the three valuation surveys

3.1 The CVM survey

This survey was part of a nationwide monthly opinion poll in May 1993 with 1229 in-person interviews with respondents drawn randomly from the entire Norwegian population. A full report is found in Halvorsen (1996). The respondents were given a scenario describing the benefits of a 50% reduction in air pollution due to reduced emissions from traffic. The benefits mentioned were: i) Reductions in risk of becoming ill from lung disease, asthma, bronchitis, allergy; and minor health effects such as a reduction in the number of days suffered with headache, tiredness, aching muscles, cold or flu. ii) A reduction in acid rain related damages to forestry, agricultural production and materials.

Table 1. The Open-Ended Contingent Valuation: Scenario of benefits from a 50% reduction in emissions from car traffic. The table shows the percentage of the population suffering from certain illnesses before and after the 50% emission reduction.

Illness	Before	After
Emphysema and other lung diseases.	3%	1.5%
Asthma and chronic bronchitis	5%	3.5%
Allergy	22%	15%
Cold, flu, headache, tiredness, aching muscles	18 days p. a.	12 days p. a.

Table 1 presents the possible health effects of a 50% reduction in air pollution described to respondents. The reductions in risks of becoming ill were calculated from a dose-response study from one of the most heavily polluted parts of Oslo (Clench-Aas et al. 1989). The respondents were told that the government would implement the desired improvement in air quality with costs covered by higher income taxes. We were careful to remind respondents of individual budget constraints and how an increase in income taxes would reduce their purchase of other goods and services. The respondents were asked an open-ended question, to state the maximum willingness to pay (WTP), in terms of increased income taxes, for a general reduction in air pollution in the Oslo area by 50%.

The sample was split in four sub-samples where the respondents were given different information about government policies, and different sequences of questions.

Split 1): Respondents in sub-samples B and D were told that the government would subsidize the introduction of electric cars to replace petrol-fuelled cars, in order to achieve the 50% reduction in air pollution. The respondents in sub-samples A and C were told that the government would use an unspecified package of tools to achieve the same reductions.

Split 2): Respondents in sub-samples A and B were initially given information on a broad set of effects of air pollution, and asked to value all benefits of a 50% air pollution reduction in the Oslo area. Sub-samples C and D were initially given information about possible adverse health effects of air pollution (such as increased risk of lung disease, asthma, allergy and bronchitis), and asked to value the 50% reduction in risk of contracting these diseases. They were then told about all other effects, and asked to value the rest of the benefits conditional on their valuation of reduced risks of getting chronic diseases. The first split was made to check the sensitivity of the answers to the framing of the suggested government action. The second split was to make it possible to control for embedding effects. To minimize unwanted ordering and embedding effects we used a "top-down" questioning approach by first asking the respondents to value all benefits. We then asked them to distribute their total WTP according to different motives, and reallocate among all budget posts. The different motives mentioned were (i) reduced risk of becoming ill for the respondent and his closest family, (ii) a reduction in other people's risk of becoming ill, and (iii) reduced damages to production and the natural environment as a result of reduced acid rain.

3.2 The Conjoint analysis survey

This study is more completely described in Sælensminde and Hammer (1994). Prior to the main study, presented here, a pilot survey was carried out in 1992 (Sælensminde and Hammer 1993), in which 179 respondents from Oslo took part. The main CA study entailed an elicitation of stated choices, with application to a particular journey. The valuation approach was of the "bottom-up" type, by first letting respondents value individual items and subsequently aggregating up the total. The respondents were faced with five different sequences of choices (called games), each involving pairwise choices between alternatives. 1,680 respondents, a representative sample of residents in Oslo and its suburbs, took part in a maximum of three games each. They were interviewed in their homes in the autumn of 1993 with the aid of a portable computer. The respondents were asked to choose between two alternative travel routes, each of which was described in terms of characteristics like vehicle time, cost, and noise and air pollution levels. The pairs were chosen such that no alternative was obviously better than the other was.

Table 2. Description of the five games and the factors included in the environmental assessment survey in Oslo/Akershus 1993.

Factor / Attribute	G 1	G 2	G 3	G 4	G 5
Cost	x	x	x	x	x
In vehicle time	x	x	x	x	x
Seat (public transportation)	x				
Walking time (car users)	x				
Noise		x		x	
Local air pollution		x		x	
Dust and dirt from road wear			x		x
CO_2*			x		x
No. resp.	897	596	289	1 179	683
Car:	580	373	196		
Public transportation:					

* CO_2 is included to give as complete an assessment as possible of environmental problems due to road traffic so as to prevent confusion between local and global environmental problems.

Table 2 shows the attributes included, and the numbers of respondents who took part in the various games. All those who had made a journey by car or by public transport first took part in game 1. They were then divided at random in two groups. One group took part in games 2 and 4, the other in games 3 and 5. Except for game 1, which always came first, the other games were presented in random order. The respondents who had made no journey took part in games 4 and 5 only.

In games 2 and 3 the respondents were to imagine that current fuels and tires were to be replaced by new ones. The new fuel was assumed to be available at all petrol stations, and give the same vehicle performance with regard to acceleration, comfort, and range as present fuels. Subjects were then faced with, and asked to choose between, two concrete journeys where the purpose and length of the journey and the means of transport were the same, while other factors such as cost, journey time and the level of environmental quality were different between the journeys. The choice situation in game 2 is shown in table 3.

Table 3. Example of conjoint choice from game 2. Interviewees are asked to choose between journey A and journey B.

Car journey A	Car journey B
Noise from road traffic reduced by 20%	Noise from road traffic reduced by 20%
Local air pollution reduced by 20%	Local air pollution reduced by 60%
In vehicle time: 30 min.	In vehicle time: 40 min.
Fuel cost for the journey: NOK 25	Fuel cost for the journey: NOK 30

In games 4 and 5, choices were to be made between different consequences of future transport policies. In other respects the choice situation in game 4 (5) was similar to that in game 2 (3). Payment was assumed to be made monthly or annually, in terms of changes in transport costs or in taxes, varied randomly among subjects.

3.3 Description of the EDP survey

A full description of this study is found in Wenstøp et al. (1994). Three expert panels were used: four persons from the State Pollution Control Authority constituted panel 1, five persons from the Directorate of Public Roads constituted panel 2, and three physicians from the National Institute of Public Health constituted panel 3. The panels were requested not to reflect possible official viewpoints of their organization, but rather act as concerned citizens with the added insights and values obtained through their professional work. Before the valuation process, the health and environmental amenities to be valued were specified in cooperation with the panels. To make the attributes more tangible, we focused on end-impacts rather than emission quantities. A selection is shown in table 4.

Table 4. List of end-impacts that were valued in the Expert Decision Panel study

Category/Area	End impact/Criteria
Adverse health effects	Person-days per year of chest irritations or coughing
	Person-days per year with headache
	Person-days per year with dizziness
	Person-days per year with nausea
	Person-days per year with colds or influenza
Irritations	Number of people complaining about noise from road traffic
	Number of people complaining about smells from road traffic
	Number of people complaining about dust and dirt from road traffic
Injuries	Number of non-fatal traffic injuries per year
	Number of fatal traffic injuries per year
Costs	Costs per year of public actions

Two specific cases were selected: a) the effects of a gradual transition to electric vehicles in Norway, and b) alternative traffic planning strategies in the city of Drammen (south of Oslo).

Two different methods were used to elicit weights, ordinal ranking of criteria as implemented in DEFINITE (Janssen 1992), and numerical tradeoffs as outlined in Keeney and Raiffa (1976). With ordinal ranking, the panel is first presented with a decision table, describing the effects of the alternative policy choices on all decision criteria. Two criteria, A and B, are selected. The panel is first asked which criterion is the more important. They must then state how much more important on a nine-point scale from "just as important" to "extremely more important". Next, a criterion C is selected and compared to the one preferred above. This procedure continues until all pairs have been compared. The implied weights given to the different attributes can then be computed. With numerical tradeoffs, the panel is asked to select one criterion as a reference criterion against which the others are compared, one at a time. For each criterion, the panel is presented with a two-by-two matrix. The rows contain the reference criterion and the selected alternative. The columns represent two situations. The matrix shows scores that are randomly chosen, without one column dominating the other. The panel is asked to adjust the numbers so that the two situations are equally preferable. The relative weights can then be computed. An example of a matrix is shown in table 5.

Table 5: Numerical trade-offs in the Expert Decision Panel study. The respondents were asked to substitute the question mark with a number to make situation A and B equally preferable

	Situation A	Situation B
Coughing days p.a.	3 000	2 000
Persons complaining about noise from traffic	100	?

The MAUT paradigm and the two evaluation methods were explained in an introductory note that was available to the expert panels before the interview sessions started. There were three interactive sessions with each panel, with no cross-information between panels. Each session used one case and one elicitation method, and lasted approximately three hours. A session would start by going through and discussing the case and elicitation method. The preference elicitation was then carried out using interactive software installed in a portable PC. During the elicitation process, the panels were supposed to discuss all matters of judgment openly and with little involvement by the session leaders. There were at least two session leaders present at each session. Questions of clarification could always be addressed to the session leaders. The panels were expected to reach unanimous judgments, if possible, or they could come up with more than one set of weights. When the weights had been produced, the implied WTPs were computed. This was possible since the criterion "annual costs of actions" was also weighted as a part of the process. A WTP associated with a criterion expresses how much the panel thinks society should pay to avoid one unit of the corresponding impact. The panels were asked to review their weights in light of the ensuing WTPs in a feedback process until they felt confident with the final result.

4 Comparing results from the three surveys

In this section we present and compare the results from the three studies. We compare the estimated WTP for five different goods that are approximately identical across the studies. The goods are: One person less complaining about respectively noise, dust and dirt from road traffic, and minor health problems; one person less with chronic lung diseases; and a 50 percent reduction in local air pollution.

4.1 Willingness to pay for one person less complaining about noise

Three different valuations of one person less complaining about noise from road traffic are presented in table 6. The results from the EDP are presented in the first row, and the results from the CA are presented in the two following rows, with the WTP from the games choosing between two different policy programs first, and then the mean WTP in the game choosing between two different fuels.

Table 6: Annual willingness to pay for one person less complaining about noise

Sample	Mean WTP ($)	Lower limit	Upper limit
Expert Panels	940	600	1480
CA - WTP for a policy program	170	0	850
CA - WTP for a private journey	1090	990	1190

From table 6 we see that the mean WTP from the CA based on choice between private journeys is higher, and the mean WTP from the CA based on choice between policy programs is lower, than the mean WTP derived from the EDP. Due to different sampling procedures in the CA and EDP studies, it is not possible to make formal statistical tests of differences in the means. Comparing the confidence regions of the estimates, however, gives an impression of the difference between two estimates. For the CA results we present ninety-five per cent confidence regions, and for the EDP study, one standard deviation. Looking at the confidence regions, the differences in mean WTP among these studies do not appear to be significant.

4.2 Willingness to pay for one person less complaining about dust and dirt

Table 7: Willingness to pay for one person less complaining about dust and dirt from road traffic

Sample	Mean WTP ($)	Lower limit	Upper limit
Expert Panels	400	220	750
CA - WTP for a policy program/	370	0	1 440
CA - WTP for a private journey	2 050	1 890	2 350

Next, we compare the WTP for one person less complaining about dust and dirt due to road traffic, in the EDP and the two applications of CA. The mean WTP over the three EDP panels is given in the first row of table 7, and the mean WTP figures from the CA are given in the two following rows. We assume that all those who are currently complaining about dust and dirt from road traffic will no longer be bothered. The first for the policy programs game, and the second for the private journey game. We find that the mean WTP from the private-journey CA is higher, and that from the policy-program CA lower, than the EDP mean, as in table 6. The mean EDP valuation differs little from that in the policy-game CA, while the private-journey CA mean is much (and significantly) higher than the two others.

4.3 Willingness to pay for one person less with minor health problems

The third good to be compared is the WTP for one person less complaining about "minor health problems", valued in the EDP and the CVM studies. In the CVM survey, the respondents were not asked to value this end-impact explicitly. It is, however, possible to decompose the total WTP from the CVM study, by the top-down procedure described in section 3.1 above. The mean decomposed WTP for one person no longer complaining about minor health problems (such as headaches, aching muscles, flu and tiredness) was then aggregated over the number of Norwegian households. To obtain an estimate of the reduction in the number of people no longer complaining, we applied the dose-response function that the respondents were facing. In the EDP, panels were asked to state a value for the entire Norwegian public of one person less complaining about headache, dizziness or nausea. These health problems were valued separately, and then aggregated to an average value attached to one less person complaining about minor health problems. To do this, we have to assume that these minor health problems are separable in utility terms. The results are presented in table 8.

Table 8: Aggregated annual willingness to pay for one person less complaining about minor health problems

Sample	Mean WTP	Lower limit	Upper limit
Expert Panels	490	320	770
CVM - Resp.w/ electric car scenario	100	80	120

The mean WTP in the EDP is almost five times the level in the CVM survey. Considering the confidence regions, the former appears to be significantly higher.

4.4 Willingness to pay for one person less suffering from chronic diseases

The valuation of one person less suffering from chronic diseases related to air pollution is studied in the EDP and CVM studies. To obtain an estimate from the CVM study, we aggregated the mean WTP for a reduction in the number of cases of chronic diseases over all households, and divided this by the estimated reduction in cases. The chronic diseases mentioned in the CVM survey were emphysema, asthma, bronchitis and allergy. The number of reduced cases of chronic lung disease was estimated applying the dose-response scenario given to the respondents. We assume that the set of people who are no longer suffering from emphysema, asthma and bronchitis and allergy are mutually exclusive. If they are not, the mean WTP in the CVM survey will increase. The good valued in the EDP was one person less suffering from chronic coughing. The results are given in table 9.

Table 9: Aggregated willingness to pay for one person less with chronic lung diseases

Sample	Mean WTP	Lower limit	Upper limit
Expert Panels	1 290	910	1 850
CVM - Resp.w/ electric car scenario	290	250	340

The mean WTP in the EDP is almost 4.5 times higher than that in the CVM survey. The difference is statistically significant. If we compare the valuation in the two studies of one person less complaining about minor health problems, relative to the valuation of one person less suffering from chronic illness, we find that these ratios are similar, namely 0.38 in the EDP survey and 0.35 in the CVM survey. This suggests that these relative valuations are approximately the same with the experts as with the public. The valuation levels are, however, systematically lower in the CVM study.

4.5 Willingness to pay for a fifty per cent reduction in emissions from cars

In table 10 we compare the mean WTP of a fifty-percent reduction in emissions from road traffic, based on results from the CVM and CA studies respectively.

Table 10: Willingness to pay for a 50 % reduction in emissions from car traffic.

Sample	Mean WTP ($)	Lower limit	Upper limit
CA - WTP for a policy program	730	70	1 400
CVM - Respondents in Oslo/Akershus	290	210	370
CVM as part of CA study	220		

The average estimate in the CA is about three times that in the two CVM surveys. This indicates that OE-CVM tend to produce lower valuation estimates than CA. Looking at confidence regions, we find, however, that the differences are not significant, due to the large variation in the CA estimate.

5 Summary and concluding remarks

From the tables 6-10 emerges a number of interesting overall conclusions:
1) There is a tendency for OE-CVM to yield values that are lower than the ones resulting from the other approaches.
2) When comparing OE-CVM to EDP in tables 8-9, the relative valuations of reductions in minor, relative to serious health problems are approximately the same, even though their levels are very different.
3) When comparing the different CA studies, we find a clear tendency for values to be greater when elicited as part of an individual choice (in terms of private journeys), than when elicited in a policy package context. This is clear both from tables 6 and 7, valuing one person less complaining about traffic noise, dust and dirt, respectively.

4) When comparing EDP to CA, average valuation results are in many cases of similar magnitudes, as found from tables 6-7.

We may distinguish between four types of explanations of the differences: a) In terms of systematic biases due to factors indicated in section 2.2. b) In terms of other systematic biases, not directly discussed in this section, among them differences in study design. c) In terms of different goods definitions and perception of goods to be valued. d) In terms of random factors and uncertainty, in EDP uncertainty among experts about appropriate valuations, and in the other studies, uncertainty about individual valuations.

It can first be noted that conclusion 1 above corresponds to that drawn in previous studies trying to compare OE-CVM on the one hand, and CA (including discrete-choice CVM) on the other (Loomis 1990; Magat, Viscusi, and Huber 1988; Seller, Chavas, and Stoll 1985; Shechter 1991). A number of explanations of these differences were offered in section 3 above, and were generally all of type b), i.e., differences in systematic bias due to other factors than those discussed directly in section 2.2. Note that these explanations all generally have the consequence that OE-CVM should tend to understate true valuations, rather than overstate. We may here underline some design issues that may have contributed to such differences. First, our OE-CVM used a top-down, and the CA studies a bottom-up, valuation sequence. Secondly, explicit considerations of the budget constraint were stressed in the OE-CVM studies, and not in the CA study. Both these design features could contribute to making stated valuations more conservative in the CVM study.

Turning to explanations of type a), implied value cues might have affected valuations in CA, as respondents here are given information about a cost that is subsequently used as a basis for eliciting the final WTP answers. Such cue biases do not occur in OE-CVM, since here no value cues are given by construction of the survey instrument. Ordering and embedding effects may be greatest for CVM surveys (at least according to CVM critics), and may tend to bias CVM valuations in the upward direction. Since our CVM valuations are much lower than those from other approaches, it appears unlikely that we have such strong upward biases in our CVM study. It is, however, conceivable that such biases could be even greater in our CA and EDP studies. In the latter, the ordering of the attributes to be valued may have led to upward bias in the WTP estimates, as the economic cost variable was one of the last attributes included. Moreover, our CA valuations may have been affected by embedding effects. From table 2, each game did not include all attributes simultaneously, which may have lead to upward bias in the valuation of a particular component of a package to be valued. See Hoehn and Randall (1989) for a discussion of such effects.

Any strategic biases with OE-CVM are as already noted, most likely to be in the downward direction. We have ruled CA to have less problems of strategic bias, while EDP may be more exposed to such biases, perhaps often in the upward direction if the panels chosen have a strategic interest in trying to increase public efforts against pollution (which our panels were likely to have). While such biases may have played a role, it is clear that they cannot alone explain the differences between the CA and EDP results, in particular as they appear from tables 6 and 7.

Econometric inference issues may have created biases in the derived results, and could contribute to explaining part of the differences in measured valuations. Inference problems are important both for CA and EDP, as we here need to make strong assumptions about the utility function, both with respect to its shape and its distribution across subjects. As shown in several previous studies, what assumptions are made here may have a considerable impact on the WTP estimates (Kealy and Turner 1993; Seller, Chavas, and Stoll 1985). Potentially, econometric inference factors may have led to upward biases in the CA and EDP valuations, although we have little control of the magnitude of these possible biases.

Turning to explanations of type c), the description of what we in this paper have denoted as minor health damages and chronic lung diseases did differ somewhat between the EDP and the CVM study. The experts may have perceived these as more serious by than the ordinary citizens who are not likely to be familiar with the relevant illnesses. This may have led to a relative upward bias in the WTP estimates from the EDP study.

In the CA and the CVM studies, another possible related source of bias is that the changes in attributes were described by changes in emissions, and not in terms of end-impacts as in the EDP study. Uncertainty about the dose-response function, both for the respondents and in the calculation of the WTP for the end-impacts in the CA and the CVM surveys, may have affected the relative estimates, but we have few clues as to the magnitude or direction of such effects.

Considering d), imperfect knowledge of public preferences among experts may have led to systematic differences in valuation between EDP on the one hand, and OE-CVM and CA on the other. It is uncertain by how much and in what direction.

Overall, the most significant finding is that OE-CVM yields lower valuations than the other approaches. We have in the discussion above tried to go some way toward explaining these differences, without drawing any unambiguous conclusion. The explanation of these differences is a difficult, but we believe crucial, issue to be addressed in subsequent research.

Many critics of CVM stress the potential upward biases inherent in this method, in particular those due to ordering and embedding effects. They maintain that CVM valuations are on the whole biased upwards. But if this were the case, CA and EDP must yield valuations that are biased upwards to an extreme degree. We view such an interpretation of the data as unrealistic, and find it more plausible that our OE-CVM results are not (at least not seriously) biased in the upward direction. Additional research, both theoretical and empirical applying different methods, is necessary for further enlightenment on these questions.

References

Adamowicz, W., J. Louviere, and M. Williams. 1994. Combining revealed and stated preference methods for valuing environmental amenities. *Journal of Environmental Economics and Management* 26:271-292.

Brookshire, D. S., R. C. d'Arge, W. D. Schulze, and M.A. Thayer. 1981. Experiments in Valuing Public Goods. *Advances in Applied Microeconomics* 1:123-172.

Carson, R.T., and R.C. Mitchell. 1995. The issue of scope in contingent valuation studies. *American Journal of Agricultural Economics* 75:1263-1267.

Clench-Aas, J., Larssen S., Bartonova A., and Johnsrud M. 1989. Virkninger av luftforurensninger fra veitrafikk på menneskers helse. Resultater fra en undersøkelse i Vålerenga/Gamlebyen-området i Oslo. Oslo: NILU.

Desvouges, W. et al. 1983. A comparison of alternative approaches for estimating recreation and related benefits of water quality improvements. Washington D.C.: USEPA.

Desvouges, W. et al. 1992. Using CV to measure non-use damages: An assessment of validity and reliability. Research Triangle Park.

Diamond, P A, and J A Hausman. 1994. Contingent Valuation: Is Some Number Better than No Number? *Journal of Economic Perspectives* 8:45-64.

Greene, W H. 1993. *Limdep 7, User's Manual.* Bellprot, New York: Econometric Software Inc.

Gregory, R, S Lichtenstein, and P Slovic. 1993. Valuing Environmental Resources: A Constructive Approach. *Journal of Risk and Uncertainty* 7:177-197.

Halvorsen, B. 1996. Ordering Effects in Contingent Valuation Surveys. Willingness to Pay for Reduced Health Damage from Air Pollution. *Environmental and Resource Economics* 8 (4).

Hanemann, W M, and B Kanninen. 1996. The Statistical Analysis of Discrete Response CV Data. Berkeley: Department of Agricultural and Resource Economics, Divison of Agricultural Resources, University of California.

Hausman, J.A., ed. 1993. *Contingent valuation: A critical assessment.* Amsterdam: North-Holland.

Heiberg, A. B., and K.-G. Hem. 1989. Use of formal methods in evaluating counter-measures to coastal water pollution. A case study of the Kristiansand Fjord, Southern Norway. In *Risk Management of Chemicals in the Environment*, edited by H. M. Seip and A. B. Heiberg: Plenum Press.

Hensher, D.A. 1994. Stated preference analysis of travel choices: the state of practice. *Transportation* 21 (2).

Hoehn, J. P., and A. Randall. 1989. A satisfactory benefit-cost indicator from contingent valuation. *Journal of Environmental Economics and Management* 14:226-247.

Janssen, R. 1992. *Multiobjective decision support for environmental managemen.* Doordrecht: Kluwer.

Kahneman, D, and J L Knetsch.1992. Valuing Public Goods: The Purchase of Moral Satisfaction. *Journal of Environmental Economics and Management* 22:57-70.

Kanninen, B. 1995. Bias in discrete response contingent valuation. *Journal of Environmental Economics and Management* 28:114-125.

Karni, R.P., P. Feigin, and A. Breiner. 1991. Multicriterion issues in energy policy making. *European Journal of Operations Research* 56:30-40.

Kealy, M. J., and R. W. Turner. 1993. A test of the equality of closed-ended and open-ended contingent valuations. *American Journal of Agricultural Economics* 75:321-331.

Keeney, R, and H Raiffa. 1976. *Decision with Multiple Objectives.* New York: John Wiley & Sons.

Kristrøm, B. 1993. Comparing continuous and discrete contingent valuation questions. *Environmental and Resource Economics* 3:63-71.

Loomis, J. B. 1990. Comparative reliability of the dichotomous choice and open-ended contingent valuation techniques. *Journal of Environmental Economics and Management* 18:78-85.

Louviere, J.J. , and D.A. Hensher. 1982. On the design and analysis of simulated or allocation experiments in travel choice modelling. *Transportation Reseach Record* 890:11-17.

Louviere, J.J., and G.G. Woodworth. 1983. Design and analysis of simulated choice or allocation experiments: an approach based on aggregated data. *Journal of Marketing Reseach* 20:350-367.

Mackenzie, J. 1993. A Comparison of Contingent Preference Models. *American Journal of Agricultural Economica* 75:593-603.

Magat, W. A., W. K. Viscusi, and J. Huber. 1988. Paired comparison and contingent valuation approaches to morbidity risk valuation. *Journal of Environmental Economics and Management* 15:395-411.

McFadden, D., and G.K. Leonard. 1993. Issues in the Contingent Valuation of Environmental Goods: Methodologies for Data Collection and Analysis. In *Contingent Valuation. A Critical assessment.*, edited by J. A. Hausman.

Mitchel, R. C., and R. T. Carson. 1989. *Using surveys to value public goods: The Contingent Valuation method.* Washington D. C.: Resources for the Future.

Roe, B, K J Boyle, and M F Teisl. 1996. Using Conjoint Analysis to Derive Estimates of Compensating Variation. *Journal of Environmental Economics and Management* 31 (2):145-159.

Seller, C., J. P. Chavas, and J. R. Stoll. 1985. Validation of empirical measures of welfare change: a comparison of nonmarket techniques. *Land Economics* 61:156-175.

Shechter, M. 1991. A comparative Study of Environmental Amenity Valuations. *Environmental and Resource Economics* 1:129-155.

Stam, A, M Kuula, and H Cesar. 1992. Transboundary Air Pollution in Europe. An Interactive Multicriteria Tradeoff Analysis. *European Journal of Operations Research* 56:249-262.

Sælensminde, K, and F Hammer. 1993. Samvalgsanalyse som metode for verdsetting av miljø - Pilotundersøkelse (Conjoint Analysis as a Method of Assessing Environmental Benefits - Pilot Study). In Norwegian with English summary. Oslo: Institute of Transport Economics.

Sælensminde, K, and F Hammer. 1994. Verdsetting av miljøgoder ved bruk av samvalgsanalyse - Hovedundersøkelse (Assessing Environmental Benefits by Means of Conjoint Analysis - Main Study). In Norwegian with English summary. Oslo: Institute of Transport Economics.

Wenstøp, Fred, and Arne J. Carlsen. 1987. Ranking hydroelectric power projects with Multi Criteria Decision Analysis. *Interfaces* 18:36-48.

Wenstøp, Fred, Arne J. Carlsen, Olvar Bergland, and Per Magnus. 1994. Valuation of Environmental Goods with Expert Panels. Sandvika: Norwegian School of Management.

The use of MCDM to evaluate trade-offs between spatial objectives

Marjan van Herwijnen[1] and Ron Janssen[1]

[1] Institute for Environmental Studies, Vrije Universiteit, De Boelelaan 1115, 1081 HV Amsterdam, The Netherlands

Keywords. Spatial information, maps, decision support

1 Introduction

In environmental management the performance of policy alternatives is often represented in maps. Such a map is a criterion map which shows the impact of an alternative for every location. The decision maker is asked to use these maps to compare alternatives and ultimately to select the one preferred. Since most people use reference maps, such as road maps, maps are familiar, and expected to represent reality. Decision makers are therefore happy to use maps to support their decisions (see for example Kohsiek *et al.* 1991).

In practice effective use of maps is a difficult task for many people and gets more difficult if the information density of the maps increases and the direct link with reality decreases (Muehrcke and Muehrcke 1992). It is known that people are able to process no more than seven stimuli at the time (Miller 1956), and that the interpretation of images is strongly dependent on map design (Bertin 1981). Comparison of map representations of alternatives is therefore an extremely difficult task which gets exponentially more difficult if more than two maps need to be compared at the same time.

Spatial evaluation methods are designed to support this task. They help the decision maker by structuring and simplifying the map representations of the alternatives. Spatial evaluation methods can be considered as a subclass of the much larger class of evaluation methods such as multicriteria analysis. They could also be considered as a specific type of spatial analysis. A large variety of methods for spatial analysis and for evaluation methods can be found in the literature. Examples of the use of multicriteria methods to support evaluation of spatial problems are more rare but can also be found in the literature (Carver 1991, Davidson *et al.* 1994, Eastman *et al.* 1993, Herwijnen *et al.* 1993, Janssen and Rietveld 1990, Pereira and Duckstein 1993). Some of these studies involve selection of an alternative, for example a location, within one map. In other studies spatial information is aggregated before evaluation, ignoring the spatial pattern of the performance of alternatives. Studies that present the performance of alternatives in maps and use evaluation methods to support the user to compare and rank these maps are relatively rare.

Spatial evaluation methods presented in this article are developed to support the comparison of mapped policy alternatives for the Green Heart of The Netherlands (see also Herwijnen *et al.* 1997). This article starts with a description of a case study on nature management in The Netherlands. In the remainder of the article the following five examples of policy questions illustrate the use of spatial evaluation methods:

1. What is the relative performance of the alternatives?
2. How does the relative performance change if comparison is linked to quality thresholds?
3. How does the relative performance change if some types of nature are considered more important than others?
4. Which of the alternatives generates the largest connected natural areas?
5. What is the best alternative?

2 Nature management in The Netherlands

Spatial evaluation methods were developed in this study to support integrated environmental and nature management of specific policy regions. The evaluation methods are integrated in the Decision Support System Nature Management (Kros *et al.* 1995). Inputs to this DSS are local and regional emission patterns of SO_2, NOx and NH_3. The system uses these emission patterns to calculate acid and nutrient deposition and to predict resulting soil quality parameters. Soil quality is used to calculate the probability that specified plant species will occur. Nature quality is evaluated using 'nature types'. Each nature type is defined by a characteristic set of plants. Model results include for each nature type the proportion of plants that are expected to occur under the different management alternatives. These results represent the potential quality of the management alternatives with a spatial resolution of 1 km^2 (Latour and Reiling 1992; Kros *et al.* 1995).

Nature management in The Netherlands uses maps to operationalise the policy objective for a specific region. This map specifies the composition and spatial pattern of the required nature types in this region. The *spatial policy objective* for the Green Heart specifies for each km^2 the required nature type (Latour *et al.* 1997). Due to the soil conditions these are all types of fen. Only areas classified as nature are included in the spatial policy objective. Some nature types, like Coppice and Holm, have been planned for large connected areas in the North East and Centre of the Green Heart. Most other nature types have been planned as small isolated units. Some of these, such as Reed and Bush land are allocated only a few km^2.

Policy alternatives are designed to provide the required environmental quality necessary for the occurrence of these nature types (Bal *et al.* 1995). The policy alternatives include instruments for reduction of emissions of SO_2, NOx and NH_3, but differ in the intensity, type and spatial pattern of these instruments. The performance of these alternatives is presented in maps. These performance maps represent the relative quality of the alternatives and are the input to the evaluation. Two performance maps for the Green Heart are shown in Figure 1. The maps show the potential quality of nature resulting from two packages of policy instruments to reduce NOx deposition. Nature quality is measured as the percentage of plant species of the nature type specified for each km^2 that can survive under the predicted conditions. If conditions for all plants of all nature types specified in the policy objective are sufficient, the alternative equals the policy objective and all scores equal 100%.

3 Relative performance

Policy question 1:   What is the relative performance of the alternatives?

The performance maps shown in Figure 1 represent for each alternative the potential nature quality. The values linked to each grid cell in these maps originally range from 0 (none of the plant species in the specified nature type can survive) to 100 (all can survive). These maps could be represented on a continuous scale as proposed by Tobler (1973). However, research has shown that too many classes reduces the readability of the map and makes it difficult, at a single glance, to get an overview and understanding of the theme mapped (Kraak and Ormeling 1996). The maps represented in Figure 1 are therefore classified in three classes. Differences among alternatives can be obscured by the use inadequate class limits. Limiting the number of classes and at the same time changing class limits may provide a better image of the differences among the alternatives. A difference index is used to determine the optimal class limits.

Each map is evaluated using an evaluation index. The index value of the map representing the policy objective is set at 100 for all evaluation criteria. Definition of the index depends on the aspect represented in the map. The performance index included in Figure 1 represents the average performance of the alternatives. This index is independent of the classification used.

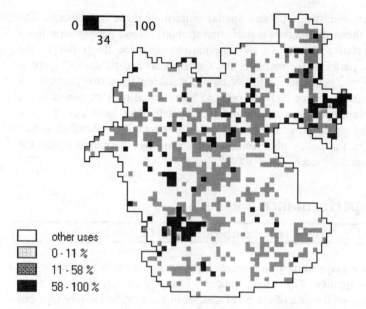

(a) Predicted performance map of Low deposition alternative

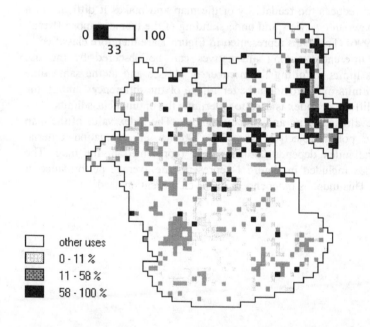

(b) Predicted performance map of High deposition alternative

Figure 1. Predicted performance of the alternatives presented in three unequal classes

4 Quality thresholds

Policy question 2:	How does the relative performance change if comparison is linked to quality thresholds?

In other words: how much nature of a certain minimum quality results from the alternatives? The answer to this question depends on the quality thresholds used. Is a nature type considered to exist if 80% of its plant species are found, or is 50% sufficient? Applying the quality thresholds on a performance map results in a total area map. By changing the quality thresholds the influence on the total area map can be examined.

Thresholds are also used to calculate indices for variation and naturalness. Variation is defined as the number of nature types specified in the policy objective that can exist in each alternative. A nature type exists if at least 3 km^2 of a nature type exceed the quality threshold. To calculate naturalness the area with specific natural qualities is calculated.

5 Priorities

Policy question 3:	How does the relative performance change if some types of nature are considered more important than others?

So far it is assumed that all km^2 and all nature types are equally important. However, decision makers can have a preference for specific rare or characteristic nature types or for protection of certain areas such as nature reserves. In this example the relative weight of each nature type is derived from the number of bird species that use this nature type as preferred habitat for at least part of their life cycle. The weights are standardised in such a way that the average weight of the total policy area is equal to 1. Linking these weights to the spatial distribution of the nature types specified in the spatial policy objective results in the priority map shown in
Figure 2.

Changes in the relative weights do not influence the performance maps (Figure 1). However changing the weights does influence the value of the evaluation indices linked to these maps. As was shown in Figure 7 the evaluation index is almost the same if all nature types are considered equally important (Low deposition: 45; High deposition: 44). If the above mentioned weights for each nature type are used these values change to 39 for the low deposition alternative and 33 for the high deposition alternative. Using these weights, the low deposition alternative is clearly the preferred alternative. Since the average weight for the whole policy area equals one, this implies that the higher weighted nature types perform badly in the high deposition alternative.

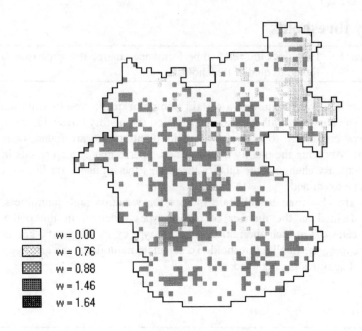

☐	w = 0.00
▦	w = 0.76
▩	w = 0.88
▣	w = 1.46
▩	w = 1.64

Figure 2. Priority map linked to number of bird species in each nature type

6 Large natural areas

Policy question 4: Which of the alternatives generates the largest connected natural areas?

At present nature is very fragmented and its quality is declining. This is leading to a decline in species diversity. It is essential to create larger connected areas to offer chances to rare species and to ensure sustainable natural qualities. It is therefore important to compare the size of connected areas of the alternatives.

Connected areas are calculated using King's neighbourhood. This implies that each grid cell can be connected to a maximum of eight other grid cells. Figure 3 shows the size of connected natural areas. In this figure no distinction is made among the nature types. This is a valid assumption for species that can survive in all different nature types. However if species are dependent on nature type, connected areas can be calculated for each nature type separately. It is also possible to generate the largest connected area for each nature type. This is relevant if thresholds are specified for target species. The index is calculated as the total natural area divided by the square root of the number of separate areas to reflect both size and fragmentation.

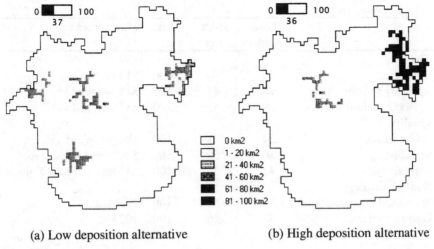

(a) Low deposition alternative   (b) High deposition alternative

Figure 3. Size of connected natural areas

7 Ranking

Policy question 5:	What is the best alternative?

Each map is evaluated using an evaluation index. This index is presented with the map. Definition of the index is dependent on the aspect represented in the map. The index value of the map representing the policy objective is in all cases equal to 100. Calculation of an evaluation index involves aggregation over space. The evaluation index can therefore be described as a non-spatial summary of the underlying map. Evaluation indices are included in an evaluation table. The evaluation table can therefore be described as a non-spatial summary of the decision problem and can be evaluated using a standard multicriteria approach (Janssen and van Herwijnen 1993).

Four classes of evaluation criteria are used: 1. size, 2. quality 3. spatial pattern and 4. costs. The evaluation criteria are included in the evaluation table (Table 1). The evaluation table shows the value of each alternative on a 0-100 scale. The last two columns show the value of 100 linked to the policy objective and the value before standardisation that corresponds to 100. For example the policy objective specifies 5 nature types. The maximum value of variation therefore corresponds to a total number of 5 nature types. The score of 80 of the Low deposition alternative therefore corresponds to the occurrence of 4 of the specified nature types. The first three categories reflect the effectiveness of the alternatives to generate natural qualities. The last category includes the cost of implementation of these alternatives. The numbers presented are tentative and represent a very rough estimate of the cost of instruments to reduce NH_3 emissions in and around the region (Herwijnen et al. 1997).

Table 1. Evaluation table

	Low dep.	High dep.	Max.	Max. represents
Size				
Total area	45	44	100	813 km^2
- priority to cons. areas	46	44	100	813 weighed km^2
- priority to bird species	39	33	100	813 weighed km^2
Quality				
performance	34	33	100	all spec. plant species
variation	80	60	100	5 different nature types
naturalness	47	35	100	513 km^2 semi-natural area
Spatial pattern				
connected areas/nat.type	55	53	100	15.1 km^2
connected natural areas	37	36	100	102 km^2
largest connected areas	46	52	100	largest area: 49.8 km^2
Cost				
Total costs	1.85	0.59	5	5 million guilders/year
Effectiveness	46.7	42.5	100	
Cost-effectiveness	25.2	72.1	∞	

All scores representing effectiveness are measured on the same scale (0-100) and can therefore be combined into an overall effectiveness index. All evaluation criteria make use of the information in the performance maps (Figure 1). This information is combined with other information such as thresholds and weights. Also evaluation criteria use different attributes of the same underlying data. Therefore the scores are, to a certain degree, interdependent especially within each category. The contribution that each of the criteria makes to the overall index is reflected in weights. These weights do not represent trade-offs, but represent the influence of the criteria in the overall objective. An effectiveness score for each alternative is calculated as the weighted sum of the scores of the evaluation criteria. The importance of the individual criteria for the overall score is dependent on the function of the region. Because the Green Heart plays an important role as a buffer between large urban areas, criteria linked to size are considered important. In this example these criteria have been given twice the weight of the quality and pattern-related criteria. These weights result in the effectiveness score presented in Table 1. The effectiveness scores show that the Low deposition alternative is more effective than the High deposition alternative. It is also shown that both alternatives are far removed from the policy objective. Only if all results are attributable to the alternatives can cost-effectiveness be calculated as the ratio of effectiveness and costs. In practice adjustments will have to be made for autonomous developments. In this case the High deposition alternative is the most cost-effective, achieving the policy objective is the least cost effective.

8 Conclusions

The results of model calculations were presented in performance maps. These maps represent the relative quality of the alternatives and are the input to the evaluation. From the performance maps the differences between the alternatives were not obvious. By looking carefully from various angles differences become clear. It was shown that the alternatives differ if priority is given to bird species and also that the alternatives differ substantially in spatial pattern. Each map was summarised using an evaluation index. These indices were included in an evaluation table. From this table it could be concluded that the differences in effectiveness between the alternatives were small. However, important differences between the alternatives were lost in aggregation. Differences that could be observed from the maps were not always reflected in the indices. Therefore maps and indices should be used in combination.

The results of this study offer many opportunities for further research. In the current study the spatial pattern of the performance maps and the policy objective is identical. Objective and performance only differ in level of achievement. As a next step performance maps with spatial patterns that differ from the policy objective could be evaluated using distance metrics. Specification of the policy objective is a complex task. Heuristics could be developed to support the decision maker in translating non-spatial objectives into an objective map. A clear reference point for the required natural qualities offers possibilities for increasing the cost-effectiveness of environmental policies. Optimisation procedures could be developed to generate optimal combinations in intensity, spatial pattern and time of emission reduction and management instruments.

Acknowledgement
The research presented in this article was commissioned by the National Institute of Public Health and Environmental Protection RIVM, Bilthoven and was included in the MAP research project Environmental quality of natural areas. The authors would like to thank Arthur van Beurden, Rob van de Velde and Joris Latour (RIVM) for their contributions to the project. The authors also thank Jos Boelens who produced the computer program and Xander Olsthoorn who provided estimates of costs.

References
Bal, D., H.M.Beije, Y.R.Hoogeveen, S.R.J.Jansen and P.J.van der Reest, 1995, *Handboek natuurdoeltypen in Nederland*. Rapport IKC Natuurbeheer nr.11, Wageningen.

Bertin, J., 1981, *Graphics and graphic information processing*. Walter de Gruyter, Berlin.

Carver, S.J., 1991, Integrating multi-criteria evaluation with geographical information systems. In: *International Journal of Geographical Information Systems*, Vol. 5, No. 3. pp. 321-339.

Davidson, D.A., S.P Theocharopoulos and R.J. Bloksma, 1994. A land evaluation project in Greece using GIS and based on Boolean and fuzzy set methodologies. In: *Int. J. Geographical Information Systems*, Vol.8, No.4, pp, 369-384.

Eastman, J.R., P.A.K. Kyem, J.Toledano and W.Jin, 1993, *GIS and Decision making*, Clarks University, Worcester/Unitar Geneva.

Herwijnen, M. van, R. Janssen and P. Nijkamp, 1993, A multi-criteria decision support model and geographic information system for sustainable development planning of the Greek islands. *Project Appraisal*, 8, pp.9-22.

Herwijnen M. van, R. Janssen, A.A. Olsthoorn, J. Boelens, 1997, *Ontwikkeling van ruimtelijke evaluatie methoden voor gebiedsgericht milieu- en natuurbeleid*. Instituut voor Milieuvraagstukken, Vrije Universiteit, Amsterdam.

Janssen, R. and M. van Herwijnen, 1993, *DEFINITE A system to support decisions on a finite set of alternatives*. Kluwer Academic Publishers, Dordrecht.

Janssen R. and P. Rietveld, 1990, Multicriteria analysis and Geographical Information Systems; an application to agricultural land use in The Netherlands. In: Scholten, H.J. and J.C.H. Stillwell (eds.), *Geographical Information Systems and Urban and Regional Planning*, Kluwer Academic Publishers, Dordrecht, pp. 129-139.

Kohsiek, L., F. van der Ven, G. Beugelink and N. Pellenbarg, 1991, *Sustainable use of groundwater; problems and threats in the European Communities*, Research report 600025001, National Institute of Public Health and Environmental Protection, Bilthoven.

Kros J., G.J. Reinds, W. de Vries, J.B. Latour en M.J.S. Bollen, 1995, *Modelling of soil acidity and nitrogen availability in natural ecosystems in response to changes in acid deposition and hydrology*. Wageningen, SC-DLO. Rapp.nr 95.

Latour, J.B. and R. Reiling, 1992, *Ecologische normen voor verzuring, vermesting en verdroging. Aanzet tot een risicobenadering*. RIVM rapport 711901003. RIVM, Bilthoven.

Latour, J.B., H. van der Vloet, R.J. van de Velde, J. van Veldhuizen, M.Ransijn en J. van der Waals, 1997, *Methodiek voor het aangeven van de milieuhaalbaarheid van gebiedsvisies: uitwerking voor het Groene Hart*. RIVM rapport in prep., RIVM, Bilthoven.

Kraak, M.J. and F.J.Ormeling, 1996, *Cartography: visualization of spatial data*. Longman, Harlow.

Miller, G.A. 1956. The magical number seven plus or minus two: some limits on our capacity for processing information. *Psycholog. Review*, Vol.63, pp. 81-97.

Muehrcke, P.C. and J.O. Muehrcke, 1992, *Map Use; reading, analysis and interpretation*. JP Publications, Wisconsin.

Pereira, J.M.C. and L. Duckstein, 1993, A multiple criteria decision-making approach to GIS-based land suitability evaluation. *International Journal of Geographical Information Systems*, Vol.7, No.5, pp. 407-424.

Tobler, W.R., 1973, Choropleth maps without class intervals? *Geographical Analysis*, 1973(5), pp. 262-265.

Interactive Decision Maps, with an Example Illustrating Ocean Waste Management Decisions

A. Lotov, Russian Academy of Sciences, Computing Center, ul. Vavilova, 40, 117967 Moscow B-333, RUSSIA

O. Chernykh, Russian Academy of Sciences, Computing Center, ul. Vavilova, 40, 117967 Moscow B-333, RUSSIA

V. Bushenkov, Russian Academy of Sciences, Computing Center, ul. Vavilova, 40, 117967 Moscow B-333, RUSSIA

Hannele Wallenius, Institute of Industrial Management, Helsinki University of Technology, Otakaari 1M, 02150 Espoo, FINLAND

Jyrki Wallenius, Helsinki School of Economics, Runeberginkatu 14-16, 00100 Helsinki, FINLAND

Abstract. Interactive Decision Maps are introduced and illustrated with an ocean waste disposal example, requiring difficult pollution-cost tradeoffs. Interactive Decision Maps are a tool for quickly displaying various decision maps for three or more decision-relevant criteria. They are based on the Generalized Reachable Sets (GRS) approach developed in Russia. Animation of decision maps is also possible. Integration of Interactive Decision Maps with Pareto Race, a free search Multiple Objective Linear Programming procedure, is proposed.

Keywords: Multiple Objective Linear Programming, Nondominated Set, Ocean Waste Management

1. Introduction

Decision maps are a well known, but rarely used multiple criteria decision support tool developed for three-criteria problems (see, for example, Haimes et al., 1990, Figure 4.2). A family of bi-criteria cross-sections of the Pareto-optimal frontier, assuming a fixed value for the third criterion is depicted. The Interactive Decision Maps (IDM) technique allows a rapid display of requested decision maps. Bi-criteria Pareto-optimal frontiers are depicted while the third criterion is not permitted to become worse than a prespecified value. A decision map which looks like a geographical map helps to understand the tradeoffs among the criteria. In case of more than three criteria, various decision maps are displayed (upon request). Animation of decision maps, i.e. display of automatically generated sequences of maps, is also possible.

The IDM technique seeks to help decision makers to identify a most preferred feasible combination of criterion values with a simple click of the mouse on the appropriate decision map. When such a combination has been identified, a point in the decision variable space which leads to this most preferred point in the criterion space is computed (the Feasible Goals Method (FGM) introduced by Lotov (1973, 1984)).

In this paper we reconsider the old problem of choosing sewage sludge disposal sites in the New York Bight with the help of the IDM/FGM technique. This problem was considered by Wallenius, Leschine, and Verdini (1987) and by Leschine, Wallenius and Verdini (1992). In conclusion, the integration of Pareto Race (Korhonen and Wallenius, 1988), a free search Multiple Objective Linear Programming procedure, with the IDM technique is proposed. An appendix contains a mathematical formulation of the IDM/FGM technique.

2. Sewage Sludge Disposal Problem

Contamination of the New York Bight has been a concern of the Environmental Protection Agency (EPA), and the neighboring municipalities for many years. Concern for water quality in the Bight region is long standing, particularly for waters in the inner portion of the Bight. Highly publicized pollution-related episodes which have occurred over the past decades have had a lasting impact on public opinion. Being concerned about the contamination of the inner Bight region, the EPA ordered in 1985 New York City and the remaining users of the inner Bight region to begin shifting their dumping operations to the 106-mile site. In this study we reexamine, following Wallenius et al. (1987) and Leschine et al. (1992), the EPA decision in a way which permits simultaneous multi-site dumping.

Three alternative disposal sites were considered in the model: the 12-mile site, the 60-mile site, and the 106-mile site. We assumed that a combination of the above sites was a possibility, such that all three sites could be used at the same time in different portions. In the model all sludge was assumed to be produced in New York City (where 52% is produced), New Jersey (41%), and Long Island (7%). Production of sludge was assumed to be constant from year to year. Two types of vessels were used for the transportation of the sludge: towed barges, and self-propelled barges. The constraint set of the model contained four parts:

    1) constraints to ensure dumping of all generated sludge;
    2) constraints of annual dumping capacity of barges;
    3) definitional constraints of amount dumped at each site;
    4) Markov constraints to model the ocean's assimilative capacity.

The following three criteria were used to evaluate different sludge disposal strategies:

- total cost of sludge disposal operation (millions of US$);
- pollution level at inshore monitoring station (pollution concentration, in percent to a given value);
- pollution level at offshore monitoring station (pollution concentration, in percent to a given value).

The decision variables included the number of self-propelled/towed barge trips from source (NY, NJ, LI) to site (12-, 60-, 106-mile sites). A formal description of the model is given in Leschine et al. (1992).

3. Interactive Decision Maps

The IDM technique is a particular form of the Generalized Reachable Sets (GRS) method (Lotov, 1973, 1984; see also Lieberman, 1991). The GRS method was developed for the exploration of nonclosed mathematical models. It consists of constructing and displaying the set of attainable output (criterion) vectors for a given feasible combination of input variables. In the MCDM context, the GRS method provides an opportunity to transform a decision problem from the space of decision variables into the criterion space. To be precise, starting with the set of feasible strategies, we construct the set of feasible combinations of criterion values. Aggregate information is provided in the form of various decision maps which are displayed on request.

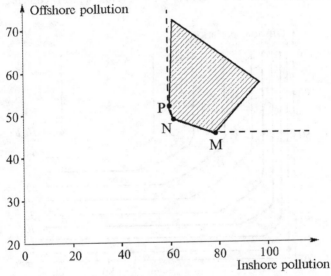

Figure 1: Feasible Set in Criterion Space

Let us consider the problem from the previous section. To begin with, let us fix the total cost. Then all feasible values of inshore pollution and of offshore pollution can be visualized on the computer screen in the form of an image (Figure 1, where total cost equals $ 15 million). Since it is preferable to decrease both inshore and offshore pollution (ceteris paribus), we are interested in the "south-western" frontier of the image (the Pareto-optimal frontier). The Pareto-optimal frontier contains those combinations of inshore and offshore pollution levels which have the following property: to decrease one objective (say, offshore pollution) one needs to increase the other (inshore pollution). The form of the Pareto-optimal frontier shows how large of a drop in the offshore pollution is connected with a certain increment in the inshore pollution, and vice versa. In Figure 1 a small decrement in the offshore pollution requires a substantial increment in the inshore pollution near point "M". And correspondingly, near point "P" just a small rise in the inshore pollution results in a sharp decrement in offshore pollution.

Note that in Figure 1 the total cost is fixed. To vary the value of the cost criterion, a decision map, consisting of a family of Pareto-optimal frontiers is constructed. The display of such maps in the framework of the IDM technique is based on approximating the Edgeworth-Pareto Hull (EPH) of the Feasible Set in Criterion Space (FSCS), i.e. of the FSCS augmented with all dominated criterion points. The shaded region in Figure 1 is an example of an FSCS for two criteria. The frontier of the EPH is depicted by dashed lines. Note that the EPH has the same Pareto-optimal frontier as the FSCS. The same is true for any number of criteria: the Pareto-optimal frontiers of the FSCS and its EPH coincide.

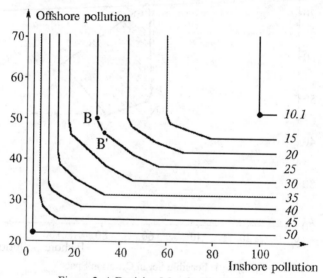

Figure 2: A Decision Map for Fixed Costs

Decision maps are constructed as collections of two-dimensional slices of the EPH. A decision map containing the Pareto-optimal frontiers between inshore and offshore pollution related to different values of total cost is shown in Figure 2. Note that Figure 1 depicts the slice corresponding to total cost equal to $ 15 million in Figure 2. With the help of the decision map, one can easily understand the relation between an increment in the total cost and the improvement in the environment (i.e., a reduction in the inshore and/or the offshore pollution). One can easily obtain different decision maps displaying the "inshore pollution vs. cost" and "offshore pollution vs. cost" tradeoffs.

An approximation of the EPH is constructed in advance in the framework of the IDM technique. This is the main mathematical and computational problem which has been solved during the development of the IDM technique. Three groups of methods for the approximation of the EPH were developed. Methods of the first group are based on direct application of the classic Fourier convolution of systems of linear inequalities and may be used in the case of linear models with a relatively small number of decision variables (see Lotov, 1996). The second group of methods can be applied for linear systems with a large number of decision variables (several thousands), but a relatively small number of criteria (three to six). The basic idea of the methods of the second group consists of constructing a sequence of polytopes iteratively approximating the FSCS. Such methods extend the idea of the NISE method earlier proposed for two-criteria problems (Cohon, 1978). Sequences of polytopes are constructed on the basis of a combination of the Fourier convolution and optimization techniques. It has been proven that the approximation methods produce optimal sequences of polytopes. The methods of the third group are related to approximating the FSCS and the EPH for nonlinear models. Details and references are given in Bushenkov et al. (1995).

The IDM technique displays collections of two-dimensional slices of the EPH in the form of decision maps. If more than three (say, five) criteria are incorporated, one needs to impose constraints on the values of the additional criteria to obtain a decision map. This can be done manually by using scroll-bars. Animation of a decision map may be performed by displaying a sequence of maps related to a sequence of constraints generated automatically. Moreover, one can display a matrix of decision maps corresponding to a collection of constraints imposed on the values of the additional criteria. These constraints may be chosen manually or generated automatically. A software system FEASIBLE GOALS for MS WINDOWS 3.1, implementing the IDM/FGM methodology has been developed.

4. Further Analysis of the Waste Disposal Problem

Let us take a closer look at Figure 2. In Figure 2 one can see the decision map in the "inshore pollution — offshore pollution" space, while the cost is changing from a minimal US $10.1 million to the maximal US $50 million. The Pareto-optimal frontiers in Figure 2 were drawn with heavy lines. They have the following important feature: they are kinked, except for the $10.1 million and $50 million frontiers. (Actually, the Pareto-optimal frontiers related to $10.1 million and $50 million consist of just one point.) The kink depends upon the cost. Above the kink, the tradeoff between inshore and offshore pollution is quite different from the tradeoff below it. Compare the distances between pairs of Pareto-optimal frontiers related to different costs. The $10.1 million and $15 million frontiers are quite apart, while the distance between the $15 million and $20 million frontiers is obviously smaller. This means that the extra $5 million investment has much more impact if the cost equals $10.1 million rather than $15 million.

The above phenomenon is explored from another angle in Figure 3 where an alternative decision map is displayed. The "cost — inshore pollution" tradeoff curves are given for several values of offshore pollution, ranging from 25% to 55% of its maximum value. One can see that every frontier contains a kink where the tradeoff changes drastically. If the offshore pollution equals 55%, the kink occurs at point A. Note that point A also belongs to another frontier which corresponds to 50% offshore pollution.

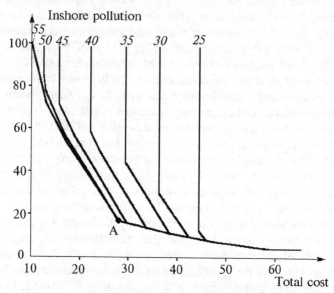

Figure 3: Total Cost Vs. Inshore Pollution

Figure 4 provides a close-up of the inshore pollution vs. total cost tradeoff, when the values of offshore pollution vary between 43% and 55%. It is interesting to note that if the offshore pollution exceeds 46%, further growth in it is practically useless: additional offshore pollution moves the frontier only marginally downwards. See point C in Figure 4.

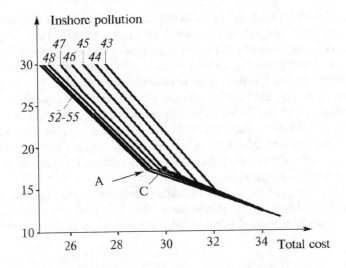

Figure 4: Inshore Pollution Vs. Cost: A Close-Up

5. Integrating IDM and Pareto Race: A Suggestion

Pareto Race was developed by Korhonen and Wallenius in the late 80's (Korhonen and Wallenius, 1988) as a dynamic and visual "free-search" type of interactive procedure for Multiple Objective Linear Programming. The procedure enables a decision maker to freely search any part of the Pareto-optimal frontier by controlling the speed and direction of motion. The criterion values are represented numerically and as bar graphs on a display. The keyboard controls include gears, an accelerator, brakes, and a steering mechanism. Using Pareto Race to explore the Pareto-optimal frontier resembles driving an automobile. The driver does not, however, have a map of the terrain that he/she explores. The decision maker usually has an idea in which direction he/she would like to move. But without a map, he/she does not always know if it is possible or worthwhile to move in a certain direction. The IDM could provide this missing map and guide the user through the search.

We describe one possible design for integrating Pareto Race and IDM. The main screen of the system is split in three main parts (Figure 5). The left/middle upper part is related to the IDM, where the decision map is displayed. As one can see, the given decision map is actually a part of Figure 2 augmented with the current point (the cross) and an additional slice related to the current point (depicted by a dashed line). The current values of inshore pollution and offshore pollution are indicated by the position of the cross. At the same time, the color (in the paper—the shading) of the field containing the cross informs the decision maker about the cost interval. The diagram relating the color of the frontier (the band) to the value of total cost is shown under the decision map. The dashed tradeoff curve which moves together with the cross over the decision map informs the decision maker about the tradeoff among inshore and offshore pollution and helps him/her decide whether it is reasonable/worthwhile to move along the chosen direction. The IDM control window is provided in the top-leftmost corner. All opportunities of the IDM technique are available. In particular, one can request another decision map related to the problem or change the number of Pareto-optimal frontiers. Moreover, a zoomed portion of the decision map may be displayed. In case of more than three (say, five) criteria, one can obtain various matrices of decision maps, animated decision maps, etc.

In the bottom part of the screen, the current criterion values are depicted in the form of horizontal bars used in Pareto Race. The right-hand side of the screen is a fictitious Pareto Race control panel supplemented with an additional PAUSE/GO window.

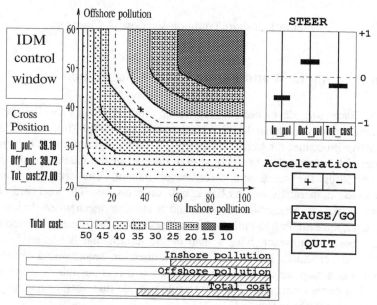

Figure 5: Integrating Pareto Race and IDM: A Suggestion

The STEER window provides an opportunity to control the direction of motion. It performs like a standard audio equalizer. 'Zero level' means that we do not want to change the search direction for this particular criterion. 'Plus one' corresponds to a maximal improvement in the underlying criterion and 'minus one' to its opposite. The ACCELERATION window provides an opportunity to increase or decrease speed. The PAUSE/GO window helps to pause the process to explore the decision map more carefully.

6. Acknowledgments

This research was conducted as part of the activities of the Russian-Finnish Technical Working Group on Operations Research which is sponsored by the Academy of Finland and the Russian Academy of Science. The authors would also like to thank the Foundation for Economic Education and the Russian Foundation for Fundamental Research (grant N95-01-00968) for financial support.

Appendix: A Mathematical Introduction to the IDM Technique

Let the decision variable x be an element of the decision space W. Let the set of feasible combinations of values of decision variables be $X \subset W$. Let the criterion vector y be an element of the linear finite-dimensional space R^m. We assume that the criterion vectors are related to decisions by a given mapping $f: W \to R^m$. Then the feasible set in the criterion space (FSCS) is defined as follows:

$$Y = \left\{ y \in R^m : y = f(x), x \in X \right\}$$

Let us suppose that we are interested, without loss of generality, in minimizing the criterion values. Then the Edgeworth-Pareto Hull (EPH) of the FSCS is defined as $Y^* = Y + R_+^m$, where R_+^m is the nonnegative cone of R^m. The Edgeworth-Pareto frontier of the FSCS is defined as

$$P(Y) = \left\{ y \in Y : \left\{ y' \in Y : y' \le y, y' \ne y \right\} = \varnothing \right\}$$

It is clear that $P(Y^*) = P(Y)$.

The IDM technique consists of constructing the EPH and interactively displaying it in the form of decision maps. Constructing set Y^* is based on approximating it by the sum of a simple body approximating set Y and the cone R_+^m. In case of a convex set Y^*, polytopes are used as the approximating bodies. Nonconvex sets Y^* are approximated as the sum of the cone R_+^m and a set of cubes.

The EPH is displayed in the form of decision maps, i.e. collections of its two-dimensional slices. Let $I^0 = \{1, 2, \ldots, m\}$. Also let $I \subset I^0$, $|I| = 2$. Finally, let $I^* = I^0 \setminus I$. Let us denote by $R(I)$ the criterion subspace with arguments from I. Then, by a two-dimensional slice of a set $V \subset R^m$ related to $z \in R(I^*)$ we mean the set

$$G(V, z) = \{u \in R(I) : (u, z) \in V\}.$$

It is important to note that in the case of the EPH, i.e. set Y^*, a slice of it contains the combinations of the values of the criteria from I which are feasible if the values of criteria from I^* aren't worse than z. Since the EPH is constructed in the framework of the IDM technique in advance, a collection of its two-dimensional slices can be depicted quite fast. Efficient algorithms for constructing the two-dimensional slices have been developed in Chernykh and Kamenev (1993).

Once a most preferred feasible point in the criterion space has been identified, it is possible to obtain the decision which will lead to this point. If an appropriate feasible point in the criterion space y' has been identified, it is regarded as the 'reference point' (Wierzbicki, 1981). An efficient decision is obtained by solving the following optimization problem

$$\min_{1 \le j \le m} \left(y'_j - y_j \right) + \sum_{j=1}^{m} \left\{ \varepsilon_j \left(y'_j - y_j \right) \right\} \Rightarrow \max,$$

while $y = f(x)$, $x \in X$, where $\varepsilon_1, \ldots, \varepsilon_m$ are small positive parameters.

References

Bushenkov V., Chernykh O., Kamenev G., and Lotov A. (1995), Multi-Dimensional Images Given by Mappings: Construction and Visualization, **Pattern Recognition and Image Analysis** 5(1), 35-56.

Chernykh O. L. and Kamenev G. K. (1993), Linear Algorithm for a Series of Parallel Two-Dimensional Slices of Multi-Dimensional Convex Polytope, **Pattern Recognition and Image Analysis** 3(2), 77.

Cohon J.(1978),*Multiobjective Programming and Planning*,Academic Press, N.Y.

Haimes Y., Tarvainen K., Shima T., and Thadathil J. (1990), *Hierarchical Multiobjective Analysis of Large-Scale Systems*, Hemisphere Publishing, N.Y.

Korhonen P., and Wallenius J. (1988), A Pareto Race, **Naval Research Logistics Quarterly** 35, 615-623.

Leschine T., Wallenius H., and Verdini W. (1992), Interactive Multiobjective Analysis and Assimilative Capacity-Based Ocean Disposal Decisions, **European Journal of Operational Research** 56, 278-289.

Lieberman E. (1991), *Multi-Objective Programming in the USSR*, Academic Press, N.Y.

Lotov A. V. (1973), An Approach to Perspective Planning in the Case of Absence of Unique Objective, in *Proc. of Conf. on Systems Approach and Perspective Planning* (Moscow, May 1972), Computing Center of the USSR Academy of Sciences, Moscow (in Russian).

Lotov A. V. (1984), Introduction into Mathematical Modeling of Economic Systems, Nauka, Moscow (in Russian).

Lotov A. V. (1996), Comment on D. J. White "A Characterization of the Feasible Set of Objective Function Vectors in Linear Multiple Objective Problems", **European Journal of Operational Research** 89, 215-220.

Wallenius H., Leschine T. M., and Verdini W. (1987), Multiple Criteria Decision Methods in Formulating Marine Pollution Policy: A Comparative Investigation, Research Paper No.126, *Proceedings of the University of Vaasa*, Finland.

Wierzbicki A. (1981), A Mathematical Basis for Satisficing Decision Making, in J. Morse (Ed.): *Organizations: Multiple Agents with Multiple Criteria*, Springer, 465-485.

Second-Generation Interactive Multi-Criteria Decision Making in Reservoir Management

Per J. Agrell[1], Antonie Stam[2], Barbara J. Lence[3]

1   Department of Operations Research, University of Copenhagen, Universitetsparken 5, DK-2100 Copenhagen Ø, DENMARK.
2   Department of Management, Terry College of Business, University of Georgia, Athens, GA 30602, USA.
3   Department of Civil and Geological Engineering, University of Manitoba, Winnipeg, Manitoba R3T 2N2, CANADA.

Abstract: Multi-purpose reservoir management is a well-known research area within Multi-Criteria Decision Making. In spite of the obvious multi-faceted decision problem, the suitability for mathematical modelling and the easily identifiable single decision maker, the many published works have apparently not affected the current operating practice, which is largely based on rules-of-thumb and policy targets. This work analyses some of the characteristics behind this from a decision-theoretical viewpoint and announces the arrival of the second generation interactive Multi-Criteria Decision Support. Based on a learning-oriented approach, designed to facilitate operator training and communication of decisions in low-key systems like spread-sheets, the proposed methodology is oriented around the problems of a contemporary dam manager. Finally, some questions are risen regarding the use of information technology and graphical aids in the communication of a decision.

Keywords. Reservoir Management, Multi-Criteria Decision Making

1. Introduction

The management of a multi-purpose reservoir, intended to satisfy flood protection, water supply and recreational requirements, exhibits three important characteristics from a decision making viewpoint. First, the reliable and effective operation of the reservoir is of primary societal and economic importance in its environment. Besides the obvious risk of flood, possibly causing loss of life and considerable damage to agricultural, residential and industrial values, there are also environmental and economic risks involved in connection to hydro- and thermo-electric power generation, dilution of municipal and industrial effluents and river navigation.

Second, due to the importance of the decision for concerned stake-holders and their often conflicting interests and the common public ownership of the reservoir, there is an apparent need to communicate the grounds for a decision and its projected implications for the stakeholders. In many cases this decision may be publicly, legally or politically challenged by dissatisfied parties, urging for their interests to be addressed. Third, the complexity of the decision is high due to the multiple objectives, seasonal and stochastic variations in the demand for and supply of water, varying inter-temporal trade-offs between objectives, ambient stream conditions, ecological and biological requirements for water quality and required releases. Effectively a real-time decision, the reservoir management must accommodate the engineering requirements necessary to achieve immediate and long-term objectives based on forecasts.

The three characteristics have each contributed to a tradition of reservoir management practice utilizing pre-determined target values and storage traces, complemented with a substantial amount of rules-of-thumb for unforeseen situations. Largely dependent on the engineering skills of the operators, the state-of-the-art is many times a stale compromise based on (at best) incomplete information regarding the multi-dimensional outcomes of the decision. Apparently, the scene has been set for improvements and suggestions from management science, an existing real-life application that both renders and deserves our full attention

The remainder of the paper will review the methodological basis for some of the earlier work in the area, compare that to the stated characteristics to explain their likely failure in practice, outline a proposed problem-oriented approach and finally briefly illustrate the proposed second-generation methodology by a reservoir modeling case study from Canada.

1.1 Reservoir Management and Multi-Criteria Decision Analysis

As an area of general academic and public interest, water resources management including reservoir management has received a fair share of attention from the management science community. An added incentive to undertake formal planning and analysis is that the investments and long-term consequences of water resource decisions are often large. For an early review of optimisation models applied to multipurpose reservoirs, see Yeh (1985). The reported applications mainly belong in either of two methodological categories; the Surrogate Worth Trade-off (SWT) Method by Haimes and Hall (1974) or Goal Programming (GP) by Ignizio (1976).

Whereas the SWT applications (e.g., Haimes *et al.*, 1975) effectively address the preferences of the individual decision makers and the selection of criteria is focused at the underlying conflicts in the problem, the method may be less than transparent to a decision-maker. In addition to the relative arbitrariness of the trade-off scale, the natural lack of commonly used computer systems may have reduced some of the practical usefulness of the approach. Referring to the characteristics of the problem, the decision maker may have been supported in the actual decision, but hardly in the communication and understanding of the problem.

Goal Programming (e.g., Can and Houck, 1984, Goulter and Castensson, 1988) may have been more easily accepted in practice, being closer to the adopted habit of using target values. However, the *a priori* goal levels and arbitrary weights on goals, possibly combined with an *a priori* lexicographic ordering of the goals, make the method sensitive to way the preferences are quantified. Also, the choice of criteria occasionally reflects computational necessity, e.g., linearity, rather than decision making importance. GP is an attractive technique to arrive at an answer already sought for, but less convincing as an argument in a public debate.

The first-generation methodology is largely based on the notion of a pre-existing preference system, either in terms of fixed trade-off ratios (SWT) or preferred points (GP). In the search for an optimal solution to an assumed problem, the methodology fails to clarify its role in the decision making environment, the significance of the criteria selected or its adaptability to future conditions. The endeavor is heavily dependent on the involvement of an analyst for the determination of technical constants, interpretation of scales and avoidance of dominated solutions. Reflecting again on the stress imposed upon the reservoir management, and the necessity to justify its actions, such dependencies are highly undesirable and likely to halt the migration of the system.

We propose the second-generation multi-criteria decision support to take its starting point in the decision making characteristics stated above. In acknowledge of the fact that formal analysis seldom is undertaken in the actual decision, which evolves as a result of negotiations with stakeholders, the need for transparency and communication in the methodology is evident. Criteria, constraints and parameters are here clearly defined and not obscured by the use of weights or scaling factors. The choice of criteria mirrors the identified conflict behind the decision problem. Although the complexity of the problem may call for a mathematical program, this is not taken for granted in the development of the support system. Rather than *eliciting* a preference structure from the decision maker, we propose *creating* one through education about the reservoir system and its inherent trade-offs. The use of a second-generation methodology is thus not in the top echelon of the decision making hierarchy, but in the level below it, where the preparation for decisions are made and the final policy is interpreted and executed. By implementing interactive support systems in familiar personal computer systems, using spreadsheets and computer graphics for the presentation of results, experimentation and learning is facilitated. In this manner, we hope that the second-generation multi-criteria decision support methodology will spread within the reservoir management community, not because of its theoretical convergence to a preferential optimum, but because of its usefulness as a tool to explore, communicate and develop system boundaries.

As a concrete example of the approach, the Shellmouth Reservoir problem (Agrell *et al.*, 1995) is used to demonstrate the ideas. Although the model is not yet operationally implemented, it may give an impression of the intended system.

2. The Shellmouth Reservoir

The Shellmouth Reservoir in Southwest Manitoba, Canada, used to regulate flows in the Assiniboine River. Originally intended for flood protection, presently the reservoir is used for a number of additional purposes, including municipal water supply for the cities of Brandon and Portage La Prairie, Manitoba, irrigation and agricultural water supply for downstream farmers on the Assiniboine River, dilution of the waste effluents from Brandon, Portage La Prairie and Winnipeg, dilution of the heated effluents from Manitoba Hydro's thermoelectric generating plant in Brandon, dilution of industrial waste effluents from facilities on the Assiniboine River, maintenance of sport fisheries in the Shellmouth Reservoir and the Assiniboine River, industrial water supply for facilities on the Assiniboine River, and recreation. A more detailed description of the reservoir is given in Agrell *et al.* (1995).

2.1　Decision Problem

The reservoir management has to strike a balance between four principal objectives: flood protection, water supply for municipal, agricultural and industrial use, water supply for thermoelectric power generation and recreational values in the Shellmouth Dam area. The flood protection objectives are concerned with the safety of the dam and a controlled rate of release. The demand-oriented objectives prompt for sufficient releases to meet realized demand at all times without violating environmental standards for instream temperature and concentration of effluents. The thermoelectric power generating plant in Brandon is primarily used Winter peak demand and dry years as a backup for more cost-efficient hydroelectric power. However, the recreational interests (fishery, boating, tourism) are in direct conflict with the power generation objective, leading to increased releases of water from the dam during Spring and high season (i.e., May-August). In our application, the main objective trade-offs result from an increase or decrease of the storage level. Whereas an increase in the storage level (up to a certain target) is beneficial from a recreational viewpoint and for safeguarding against expected or unexpected drought, a decrease in storage (i.e., an increased release) would enable higher thermal power generation, and meet downstream flood control, dilution and water demands. Subsequently, the reservoir manager desires a system enabling per-calculation and communication of projected consequences of a release plan.

3. The MCDM Model for Reservoir Management

The multi-criteria decision making framework for dam management presented here draws on Agrell *et al.* (1995), improving upon a previous models by Lence *et al.* (1992) and Yang *et al.* (1993).

3.1　Assumptions

Suppose that 1) the planning period is n years and let each year be sub-divided into m discrete planning periods (e.g., weeks or months), 2) there are K river reaches, $k = 1, ..., K$, downstream from the reservoir, 3) thermoelectric power plants are located in a subset of the river reaches $K_T \subseteq \{1, ..., K\}$, 4) hydrologic forecasts for evaporation, inflows and temperatures exist for the entire horizon, 5) tributary inflows into a river reach is available only at the end point of the reach, 6) the total demand for water in the reach is withdrawn at the very beginning of the reach and 7) water used for cooling of thermoelectric power plants is returned to the river.

This work is based on three types of annual hydrologic conditions; wet, average and dry years, from historical data and then optimises the storage for each type of year over the planning period horizon. The resulting policy can successfully be applied under the plausible condition that the decision maker can roughly categorise the upcoming year in terms of the three types. Thus, the decision vector consists of periodic storage levels to be determined for each storage period $i = 2, ..., m$, S_i^W during wet years, S_i^A during average years and S_i^D during dry years. To provide for continuity, the first period of each year, regardless of type, is excluded from the storage policy.

The relationship between river flow and maximum allowable power generation, Γ_{ij} (in GWh), is estimated by a linear function in Lence *et al.* (1992):

$$\Gamma_{ij}^k = a_i^k + b_i^k Q_{ij}^k, \quad i = 1,...,m; \ j = 1,...,n; \ k \in K_T, \tag{1}$$

where a_i^k and b_i^k are the intercept and the slope of the function, respectively and Q_{ij}^k is the total water volume available (in $10^6 \ \text{m}^3$) for cooling the thermoelectric power plant in river reach k during period i of year j.

4. Formulation of Criteria

Based on the perceived decision problem, we identify nine criteria in four groups to capture the conflict. Pertaining to operational safety are criteria for flood protection and release stability. The municipal, industrial and agricultural water demand is represented in one aggregate criterion. The energy-production related demand for water is present through one criterion, and the recreational interest is represented through four criteria, each with a specific purpose. Of course, the number of criteria in the groups has no implication regarding the relative importance of the group.

Flood Protection: We include a normalised threshold violation of a known channel capacity, $\hat{R}$ (in $10^6 \ \text{m}^3$) as a flood protection criterion to be minimised, rather than as a "hard" constraint. Denote the release during period i of year j by R_{ij}. Since large violations should be avoided at any time, we use a normalised criterion that minimises the maximum threshold violation $R_{ij} - \hat{R}$ over all i, j:

$$\text{minimise} \ y_1 = \frac{100}{\hat{R}} \max_{i,j} \left\{ R_{ij} - \hat{R}, 0 \right\}. \tag{2}$$

Dam Safety: The concerns of dam operation are primarily related to dam safety and operational stability. The structural safety of the dam is assured by "hard" constraints on the physical upper and lower reservoir levels. We interpret operational stability as a measure of change in release, ΔR_{ij}:

$$\Delta R_{ij} = \begin{cases} R_{ij} - R_{i-1,j}, & i = 2,...,m, j = 1,...,n, \\ R_{1,j} - R_{n,j-1}, & j = 2,...n, \\ 0, & i = 1, j = 1 \end{cases} \tag{3}$$

where R_{ij} denotes the release in period i, year j. The dam safety criterion is formulated as a minimisation of the sum of release changes:

$$\text{minimise} \ y_2 = \sum_{i=1}^{m} \sum_{j=1}^{n} \Delta R_{ij} \tag{4}$$

Municipal, Industrial and Agricultural Water Supply: As the ability to supply consumers with water is an important criterion, we include the fraction of unsatisfied demand as a performance measure, minimising the maximum water deficit:

$$\text{minimise } y_3 = \max_{i,j}\left\{1 - R_{ij}\left\{\sum_{k=1}^{K} F_{ij}^k\right\}^{-1}, 0\right\} \cdot 100,\tag{5}$$

where F_{ij}^k is the total net water requirement in river reach k, during period i of year j. If all demand is satisfied at all time, $y_3 = 0$.

Power Generation: The deviation of the power production, G_{ij}^k, at a thermoelectric plant in reach k during period i of year j, from its desired production target, (in GWh), is of interest to management. For each thermoelectric power plant in reach $k \in K_T$ under separate ownership, we minimize the largest deviation of G_{ij} from G_{ij}:

$$\text{minimise } y_4^k = \max_{i,j}\left\{G_{ij}^k - G_{ij}^k, 0\right\}, \qquad k \in K_T.\tag{6}$$

The generation of energy by means of thermoelectric plants requires that water be released from the reservoir. Thus, the release decision directly constrains output of the thermoelectric power plants.

Recreation: We define criteria as the (annual) average (y_6, y_8) and maximum deviation (y_5, y_7) from the storage level targets, SU_i, SL_i (in 10^6 m) established for fishery, recreation and tourism in the dam area.

$$\text{minimise } y_5 = \frac{100}{SU_i} \max_{i,j}\left\{S_{ij} - SU_i, 0\right\},\tag{7}$$

$$\text{minimise } y_6 = \frac{1}{mn}\sum_{i=1}^{m}\sum_{j=1}^{n}\frac{100}{SU_i}\max\left\{S_{ij} - SU_i, 0\right\},\tag{8}$$

$$\text{minimise } y_7 = \frac{100}{SL_i}\max_{i,j}\left\{SL_i - S_{ij}, 0\right\},\tag{9}$$

$$\text{minimise } y_8 = \frac{1}{mn}\sum_{i=1}^{m}\sum_{j=1}^{n}\frac{100}{SL_i}\max\left\{SL_i - S_{ij}, 0\right\},\tag{10}$$

$$\text{minimise } y_9 = \sum_{i=1}^{m}\sum_{j=1}^{n}\left|S_{ij} - \tfrac{1}{2}\left(SU_i + SL_i\right)\right|.\tag{11}$$

4.1   The Constraint Set

To simplify the notation, let S_{ij} denote the reservoir storage level by the end of period i of year j, where j can be either a dry, wet or average year. The within-year and between-year storage continuity equations are provided in (12) and (13), respectively:

$$S_{ij} - S_{i-1,j} + R_{ij} = I_{ij} - V_{ij}, \quad i = 2,...,m; \; j = 1,...,n, \tag{12}$$

$$S_{1j} - S_{m,j-1} + R_{ij} = I_{1j} - V_{1j}, \qquad j = 2,...,n, \tag{13}$$

where I_{ij} and V_{ij} are the inflow and evaporation, respectively, in the reservoir, during period i of year j. The limits on release based on ecological and flood control requirements are given by:

$$R_{min} \leq R_{ij} \leq R_{max}, \; i = 1,...,m; \; j = 1,...,n. \tag{14}$$

Due to the elevation of the dam outlet, the storage level has a lower bound of S_{min}, which is the dead pool volume. The structural upper storage limit to prevent dam collapse is given by S_{max}:

$$S_{min} \leq S_{ij} \leq S_{max}, \quad i = 1,...,m; \; j = 1,...,n. \tag{15}$$

Following the assumptions regarding tributary inflow and transit of excess release, the transit release and net water requirement for downstream reaches are given by (16) and (17), respectively:

$$TR_{ij}^{k} = \max\left\{ TR_{ij}^{k-1} + TI_{ij}^{k-1} - D_{ij}^{k}, 0 \right\}, \quad i = 1,...,m; \; j = 1,...,n; \; k = 2,...,K, \tag{16}$$

$$F_{ij}^{k} = \max\left\{ D_{ij}^{k} - TR_{ij}^{k-1} - TI_{ij}^{k-1}, \varepsilon_{F} \right\}, \; i = 1,...,m; \; j = 1,...,n; \; k = 2,...,K, \tag{17}$$

where D_{ij} and TI_{ij} are the total water demand in and total tributary inflow to river reach k, respectively, during period i of year j, and $\varepsilon_{F} > 0$ is a small number to assure the feasibility of (10). Conditions (16) and (17) are adjusted somewhat for the first reach, since in our model the first reach enjoys no tributary inflows and the release from the reservoir substitutes for the transit release.

The magnitude of the discharge at the site of each thermoelectric power plant, on reach k, is obtained as:

$$Q_{ij}^{k} = TR_{ij}^{k-1} \qquad i = 1,...,m; \; j = 1,...,n; \; k > 1; \; k \in K_{T}. \tag{18}$$

The attainable power generation is bounded (1):

$$G_{ij}^{k} \leq a_{i}^{k} + b_{i}^{k} Q_{ij}^{k}, \quad i = 1,...,m; \; j = 1,...,n; \; k \in K_{T}, \tag{19}$$

and non-negativity applies to all variables.

5. Multi-Criteria Methodology

An MCDM problem, minimizing q criteria functions, can be written as:

$$\min \{ \mathbf{y} \} \tag{20}$$

subject to $\mathbf{x} \in X$,

where the y_i are criterion functions, $\mathbf{y} \in Y \subseteq \mathfrak{R}^q$, Y is the criterion (outcome) space, $\mathbf{x}$ is any vector of decision variables and X is the feasible region in decision space.

The ideal $(\mathbf{y}^*)$ and nadir $(\mathbf{y}_*)$ vectors are used in the decision analysis to indicate the relevant ranges of solutions that can be expected during the interactive process of determining the most preferred solution. The nadir vector is an approximate upper bound of the non-dominated set, estimated by collecting the maximum row values in the payoff table for the problem.

We now continue to select an interactive multi-objective linear programming (interactive MOLP) method to solve our mathematical problem.

5.1   Multi-Criteria Solution Procedure

A variety of MOLP solution procedures has been proposed (see, e.g., Evans, 1984; Shin and Ravindran, 1991; Gardiner and Steuer, 1994). We chose the Interactive Weighted Tchebycheff Procedure (IWTP) of Steuer and Choo (1983), due to its desirable mathematical and decision-support related properties. The IWTP is attractive computationally, since it can be implemented using existing commercial optimisation software, since it is guaranteed to provide the decision maker with non-dominated solutions at all times, and since it is user-interactive and provides flexible decision-support for multi-criteria programming problems. A comparative experimental study of interactive MOLP methods by Buchanan and Daellenbach (1987) found the IWTP to be favoured over several competing methods, from the user's viewpoint.

5.2   Implementation

A planning period of nine years is chosen, and the planning scenario that is investigated consists of three cycles of a wet, an average and a dry year. The three-cycle planning scenario used here is examined by Lence *et al.* (1992) and may be considered as a relatively difficult hydrologic scenario. We solve the underlying model with 1,150 constraints and 800 variables using GAMS (Brooke *et al.*, 1988). The MCDM shell is coded in Visual Basic under MS Excel and the time needed to solve each sub-problem on a 133 MHz Pentium PC averages 12 seconds. To assure that the decision maker receives maximum support, storage traces were presented for each candidate solution.

6. Illustration

The decision maker for this example session, experienced in reservoir management, is familiar with the reservoir and the preferences of the dam management.

6.1 Decision Making Experiment

Initially, the ideal vector $\mathbf{y}^*$ and the nadir vector $\mathbf{y}_*$ determined from the initial pay-off table in Table 1 and the selfish solutions S1-1 to S1-9, are discussed with the decision maker. The decision maker recognizes the conflict between the release rate criterion, y_1, recreational criteria, (y_5, y_7, y_8), and power generation, y_4. However, the conflict between the dam stability criterion, y_2, criteria related to the upper storage target (y_5, y_6), and the reservoir stability criterion, y_9, in Table 1 from solutions S1-2 and S1-9 is less immediate. It is expected that a reservoir policy minimising release change would result in a highly varying storage trace, given the variations in inflow.

Table 1. Criterion Vectors S1-1 – S1-9, Selfish Optimization, Iteration 1.

Crit.	Vector									Ideal	Nadir
	S1-1	S1-2	S1-3	S1-4	S1-5	S1-6	S1-7	S1-8	S1-9	Ideal	Nadir
y_1	0%	0%	0%	0%	10%	10%	6%	6%	10%	0%	10%
y_2	1 145	949	1 142	1 224	1 409	1 459	1 203	1 353	1 318	949	1 459
y_3	0%	0%	0%	0%	0%	0%	0%	0%	0%	0%	0%
y_4	52	95	52	38	64	113	56	75	76	38.0	112.8
y_5	90%	90%	90%	90%	60%	60%	90%	81%	90%	60%	90%
y_6	3.7%	5.0%	3.9%	3.6%	2.7%	0.6%	3.9%	4.0%	3.6%	0.6%	5.0%
y_7	14%	18%	11%	14%	22%	45%	0%	0%	25%	0%	45%
y_8	0.3%	0.8%	0.2%	0.5%	1.3%	5.3%	0.0%	0.0%	0.4%	0.0%	5.3%
y_9	1 138	1 937	1 257	1 294	1 307	2 627	1 068	1 145	315	315	2 627

Notice that all vectors presented in Table 2, the result of the first ITWP iteration, are different from the selfish solutions of Table 1. One desirable property of the IWTP is the ability to easily generate non-extreme non-dominated points. The point is here to expose the decision maker to diverse solutions, suggest trade-offs which may have been blocked by early imposed constraints and encourage experimentation with the model.

Table 2. Filtered Sample of Non-Dominated Criterion Vectors 1-1 – 1-7, Iteration 1.

Criterion (all *min*)	Vector						
	1-1	1-2	1-3	1-4	1-5	1-6	1-7
y_1 Flooding	6%	10%	3%	10%	0%	9%	0%
y_2 Stability	1 241	1 453	1 319	1 200	1 313	1 419	1 057
y_3 Supply shortage	0%	0%	0%	0%	0%	0%	0
y_4 Max power shortage	58	113	96	88	55	75	53
y_5 Max SU deviation	81%	60%	65%	90%	73%	60%	90%
y_6 Ave SU deviation	4%	1%	2%	4%	3%	2%	5%
y_7 Max SL deviation	0%	44%	30%	25%	18%	25%	13%
y_8 Ave lower SL deviation	0%	5%	3%	0%	0%	2%	0.16%
y_9 Midtarget deviation	1 158	2 627	2 127	354	940	1 527	1 669

Through the iterations, the decision maker tests different bounds for flooding before dismissing the options, thereby learning an important but less known system property. By such experimentation and preference learning, the dam manager will be able to counter requests for increased release rates, e.g., to meet water demand. The model also highlights the inter-temporal effects of dry-year releases during dry years, especially for recreation in the case of a consecutive dry year.

In order to demonstrate the gradual evolving of preferred solutions and to compare the solutions from the presented model with other works, we finally compare the first and final solutions (1-5, 3-1) and a corresponding GP-solution by Latheef (1991). Figure 2 depicts the storage traces associated with the three vectors and Table 3 gives the criterion values for the corresponding vectors. By comparing the criterion values for the final solution and Latheef (1991), the former is clearly preferred.

Table 3. Criterion Vectors for Solutions 1-5, 3-1 and Latheef (1991).

		Vector		
Criterion (all *min*)		1-5	3-1	Latheef (1991)
y	Flooding	0%	0%	0%
y_2	Stability	1 313	1381	2 058
y_3	Supply shortage	0%	0%	0%
y_4	Max power shortage	55	63	89
y_5	Max SU deviation	73%	76%	72%
y_6	Ave SU deviation	3%	3.5%	3.7%
y_7	Max SL deviation	18%	9%	18%
y_8	Ave lower SL deviation	0%	0.1%	0.2%
y_9	Midtarget deviation	940	940	1 596

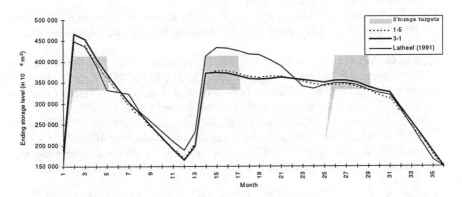

Figure 1. Storage Policies for Solutions 1-5, 2-5, 3-1 and Latheef (1991).

7. Conclusion

If multi-criteria decision support is ever to be widely adopted by practitioners, considerable work has to be spent on the pre-analysis of the decision, the decision-making situation, the information required by the user and the quality and accessibility of the feed-back. Reservoir management is an example of an area which exhibits a potential market for applied work, providing that the system matches the practical requirements of accountability, transparency and versatility. The condition that most previous research in the area belongs to what we call a first-generation, search-oriented methodology, has inspired the formulation of a comprehensive, yet manageable, second-generation methodology.

Corner stones of the methodology are simple and natural: the decision, the decision maker, his context, expectations and authority. By verbally clarifying the actual decision, the crucial conflict, the relevant criteria and system boundaries, a mathematical program is compiled. Using a non-presumptions algorithm (Steuer and Choo, 1983), pedagogical information and a familiar, low-key implementation using spread-sheets, the decision maker can simultaneous learn about and explore the decision. The overall ambition of the tool is consequently downplayed to be a rigorous support, rather than an oracle.

References

Agrell, P. J., B. J. Lence and A. Stam. (1995) *An Interactive Multi-Criteria Decision Model for Reservoir Management: the Shellmouth Reservoir Case.* Working Paper WP-95-090, International Institute for Applied Systems Analysis, Laxenburg, Austria.

Brooke, A., D. Kendrick and A. Meeraus (1988) *GAMS: A User's Guide.* Scientific Press, Redwood City, CA, USA.

Buchanan, J. T. and H. G. Daellenbach (1987) A Comparative Evaluation of Interactive Solution Methods for Multiple Objective Decision Models. *European Journal of Operational Research* 29(3), 353-359.

Can, E. K. and M. H. Houck, Real Time Reservoir Operations by Goal Programming. *ASCE Journal of Water Resources Planning and Management* 110(3), 1984, 297-309.

Evans, G. W. (1984) An Overview of Techniques for Solving Multiobjective Mathematical Programs. *Management Science* 30(11), 1268-1282.

Gardiner, L. K. and R. E. Steuer (1994) Unified Interactive Multiple Objective Programming. *European Journal of Operational Research* 74(3), 391-406.

Goulter, I. C. and R. Castensson (1988) Multi-Objective Analysis of Boating and Fishlife in Lake Sommen, Sweden. *Water Resources Development* 4(3), 191-198.

Haimes, Y. Y. and W. Hall (1974) Multiobjectives in Water Resource Systems Analysis: The Surrogate Worth Trade-Off Method. *Water Resources Research* 10, 615-624.

Haimes, Y. Y., W. A. Hall and W. A. Freedman (1975) *Multiobjective Opimization in Water Resources Systems.* Elsevier, Amsterdam, the Netherlands.

Ignizio, J. P. (1976) *Goal Programming and Extensions.* Lexington, New York.

Latheef, M. I. (1991) *Application of Optimization Techniques for Reservoir Management in Conjunction with Thermoelectric Power Generation.* M.S. Thesis, University of Manitoba, Winnipeg, Manitoba, Canada.

Lence, B. J., M. I. Latheef and D. H. Burn (1992) Reservoir Management and Thermoelectric Power Generation. *ASCE Journal of Water Resources Planning and Management* **118**(4), 388-405.

Shin, W. S. and A. Ravindran (1991) Interactive Multiple Objective Optimization: Survey I - Continuous Case. *Computers & Operations Research* **18**(1), 97-114.

Steuer, R. E. and E. U. Choo (1983) An Interactive Weighted Tchebycheff Procedure for Multiple Objective Programming. *Mathematical Programming* **26**(1), 326-344.

Yang, Y., D. H. Burn and B. J. Lence (1992) Development of a Framework for the Selection of a Reservoir Operating Policy. *Canadian Journal of Civil Engineering* **19**, 865-874.

Yeh, W. W-G. (1985) Reservoir Management and Operations Models: A State of the Art Review. *Water Resources Research* **21**(12), 1797-1818.

Carbon Sequestration and Timber Production: A Bi-Criteria Optimisation Problem

Valeria Ríos[1], Luis Díaz-Balteiro[2] and Carlos Romero[1]

[1] Departamento de Economía y Gestión, E.T.S. Ingenieros de Montes, Avenida Complutense s/n, 28040 Madrid, Spain
[2] Departamento de Economía y Ciencias Sociales Agrarias, E.T.S. Ingenieros Agrónomos, Avenida Complutense s/n, 28040 Madrid, Spain

Abstract. This paper presents a multi-criteria approach to determine the optimal forest rotation age when timber production as well as carbon uptake are considered. With this purpose a utility function with two criteria (net present value and carbon uptake) is optimised over the corresponding production possibility frontier. The unknown optimum is surrogated by several best-compromise solutions which are interpreted in utility terms. The theoretical framework is applied to a coastal forest in British Columbia.

Key words: Forest management, compromise programming, environmental economics.

1 Introduction

Nowadays, it is well accepted that carbon dioxide (CO_2) is the gas most responsible for the greenhouse effect. Global emissions of CO_2 have dramatically increased in the last few years. Since trees absorb CO_2 during growth, storing it as harmless carbon, some authors have recently argued on the importance of considering forests as carbon pools (e.g. Lewis et al. 1996).

Under this context, the methods usually used to determine the optimal forest rotation age are not applicable. For instance, Hartman's approach (1976) requires a flow of valuable services (chiefly recreational activities) measured in monetary units in the analysis and consequently, cannot be used. In cases like ours, we are dealing with a public good (carbon stored in the timber biomass) which is very difficult to measure in monetary units.

This paper presents an approach to determine the optimal forest rotation age when considering timber production and carbon uptake. With this approach the optimal rotation age is given by the optimisation of a utility function with two arguments (timber and carbon) over the corresponding production possibility frontier or transformation curve.

2 Analytical Framework

Let us introduce the following social welfare function:

u (W, S)   (1)

Where: W= net worth associated to timber production (e.g. the net present value (NPV) of the underlying investment) and S= carbon uptake. W and S can be represented by the following functions:

$$W = W(t|\Omega) \qquad \Omega = \{P, i, ...\} \qquad (2)$$

$$S = S(t|\Omega) \qquad \Omega' = \{\gamma, \beta, ...\} \qquad (3)$$

where t is the forest rotation age, Ω and Ω' two sets of exogenous parameters affecting the values of W and S with the following meaning: P= timber price, i= discount rate, ..., and γ = proportion of carbon in timber biomass and β = fraction of timber harvested whose carbon content is released into the atmosphere. If a timber growth curve V= f(t) -which relates timber volume V with rotation age t- is combined with expressions (2) and (3) a production possibility frontier or transformation curve in the W-S space is obtained:

$$T(W,S) = K \qquad (4)$$

It is assumed that the transformation curve has a typical concave shape what implies diminishing marginal rates of transformation. This assumption will be empirically corroborated in Section 4. From this setting the following optima can be derived:

a) **The private optimum**; that is the point where the net worth W reaches a maximum. This optimum W* is obtained by maximising (2) and it is graphically represented by the point where T(W,S) = K cuts the W- axis (see figure 1). The corresponding rotation age will be represented by t_F.

b) **The environmental optimum**; that is, the point where the carbon uptake S reaches a maximum. This optimum S* is obtained by maximising (3) and it is graphically represented by the point where T(W,S) = K cuts the S- axis (see figure 1). The corresponding rotation age will be represented by t_E.

c) **The social optimum**; that is, the point where the social welfare function u(W,S) reaches a maximum over the feasible frontier T(W,S) = K as the following optimisation problem indicates:

Max   u (W, S)

$$\qquad (5)$$

s.t.   T (W, S) = K

Graphically solution of (5) corresponds to the point of tangency between the family of iso-contours u(W,S) = λ and the production possibility frontier T(W,S) = K as shown in the figure 1. The corresponding rotation age will be represented by t_S.

Therefore, the presence of an externality, positive in this case generates a divergence between the private and social optima (see e.g., Turvey, 1963). This divergence provides a market failure what implies an inefficient allocation of resources. This inefficiency can be removed by resorting to a Pigovian premium or subsidy as it was initially proposed by Pigou (1932). To determine this subsidy is necessary the previous determination of the social optimum what requires a reliable estimation of u(W,S). Given the conceptual and operational difficulties in determining a reliable representation of a social welfare function, a sensible surrogate of the true but unknown social optimum will be approximated.

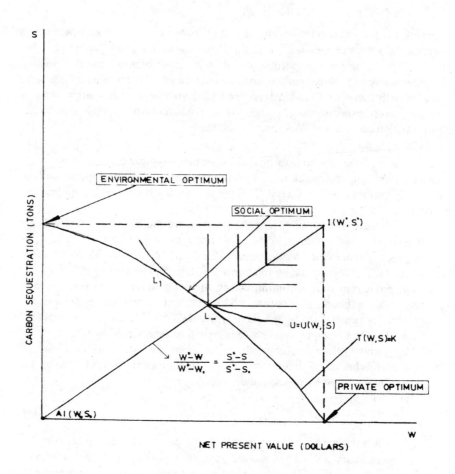

FIGURE 1. PRODUCTION POSSIBILITY FRONTIER AND COMPROMISE PROGRAMMING SETTING IN THE W-S SPACE.

3 A Surrogate of the Social Optimum

Within a compromise programming (CP) approach the private and environmental optima represent the ideal point. Hence, the following compromise model can be formulated (Yu 1973, 1985, Zeleny 1974, 1982):

$$\text{Min } L_\pi = \left[\alpha_1^\pi \cdot \left(W^* - W \right)^\pi + \alpha_2^\pi \cdot \left(S^* - S \right)^\pi \right]^{\frac{1}{\pi}} \tag{6}$$

$$\text{s.t. } T\left(W, S \right) = K$$

where W and S are given by (2) and (3), respectively; π is the metric which define the distance functions and weights α_1 and α_2 play two different roles: (1) normalise both W and S magnitudes to make their aggregation possible and 2) measure the social preferences for W and S, when utility functions are not considered in the analysis. In this paper α_1 and α_2 will only be used for normalising purposes, since utility functions u(W, S) involve preferential weights. It has been proven elsewhere (Ballestero and Romero 1993) that from a shadow price perspective, the best weights α_1 and α_2 for normalising purposes is given by:

$$\alpha_1 = \frac{1}{W^* - W_*} \qquad \alpha_2 = \frac{1}{S^* - S_*} \tag{7}$$

where W_* and S_* are the anti-ideal or nadir values.

By solving (6) for metrics $\pi = 1$ and $\pi = \infty$, a compromise set, such as an arc $[L_1, L_\infty]$ of the transformation curve nearest to the ideal point will be obtained (see figure 1). The following optimisation problem is obtained making $\pi = 1$ in (6), substituting α_1 and α_2 by their values in (7) and after simple algebraic manipulation:

$$\text{MAX} \quad \frac{W}{W^* - W_*} + \frac{S}{S^* - S_*} \tag{8}$$

$$\text{s.t.} \quad T\left(W, S \right) = K$$

By solving (8), the L_1 bound of the compromise set is obtained. This solution provides a compromise between carbon and timber uptake as well as the corresponding forest rotation age, represented by t_1. This solution implies a technological optimum where the normalised sum of timber and carbon benefits are maximised. It is interesting to note that the L_1 bound implies the maximisation of a weighted additive social welfare function u (α_1 W + α_2 S).

The L_∞ bound coincides with the point were path α_1 (W*-W) = α_2 (S*-S) intercepts the transformation curve T(W, S) = K (see Ballestero and Romero 1991). For weighting (7), the L_∞ bound is obtained by solving the following system of equations:

$$\frac{W^* - W}{W^* - W_*} - \frac{S^* - S}{S^* - S_*} = 0 \tag{9}$$

$$\text{s.t.} \quad T\left(W, S \right) = K$$

Note that points on path $(W^*- W) / (W^*- W^*) = (S^*- S) / (S^*- S^*)$ represent balanced allocations between the outputs of timber and carbon uptake. In other words, the forest rotation age, t_∞ corresponding to this solution implies the equality of the weighted discrepancy of the two outputs provided by the forest. It should also be mentioned that the L_∞ bound implies the maximisation of an L-shaped utility function where the maximum deviation is minimised, as represented in figure 1.

To validate the compromise set obtained as the tangency zone for the unknown social welfare function u(W,S) the following theorem demonstrated elsewhere (Ballestero and Romero 1991) and adapted to our context is used:

Theorem I. With any social welfare function u(W, S), the necessary and sufficient condition under which the maximum of u always belongs to the compromise set on every T(W, S)=K is:

Marginal rate of substitution $= \mathrm{MRS}(W,S) = \dfrac{u_W}{u_S} = \dfrac{\alpha_1}{\alpha_2} = \dfrac{S^*-S_*}{W^*-W_*}$ on the L_∞

path, $\dfrac{W^*-W}{W^*-W_*} = \dfrac{S^*-S}{S^*-S_*}$, being u_W and u_S the corresponding partial derivatives.

The condition underpinning Theorem I seems sensible since it simply implies a behaviour coherent with the diminishing MRS law. Notice that we do not state a rigid assumption such as "society's preferences for the balanced position $\alpha_1 (W^*-W) = \alpha_2 (S^*-S)$ are so marked that will represent the social optimum under any circumstances". This would be a restrictive assumption, leading the equilibrium and rotation age to the solution given by the L_∞ point. However, we only assumed conservative behaviour when society is in a balanced position. According to Theorem I, this conservative behaviour is sufficient to determine that the social optimum lies on the compromise set. More details on the economic meaning of Theorem I can be seen in Ballestero and Romero (1994). Morón et al. (1996) demonstrate that there are large families of utility functions holding the condition that underpins Theorem I. Thus the compromise set $[L_1 , L_\infty]$ derived in our problem is reinforced as a good surrogate of the social optimum.

The theoretical framework presented in Sections 2 and 3 will be operationalised in the next section.

4 A Basic Model

The most simple expression to measure the net worth associated to timber production is the following:

$$W = \frac{P \cdot f(t) \cdot e^{-it}}{1 - e^{-it}} \tag{10}$$

All the symbols of expression (10) were defined in Section 2. This expression is known as Faustmann formula and corresponds to the NPV of timber harvest by considering infinite plantation cycles (e.g. Johansson and Löfgren 1985).

Carbon uptake throughout forest rotation can be measured as follows:

$$S \ = \ \gamma \cdot f(t) - \gamma \cdot \beta \cdot f(t) \ = \ \gamma \cdot (1 - \beta) \cdot f(t) \tag{11}$$

Again all the symbols of expression (11) were defined in Section 2.

In order to equalise the private and social optima a premium or subsidy of P_c dollars per ton of CO_2 captured or of γP_c dollars per m^3 of timber added to the growing stock is established. The net present value of the would be given by:

$$\frac{P \cdot f(t) \cdot e^{-it}}{1 - e^{-it}} \ + \ \gamma \ P_c \ \frac{\int_0^t f'(t) \ e^{-it} \ dt}{1 - e^{-it}} \tag{12}$$

The first term of (12) represents timber benefits of the forest owner provided by the market while the second term represents carbon uptake benefits of the forest owner paid by society as a compensation for the positive externality generated by the forest. By maximising (12), the following first-order condition is obtained (see van Kooten $et\ al.$ 1995 for a similar mathematical derivation):

$$(P + \gamma \ Pc) \ \frac{f'(t)}{f(t)} + i\gamma \ Pc = \frac{i}{1 - e^{-it}} \left[P + \gamma \ Pc + \frac{i\gamma \ Pc}{f(t)} \int_0^t f(t) \ e^{-it} \ dt \right] \tag{13}$$

By substituting in (13), γ, P and i by their values and t by t_1 and t_∞ respectively, the corresponding optimal Pigovian subsidies P_c ($\pi = 1$) and P_c ($\pi = \infty$) are obtained. The unknown optimum subsidy lies on the closed interval [P_c ($\pi = 1$), P_c ($\pi = \infty$)] when the MRS between both outputs behaves accordingly to Theorem I.

The main ideas developed in this paper will be illustrated with the help of the following data adapted from van Kooten $et\ al.$ (1995) for a coastal forest in British Columbia:

$\gamma = 0.2$ tons of CO_2/m^3 of timber; $\beta = 0$; $i = 0.05$; $P = 25$ dollars/m^3
$f(t) = 0.000573 \cdot t^{3.7819} \cdot e^{-0.030965t}$

By substituting the above data in (10) and (11) the following expressions are obtained:

$$W = \frac{0.0143 \cdot t^{3.7819} \cdot e^{-0.080965 \cdot t}}{1 - e^{-0.05 \cdot t}} \tag{14}$$

$$S = 0.01146 \cdot t^{3.7819} \cdot e^{-0.030965 \cdot t} \tag{15}$$

By maximising (14) and (15) separately and then computing the value of each objective at each of the optimal solutions the following pay-off matrix is obtained:

Table 1. Pay-off matrix for net present value and carbon uptake

	Net present value (W) dollars / ha	Carbon uptake (S) ton / ha
Net present value	751	46
Carbon uptake	57	204

The first row of the above matrix represents the private optimum (Faustmann solution) with a rotation age t_F of 43 years, while the second row represents the environmental optimum with a rotation age t_E of 122 years (see Table 2 and figure 2).

To approximate the frontier $T(W,S) = K$, we can maximise (14) by considering (15) as a parametric restraint (i.e.. the constraint method). In this way a set of fifteen efficient mixes in the W-S space were obtained. From this cloud of points, the following frontier or transformation curve was obtained by statistical fitting:

$$T(W,S) = 79.98 \cdot W + 369.80 \cdot S - 0.04 \cdot W^2 - 0.86 \cdot S^2 - 0.37 \cdot W \cdot S = 40000 \quad (16)$$

According to (8), the L_1 bound of the compromise set will be obtained by solving the following optimisation problem:

$$\text{Max} \quad \frac{W}{751-57} + \frac{S}{204-46}$$

s.t. $W = \text{expression (14)}; \quad S = \text{expression (15)}$ $\qquad\qquad (17)$

expression (16)

Solution to model (17) is shown in row 3 of Table 2 and in figure 2.

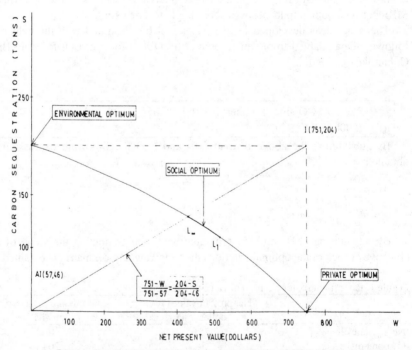

FIGURE 2. PRIVATE, ENVIRONMENTAL AND SOCIAL OPTIMA (COASTAL FOREST IN BRITISH COLUMBIA).

According to (9) the L_∞ bound of the compromise set will be obtained by solving the following system of equations:

$$\frac{751 - W}{751 - 57} = \frac{204 - S}{204 - 46}$$

s.t. $W = \mathrm{expression}\ (14);\quad S = \mathrm{expression}\ (15)$ (18)

        $\mathrm{expression}\ (16)$

Solution to model (18) is shown in the last row of Table 2 and in figure 2.

Table 2. Private, environmental and social optima

	Net present value dollars / ha	Carbon uptake tons / ha	Timber production m^3 / ha	Forest rotation age years
Private optimum	751	46	231	43 (t_F)
Environmental optimum	57	204	1019	122 (t_E)
Social optimum (L_1 bound)	592	102	509	62 (t_1)
Social optimum (L_∞ bound)	441	134	669	73 (t_∞)

Therefore, the social optimum rotation age is comprised between 62 and 73 years. It seems that the inclusion of the positive externality generated by carbon sequestration increase rotation age with respect to traditional Faustmann solutions. By applying our data to equation (13), an optimal Pigovian subsidy comprised between 132 dollars / ton CO_2 for the L_1 solution and 252 dollars / ton of CO_2 for the L_∞ solution was obtained. This subsidy is equivalent to 26-50 dollars / m^3 of timber added to the growing stock.

Acknowledgements

Comments raised by David Ríos-Insua and by two referees are appreciated. English language was checked by Christine Méndez. This research was supported by "Comision Interministerial de Ciencia y Tecnología" and by "Consejería de Educación y Cultura, Comunidad de Madrid"

6. References

BALLESTERO, E. and ROMERO C. (1991), "A theorem connecting utility function optimization and compromise programming", *Operations Research Letters* **10**, 421-427.

BALLESTERO, E. and ROMERO C. (1993), "Weighting in Compromise programming: a theorem on shadow prices", *Operations Research Letters* **13**, 325-329.

BALLESTERO, E. and ROMERO C. (1994), "Utility optimization when the utility function is virtually unknown", *Theory and Decision* **37**, 233-243.

HARTMAN, R. (1976), "The harvesting decision when a standing forest has a value, Economic Inquiry " **14**, 52-58.

JOHANSSON, P-O. and LÖFGREN, K-G. (1985) *The Economics of Forestry and Natural Resources*. Basil Blackwell, Oxford.

LEWIS, D.K., TURNER, D.P. and WINJUM J.P. (1996), "An inventory-based procedure to estimate economic costs of forest management on a regional scale to conserve and sequester atmospheric carbon", *Ecological Economics* **16**, 35-49.

MORÓN, M.A., ROMERO C. and RUIZ DEL PORTAL F.R. (1996), "Generating well-behaved utility functions for compromise programming", *Journal of Optimization Theory and Applications* **91**, 643-649.

PIGOU, A.C. (1932) *The Economics of Welfare*, 4th de. Macmillan, London.

TURVEY, R. (1963), "On divergences between social and private cost", *Economica* **30**, 309-313.

VAN KOOTEN, G.C., BINKLEY C.S. and DELCOURT G. (1995), "Effect of carbon taxes and subsidies on optimal forest rotation age and supply of carbon services", *American Journal of Agricultural Economics* **77**, 365-374.

YU, P-L. (1973), "A class of solutions for group decision problems", *Management Science* **19**, 936-946.

YU, P-L. (1985), *Multiple-Criteria Decision Making. Concepts, Techniques and Applications*. Plenum Press, New York.

ZELENY, M. (1974), "A concept of compromise solutions and the method of the displaced ideal", *Computers and Operations Research* **1**, 479-496.

ZELENY, M. (1982), *Multiple Criteria Decision Making*. McGraw-Hill, New York.

Part 7
General Applications of MCDA

Integrated MCDA: A Simulation Case Study

Valerie Belton and Mark Elder
Management Science, University of Strathclyde, Glasgow, Scotland[1]

Abstract. Despite the fact that the origins of the MCDM school lie in extensions of linear programming to problems with multiple objectives, multicriteria analysis, in particular methods for choosing between discrete alternatives, tend to be viewed by Operational Research/Management Science (OR/MS) practitioners as distinct from other methods of OR/MS. In this paper we argue that MCDA should be viewed as an integral part of all problem solving methodologies, not as a set of tools applicable only to certain categories of problem. We will illustrate our argument by reference to a problem faced by the management of a regional UK airport, namely the allocation of check-in desks to airlines. In this case study simulation and multicriteria analysis have been integrated, using the softwares Simul8 and V•I•S•A , to provide a decision support tool for the airport Planning Manager. The DSS will be used in the short term to allocate check-in time slots in a way which achieves effective management of queues and utilization of the desks; in the longer term it will inform the allocation of take-off times.

Keywords, Simulation, Multi-attribute value theory, Integrating methods

---

[1] *Address for correspondence*
Dr Valerie Belton, Management Science, University of Strathclyde,
40 George Street, Glasgow G1 1 QE, UK
Tel: 44 141 548 3615, Fax: 44 141 552 6686, Email: val@mansci.strath.ac.uk

1 Introduction

Amongst the many influences on and contributions to the development of multicriteria analysis, a term which we use to embrace all multicriteria approaches, it is possible to identify three clear threads - the MCDM/MOLP school having its origins in extensions of linear programming to problems with multiple objectives, the US dominated field of decision theory, incorporating multiattribute value/utility theory, and the French led school of outranking methods. Despite the fact that, at least, the first two of these represent apparent attempts to broaden the scope of single objective modelling approaches to cope with the added complexity which is characterised in practice by multiple, conflicting objectives, multicriteria analysis, in particular methods for choosing between discrete alternatives (which we will refer to as MCDA methods), is generally viewed by Operational Research / Management Science (OR/MS) practitioners as distinct from other methods of OR/MS. In this paper we argue that multicriteria analysis should be viewed as an integral part of all problem solving methodologies, not as a set of tools applicable only to certain categories of problem. In the following section we expand on this argument and comment briefly on the extent to which integration is reflected in the literature. In section 3 we describe a case study in which simulation and multicriteria analysis are used in an integrated way to examine the problem of allocating check-in desks to airlines at a UK regional airport. In section 4 we reflect on the lessons from this experience and suggest avenues for further investigation.

2 Multicriteria Analysis and OR/MS

In considering the extent of integration of multicriteria analysis and OR/MS it is helpful to distinguish between MOLP methods for the exploration of a continuous solution space in which alternatives are implicitly defined by a set of constraints (MCDM), and methods for the evaluation of and choice between a finite set of discretely defined alternatives (MCDA). It is the second group of approaches, comprising multi-attribute value theory, AHP, outranking methods, etc., which are of particular interest in this paper.

MOLP and OR/MS
We have already noted that the MCDM school has its origins in the extensions of linear programming to problems with multiple objectives, thus, in one sense we can view MOLP as the integration of linear programming and multicriteria analysis. There are many reported applications of MOLP, in particular of goal programming. However, rather than being viewed as an integrated part of optimisation, we feel that MCDM has developed as a separate area of study, as something that is distinguished from single objective optimisation rather than a natural extension of it. This is a concern which is echoed in current discussions

in the multicriteria community, as highlighted by Korhonen (1997) in the Opinion Makers Section of the newsletter of the European Working Group (EWG) for "Multicriteria Aid for Decisions".

MCDA and OR/MS

A cursory examination of the literature and practice of OR/MS suggests that MCDA is viewed as a stand-alone body of theory, or set of tools, in the same way as, say, simulation or forecasting methods. This sense of isolation is reflected in published accounts of both practical applications and theoretical developments. The latter tend to focus on introspective issues and the majority of reported practical interventions are accounts of "stand-alone" applications of MCDA methods. This is supported by a review of the contents of the Journal of Multi-Criteria Analysis and the published proceedings of these Conferences (International Special Interest Group on MCDM). Only a small number of studies draw on, or incorporate MCDA within the broader repertoire of OR/MS approaches to problem structuring and analysis. Examples are: the paper by Spengler and Penkuhn (1996) describing a DSS which combines a flowsheet-based simulation with multicriteria analysis: work by Macharis (1997) which incorporates a multicriteria choice rule within a system dynamics analysis of transport policy: and the study by Gravel et al (1991) using multicriteria analysis to evaluate production plans developed using simulation. It is interesting to note that the isolation is bi-directional; MCDA analysts do not seek to make use of other OR/MS methodologies but neither do OR/MS practitioners specialising in other methodologies draw extensively on MCDA methods. This issue is addressed by Belton and Pictet (1997) in a later Opinion Makers Section of the EWG newsletter, commenting on the views expressed by Korhonen (1997) and others.

Clearly there is an important role for MCDA in "stand-alone" applications and we do not wish to detract in any way from this. Many decisions present themselves as a need to choose between a number of well specified alternative courses of action - tender evaluation, personnel selection and equipment selection to give just a few examples. However, we feel that MCDA can and should be much more widely used by OR/MS practitioners in conjunction with other analytic tools.

Ways of combining MCDA and OR/MS

There are many ways in which the use of MCDA in conjunction with other OR/MS methods can lead to mutual enhancement. Exploring these possibilities, with the aim of expanding the influence of MCDA and its acceptance as a valuable tool for OR/MS practitioners, has been a particular interest of one of the authors. This work has identified a number of ways in which MCDA can work synergistically with other approaches: Belton, Ackermann and Shepherd (1995) discuss the use of the SODA methodology together with MCDA to give an integrated approach from problem structuring through to evaluation: Belton and

Elder (1996) describe a visual interactive DSS for production scheduling in which multicriteria analysis is embedded in the scheduling algorithm: Belton and Vickers (1993) discuss the parallels between MCDA and DEA and put forward the suggestion that an MCDA interpretation of DEA can facilitate understanding. We refer to these three "combinations" as *integrating, embedding* and *exploring parallels*. In this paper we explore further possibilities for the *integration* of MCDA and other approaches, focusing here on simulation.

3 Background to the Case Study

Most airports in the world face the problem described here - that of allocating check-in desks to flights and airlines. We describe ongoing work to assist the planning manager of a major UK airport (the 5th busiest in the UK).

Significance of the problem
Like many decisions facing any management this decision is influenced by many factors: it is affected by, and itself influences, other decisions. The average time passengers queue to check-in for each flight along with the worst case of queuing time for each flight are key factors which drive the decision. However, there are other considerations for example: the length of the queue (which is only partially correlated with queuing time): the space allocated to each airline: and the details of queuing times for different types of flight such as business / holiday, and domestic / international / long haul flights. The length of time prior to takeoff that a check-in can be opened is also relevant. The aim of the planning manager is to allocate flights to check-in desks in a way, which achieves effective management of queues and utilisation of the desks. In doing this he aims to meet the airports own performance measures, which are aligned with meeting the objectives of the airlines and passengers.

At one level this is a weekly decision about which flights will use which check-in desks; however, it also has a longer-term impact on the allocation of takeoff times to flights. Airlines state their preferences for takeoff times at a bi-annual conference at which all international airlines and airports get together to coordinate plans for timetables which will be compatible with both the airlines' commercial demands and the airport capacities. Runway capacity is clearly an important factor in determining capacity, but the capacity of the check-in hall is in fact much more significant for our decision maker. For example, the check-in facilities would have some difficulty in coping if more than four 747's (or aircraft of equivalent size) were scheduled to leave within half an hour, although this presents no problems with respect to runway capacity.

The decision is taken weekly by a planner who is directly responsible to a member of the airport's senior management team. It is this member of the management team who has to deal with subsequent complaints, for example: complaints from airlines unhappy with their allocation: complaints from passengers about the length of queues: complaints from passengers about missing flights because of queuing times: or complaints about the overcrowding of the check-in hall. The airport and its parent company place a very high emphasis on customer service and getting this decision right is seen as a key element in the process of increasing the use of the airport (and thus its profitability). Consequently, the responsible member of the management team monitors the situation very closely.

The physical environment
Each check-in desk is identical. The sign on each check-in desk is a video screen, which can be easily and automatically changed to show the currently allocated airline and, if appropriate, flight number. The schedule for the coming week is programmed once the allocation decision has been taken.

There are 34 check-in desks and around 800 passenger carrying flights leave the airport in a peak week. Around 40 airlines operate from the airport and of these only 2 or 3 have the need for a permanently allocated desk. Others operate only a few flights each day, or in some cases 1 or 2 flights per week. A holiday charter flight to a long-haul destination, with a large number of passengers, needs to open one check-in up to five hours before takeoff time. For the period between 3 to 2 hours prior to departure there may be 4 desks allocated to the flight, the number reducing as the flight time approaches. The consequence of this pattern of demand is that most check-in desks must be reallocated continually.

Some airlines are prepared to have several of their flights sharing one or more check-ins, but others prefer to dedicate check-ins to particular flights. When an airline is allocated several check-in desks at one time they have a strong preference for all of these to be adjacent. Some airlines have a strong preference to keep the same check-in desks for the entire week, or at least to have desks in a group which passengers will come to recognize as approximately the area to use for that airline. Some airlines are prepared to pay extra to be allocated more desks than they really deserve in order to reduce queues for their passengers or, in the case of one airline, simply to make it appear that they are a larger airline than is actually the case!

Current decision process
The decision on how to allocate check-in desks is taken by the same person each week drawing on past experience. Prior to this study the only tool available to the planner was a spreadsheet into which he entered proposed check-in allocations taking into account scheduled take-off times. The spreadsheet is illustrated in Figure 1: each row corresponds to a check-in desk and each column to a 30 minute time interval. The planner enters into each cell in the matrix the airline to be allocated to a given desk at the specified time. It is simply a visual presentation tool; no analysis is provided. At certain times of year there will be few changes from one week to the next, but changing schedules necessitate some

change in most weeks. When the planner arrives at an allocation he is satisfied with the spreadsheet is emailed to the airlines and other interested parties. Airlines can quickly see their allocation and are quick to complain if they believe this to be unfair.

However, without the support of an analytic technique it is difficult to assess the likely impact of an allocation of check-in desks on the queuing times or queue lengths experienced by passengers. Thus it is difficult to judge if airlines are justified in complaining and difficult to assess the impact of a change in the allocation resulting from negotiation. Of course, the true effectiveness of a schedule can only be assessed retrospectively at the end of the week due to the stochastic nature of the problem; however, the aim of this study is to provide a tool which will give an indication of performance. This will assist the planner in making the initial allocation and in subsequent negotiation with the airlines.

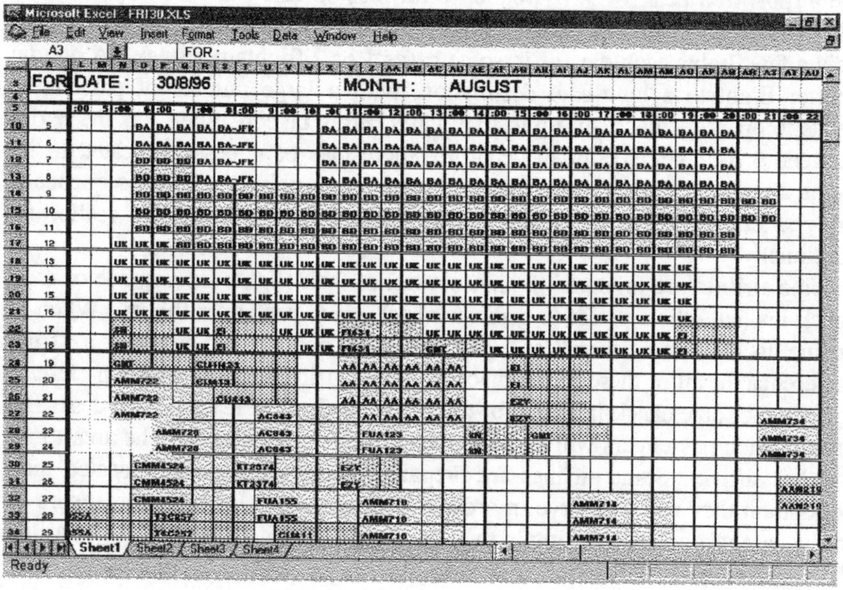

Figure 1 - illustrative allocation of airlines to check-in desks

When asked to help with this problem we felt that a discrete event visual interactive simulation model would be an appropriate analytic tool. Our aim was not to provide a tool which did the planner's job for him, but one which facilitated and supported him in this task. Given a particular allocation schedule the simulation would utilise statistical distributions for arrival and service times of passengers to predict queuing patterns. The visual representation would facilitate communication with the airlines and other interested parties about the impact of particular allocations.

A Note about Simulation

Developing a simulation model nowadays is made easy by the availability of many high-level, visual interactive software systems. The developer no longer has to concern themselves with the details of event-by-event coding; the days of models which were difficult and time consuming to build and cumbersome to use are now past. It is now possible for the developer to work with a client interactively developing a model which can be run immediately. The modeller can concentrate on helping the client rather than on building the model. An initial model can be built very quickly, to be refined at a later stage if appropriate.

A simulation is comprised of a logical model of the system being investigated, incorporating as many rules governing the behaviour of the system as are necessary to give an acceptable representation. The model also captures the stochastic nature of the system, representing variability by theoretical or empirical statistical distributions. Typically the user of a simulation model wants to test a number of possible scenarios and to gather information on the performance of the system in each instance. The simulation model is run a number of times for each scenario to generate a distribution of performance measures.

The Model

The simulation model was built using the simulation software system "SIMUL8" (Elder, 1995), to provide information for the planner on the expected performance of the system under different allocation schedules. It models the arrival of passengers at the airport, their decision about which queue to join and the time taken to check-in each passenger or group of passengers. The number of passengers, arrival times and check-in times, are generated randomly utilising available data for each flight type, i.e. business / leisure and domestic / international / long haul. The main simulation window, built together with the planner, depicts the physical layout of the check-in hall and the flow of passengers through it. A view of the model, annotated to assist the reader, is shown in Figure 2. When the model is running the display shows the length of each of the queues and the planner can see immediately if a particular allocation results in excessively long queues. This facility is important in helping the user to understand and informally validate the model, thereby developing ownership of the model and building confidence in the results it provides (Belton and Elder, 1994).

The planner inputs to the model a particular allocation schedule of airlines and flights to check-in desks, such as the one illustrated in figure 1. The simulation takes this information from the spreadsheet and executes multiple runs to provide predictions of the following performance measures:

- Number of flights for which any passenger queues more than 12 minutes
- Number of flights for which more than 3% of passengers queue more than 12 minutes
- Number of flights for which the average queuing time is greater than 4 minutes
- Number of flights for which any queue length exceeds 10 passengers
- Percentage of airline B's passengers who queue for more than 6 minutes
- Percentage of airline M's passengers who queue for more than 6 minutes
- Percentage of airline S's passengers who queue for more than 9 minutes

These measures were derived in discussion with the planner; the first two are ones which are specified by the parent company as corporate objectives; the last three relate to specific operators and on the basis of the planner's experience are indicators of the likely level of complaints from the airlines; the remaining two measures are regarded as good indicators of general performance. At the time of writing the model is still under development and as the planner gains more experience with its use he may wish to add to, or change these. Such changes are easily incorporated.

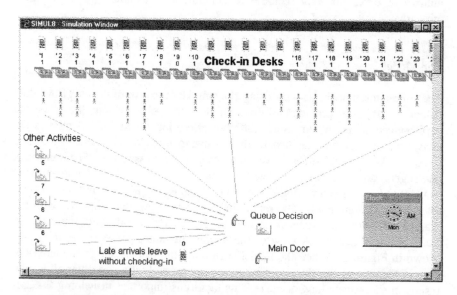

Figure 2 Main screen of the simulation model

The model also provides the planner with the following information which helps him to see how a given plan might easily be improved:

- The 10 flights which have the longest average queuing times
- The 10 flights which have the longest maximum queuing time

The planner can change the allocation schedule and re-run the simulation whenever he thinks he has a new schedule, which is worth evaluating.

The simulation model was initially intended as a device to be used by the planner to find a good schedule. It was only later, when use of the model caused the investigation team to realise that the concept of a "good schedule" was a complex multicriteria issue, that the integration with MCDA became important.

Integration of Simulation and MCDA

From the discussion in the previous section of this paper, it is clear that the problem is not one for which there is an "optimal" solution. The planners are interested in multiple measures of performance, which cannot easily be reduced to a single dimension. Having been provided with more information to use as a basis for deciding how to allocate check-in desks, the planners are now faced with the multicriteria problem of identifying a preferred allocation. We decided to extend the system to draw on existing software for MCDA by integrating the simulation system (SIMUL8) with the multi-criteria decision support system, V•I•S•A.

V•I•S•A is a system which supports the use of a multiattribute value function for multicriteria decision support. As with SIMUL8, the focus is on providing visual interactive facilities for problem representation and analysis. V•I•S•A utilises simple, easy to understand, visual displays to reflect back information and act as a catalyst for learning about the problem and about ones own and others' values. The visual interactive interface provides a powerful vehicle for exploring the implications of uncertainty about values and priorities.

The alternatives to be considered here are the different allocation schedules suggested by the planners and the criteria for the evaluation of these schedules are initially taken to be the factors outlined above. Information about the performance of each allocation schedule is passed automatically to the multi-criteria model as the scenario is run through the simulation.

The planners spend about one day each week drawing up the schedule for the following week; the following section illustrates the analysis of six schedules which they felt were worth considering further for that particular week.

The Multicriteria Analysis

Figure 3 illustrates the initial V•I•S•A model of the problem used to evaluate six proposed schedules.

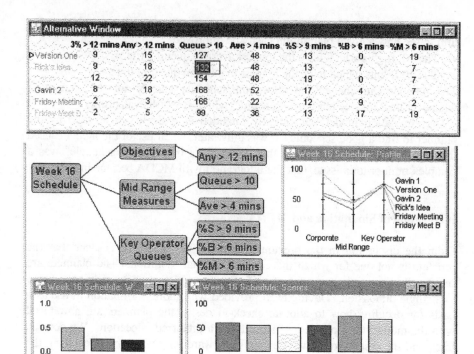

Figure 3   **Illustration of the V•I•S•A model**

The seven measures outlined earlier are grouped into the three families indicated to give the value tree shown. The Alternatives Window shows the actual performance of each of the six test schedules against these measures; this data is imported directly from the simulation model. These performances are converted, within the V•I•S•A model, into value scores on a 0 to 100 scale (where 0 and 100 are defined as worst and best possible performance against each factor). Each of the criteria in the value tree are weighted to reflect acceptable trade-offs to the decision maker - the cumulative weights reflecting the importance of the three top-level criteria are shown in the window on the bottom left. The window on the bottom right shows the overall score for each of the six schedules and the window above that shows how that score is made up from performance on the three top-level criteria. As can be seen from Figure 3, in this illustration the schedule named Friday Meeting is rated most highly overall, with Friday meeting B close behind. The profile graph shows that these two schedules perform substantially better than the others against "Corporate Objectives", they are also the top two performers against "Mid-Range Measures". Poorer performance against "Key Operator Queues", particularly in the case of Friday Meeting B, is mitigated by the lower weight given to that criterion. The planners can use the software to interactively investigate the effect of changes to the criteria weights on the evaluation of the schedules. In this example, the weight on "Corporate Objectives" has to be significantly reduced to change the overall evaluation.

When the planner creates a new schedule he wishes to consider, the information is picked up from the spreadsheet by the simulation package, which executes multiple runs and automatically creates a new alternative, together with performance measures, in V•I•S•A. However, the multicriteria analysis is not restricted to the measures generated by the simulation model; other factors may be introduced into the value tree and performance measures entered directly into V•I•S•A. For example, the planner may wish to take into account his subjective expectation of the likelihood that a schedule will be accepted by the airlines without complaint.

4 Reflections

This study is still ongoing and it is likely that both the simulation model and the multicriteria evaluation will be developed further in consultation with the client, a process that is made easy by the nature of the two softwares used. However, even at this stage of the process we feel that a number of important lessons have been learned.

The linking of simulation and multicriteria analysis provides a decision support tool which spans the five categories of learning we have described elsewhere (Belton and Elder, 1994) as illustrated in Figure 4.

Simulation provides the link between decision space and solution space (discovery), whilst multicriteria analysis provides the link between solution space and value space (clarification and explication). The overall analysis promotes understanding and creativity by allowing the user rapidly to try out new ways of doing things and immediately to see the consequences, which may lead the decision maker to change their view.

This case study began with the simulation model and the multicriteria analysis was incorporated at a later stage. It may have been the case that a consideration of the multicriteria aspect of the problem, in particular the elicitation of relevant criteria, could have usefully informed the building of the simulation model. We anticipate that working with the V•I•S•A model will prompt the planners to think of other factors they would like to include in the evaluation, prompting an update of the simulation model in order to collect the appropriate performance measures.

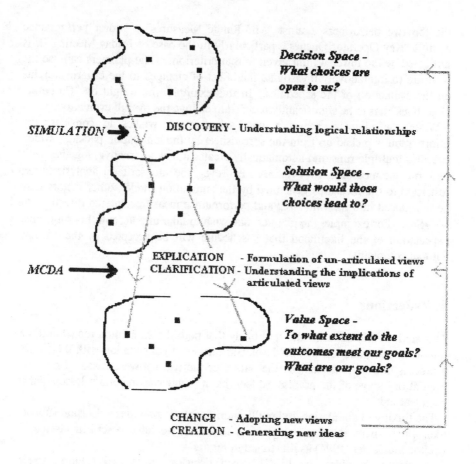

Decision Space -
What choices are
open to us?

SIMULATION ➔ DISCOVERY - Understanding logical relationships

Solution Space -
What would those
choices lead to?

MCDA ➔ EXPLICATION - Formulation of un-articulated views
CLARIFICATION - Understanding the implications of
articulated views

Value Space -
To what extent do the
outcomes meet our goals?
What are our goals?

CHANGE - Adopting new views
CREATION - Generating new ideas

Figure 4:   A Model of Learning

We find it rather surprising that the integration of simulation and multicriteria analysis has not occurred more often. A simulation model more often than not provides the user with multiple performance measures which are typically handled intuitively by the decision makers: MCDA is a natural extension of the analysis which can draw together all the elements and enable the decision makers to fully appreciate the significance of the analysis. One barrier to integration may be software related - although the link is easy to conceptualise it has not been easy to implement using two separate software tools. However, this barrier has not been insurmountable in the past and it would not be difficult for simulation software suppliers to integrate a tool, or tools, for MCDA. It would, perhaps, be instructive for the MCDA community to reflect on why this has not happened - but we will leave this as an issue for debate.

REFERENCES

BELTON, V., ACKERMANN, F. and SHEPHERD, I. (1997) "Integrated support from problem structuring through to alternative evaluation using COPE and V•I•S•A" Journal of Multiple Criteria Analysis

BELTON, V. and ELDER, M.D. (1994). "Decision Support Systems - Learning from Visual Interactive Modelling", Decision Support Systems (12), 355-364

BELTON, V. and ELDER, M.D. (1996) "Exploring a Multicriteria Approach to Production Scheduling", Journal of the Operational Research Society (47), 162-174

BELTON, V. and PICTET, J. (1997) Opinion Makers Section, Newsletter of the European Wirkin Group on "Multicriteria Aid for Decisions", Series 2, Number 11

BELTON, V. and VICKERS, S.P. (1993) "Demystifying DEA: A Visual Interactive Approach Based on Multicriteria Analysis", Journal of the Operational Research Society (44), 883-896

GRAVEL, M., MARTEL,J.M., NADEAU, R., PRICE, W. and TREMBLAY, R. (1991) "A Multicriterion View of Optimal Resource Allocation in Job-Shop Production" presented to IFORS SPC1 on Decision Support Systems, Bruges.

KORHONEN, P. (1997), Opinion Makers Section, Newsletter of the European Wirkin Group on "Multicriteria Aid for Decisions", Series 2, Number 10

MACHARIS, C. (1997) "Hybrid Mmodelling: System Dynamics Combined with Multicriteria Analysis", Vrije Universitet Brussels, Centre for Statistics and Operational Research, Working Paper.

SPENGLER, T. and PENKUHN, T. (1996) "KOSIMEUS - A Combination of a Flowsheeting Program and a Multicriteria Decision Support System" accepted for publication in the Proceedings of SAMBA II Workshop (State of the Art Computer Programs for Material Balancing), Wuppertal Institut

Viewing a Plant Layout Problem as a Multiobjective Case to Enhance Manufacturing Flexibility

H.T. Phong[1] and M.T. Tabucanon[2]

[1] School of Industrial Management
HoChiMinh City University of Technology (HUT),
268 Ly Thuong Kiet St., Dist. 10, HoChiMinh City, Vietnam.
[2] Industrial Systems Engineering Program, School of Advanced Technologies,
Asian Institute of Technology.
P.O. Box 4, Klong Luang, Pathumthani 12120, Thailand.
Email: phong.ho@bdvn.vnmail.vnd.net , mtt@ait.ac.th

Abstract. Traditionally, plant layout is viewed as a single-objective optimization problem, oftentimes minimizing total transport distance of in-process materials. This is no longer valid if we were to remain efficient in the contemporary competitive environment. Three objectives which are important in setting up the layout of a plant are material handling cost, physical closeness of machines and other facilities, and production routing flexibility especially in a Flexible Manufacturing Systems Environment. The three-objective nonlinear programming model is solved using Genetic Algorithms.

Keywords. Multiobjective Nonlinear Programming, Genetic Algorithms, Plant Layout

1 Introduction

Going over the operations research and industrial engineering literature would reveal that most research works have paid attention to operational problems - planning, scheduling, sequencing, control and distribution. The problem of design whether of product or of a facility has not been adequately analyzes. In multiobjective decision analysis of these operational problems, one should not lose sight of the fact that conflict of objectives could be partly or fully due to poor design. This conflict could probably be avoided partially or completely if the design was better made taking into account the issues and problems encountered at the operational phase. For example, manufacturing flexibility which is the focus of attention of many researchers, could be hindered or restricted by the *non-flexible* nature of the plant layout. The layout of the plant must be conducive to a flexible manufacturing operation.

The Concurrent Engineering (CE) paradigm is an excellent development in design. But hitherto, the thrust of application has been product (consumer and capital goods) design. Kusiak (1993) summarized CE into mainly two approaches: human oriented and computer-oriented approach. Smith and Browne (1993) pointed out that the design can be understood in terms of five concepts prominent in design thinking and practice: goals, constraints, alternatives, representations and solutions. Dowlatshahi (1992) also developed a five-step algorithm containing attribute-based utility values for CE design. Thurston (1990), Thurston and Locascio (1993) developed an integrated approach to cover the CE concept by applying rule-based system with multiple utility assessment.

Concurrent Engineering, as a concept, ought to be propagated in facility design as well, particularly in plant layout. It is along this vein that this paper is attempting to relate by way of viewing a plant layout problem as a multiobjective optimization case to enhance the flexibility of manufacturing.

There have been attempts to tackle plan layout as a multicriteria decision making problem. Francis and White (1974) and Rosenblatt (1979) proposed considering two objectives, that for minimizing material handling cost (material handling cost is proportional with total workflow) and maximizing closeness ratings. Fortenberry and Cox (1985) simplified the problem as a single criterion problem that is minimizing the cost when distance has been weighted by closeness rating scores.

Malakooti and D'Souza (1987) and Malakooti (1989) introduced the multiobjective aspect for the Quadratic Assignment Problem (QAP) and used a heuristic method to generate efficient points associated with given weights. Urban (1987) used the same criteria as proposed by Rosenblatt (1979) and Fortenberry and Cox (1985) but used the modified formulation of additive models for work flows and closeness rating scores.

Using Expert System Approach (ESA) with priorities was discussed by Malakooti and Tsurushima (1989). They proposed a procedure that allow the decision maker to interact with the system. On the other hand, Heragu and Kusiak (1988) stated that the PLP in Flexible Manufacturing Systems (FMS) can not be formulated as QAP because the machine sizes are varied from machine to machine; they proposed heuristic algorithms to solve the problem for single row layout. ESA was also used by Heragu and Kusiak (1990) for PLP.

Abdou and Dutta (1990) used ESA to solve the Facility Layout problem considering factors that effect the layouts including qualitative criteria that are difficult to quantify such as safety, noise, aesthetics, ease of supervision, floor space utilization, layout flexibility, ease of expansion, organization structure.

Partovi and Burton (1992) used five-level rating to evaluate the criteria, the work based on the Analytic Hierarchy Process (AHP) of Saaty (1980). An attempt to solve Facility Layout problem by heuristic improvement procedure was proposed by Harmonosky and Tothero (1992); a practical example of large plant with 36 departments was treated. Another application of AHP to Facility Layout problem was proposed by Shang (1993), who tackled the qualitative aspects of Facility Layout problem by AHP, obtaining the relative weights for each qualitative criterion and the QAP is formulated.

2 Design Objectives for Plant Layout Problem

2.1 Material Handling Cost

The first criterion is minimizing the material handling cost. This is equivalent to minimizing total transport/travel time between machines in the layout. Most of the early works have considered that the material handling cost between machine i and machine j is proportioned with $f_{ij}d_{ij}$ where f_{ij} is work flow from machine i to machine j and d_{ij} is physical distance between the two machines.

The distance d_{ij} is considered as rectilinear distance if the material handling system to be used is Automated Guided Vehicles (AGVs) or Euclidean distance if the material handling system is robot. If robot is placed at point O, and machines are placed at points A, B, C etc. there are some specific geometrical relationships (Fig. 1):

$$(x_A - x_O)^2 + (y_A - y_O)^2 = (x_B - x_O)^2 + (y_B - y_O)^2 \qquad (1)$$

(A and B are on the same circle)

$$(x_A - x_O)^2 + (y_A - y_O)^2 = (x_C - x_O)^2 + (y_C - y_O)^2 \qquad (2)$$

(A and C are on the same circle)

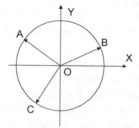

Fig. 1. Geometrical relationship if machines are served by robot

Let us look at an example in Fig. 2, the layout B is considered better than layout A in terms of having lower total travel time. We shall use the measurement of travel time as $(f_{ij} \cdot d_{ij})$ in computing the material handling costs taking into account both rectilinear and Euclidean distances.

2.2 Total Closeness Rating Scores

The second objective is to maximize the total closeness rating scores (TCR). It is an attempt to quantify the qualitative measurements of relationships among facilities. We consider the use ratings r_{ij} as A, E, I, O, U and X to indicate, respectively, that the necessity for two facilities to be located close to each other (or adjacent) is absolute, essential, important, ordinary unimportant and undesirable. The example in Fig. 3 shows that layout B is better than layout A in terms of having more closeness rating scores. According to this definition, numerical values are assigned to the ratings so that they have the ranks A > E > I > O > U > X. We suggest using the following numerical values: A = 1250, E = 250, I = 50, O = 10, U = 0 and X = -1250 (Francis et. al., 1992).

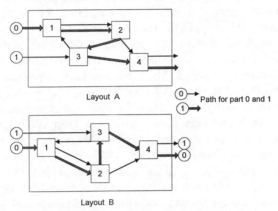

Fig. 2. Screening layouts by travel time objective.

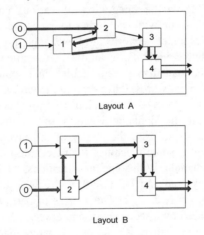

Fig. 3. Screening layouts by closeness rating score objective.

The negative value assigned to X will better separate the machines which are undesirable to be close to each other (Fortenberry and Cox, 1985). Of course, one can choose another numerical values set so that the differences between different rating score are large enough. This measure is a traditional way to treat the relationships among facilities and the setting is subjective depending upon the DMs and analysts. However, we use this concept with some considerations on the exchanges of available machines which can perform the same operations. If machines A and B are able to perform a task C then A and B should be placed close to each other. We suggest a measure for total closeness rating as the ratio between rating score and the distance between two machines as (r_{ij}/d_{ij}).

2.3. Flexibility of the Layout

The third objective is maximizing the flexibility of the layout. Several works have been done in an attempt to identify such measure for the flexibility of manufacturing system, especially, FMS. The need for flexibility is growing due to the change of the nature of competition in the market. Flexibility is the ability of the system to quickly adjust to any changes in relevant factors like product, process, loads and machine failures. Many authors have attempted to attain a qualitative understanding of flexibility of manufacturing system. One measure of flexibility is based on counting of operation for routing and process center. The concept of *entropy* in information theory have also been applied to measure various types of flexibility such as routing flexibility as proposed by Yao (1985), and loading and operation flexibility (Kumar, 1987, 1988). Another attempt to define flexibility is within *task set* (Brill and Mandelbaum, 1990). The task set can be defined in terms of an industry, a particular product, a given set of operations, etc. The authors compare a specific task set with a sample space and develops the concepts of flexibility using a measure-theoretic approach.

An advantage of FMS is the integration of various components. Any two components can communicate with each other to exchange "information" as well as "material" through computer networks or to exchange in-process products using relevant material handling devices. For example, if a machine is down or if it is felt that it carries a heavier load than it can handle, then some of the jobs can be diverted to others, or in other words, it has *routing flexibility*.

The main thrust of the paper is to develop a comprehensive measure for the layout adaptability with the dynamic changes of the manufacturing system. Specifically, the aim is to measure the ability of the layout to offer a changed routing in case that one or some machines (work centers) failed or faced with a heavier load that it can handle. With some modifications, we follow the production task model proposed by Yao and Pei (1990). A manufacturing task consists of several types of parts t, $t = 1, 2, ..., T$. The lot size of part type t is $Q(t)$, for all t. Each part will be produced by a sequence of operations, which is indexed by i, $i = 1, ..., I_t$. For each operation i, there is a set of alternative machines or work center $M(i_j)$, indexing machines by m, $m = i_1, ..., i_N$.

The flexibility of parts routing is due to the existence of multiple machines to process an operation. It means that the flexibility of layout has so far been contributed by the contribution through the alternative machines only. This is the restriction compared with the work of Yao and Pei (1990) is illustrated in Fig. 4.

The flexibility of a system includes two components: *availability* (or reliability, capacity to perform operation i) corresponding with individual machine m denoted by $av(i_m)$, and *reachability*, the ability of the system to reach to another machine when the current machine is down.

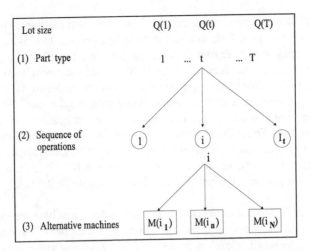

Fig. 4. The structure of the processing requirements of a part in FMS.
(Modified from source: Yao and Pei, 1990)

Availability is defined based on the reliability of machine. Let λ_m be the failure rate of the machine and let μ_m be the repair rate. Then the probability (or the average proportion time) that the machine is functioning as per the alternating renewal process (Ross, 1985), is

$$r_m = \frac{\mu_m}{(\lambda_m + \mu_m)} \tag{3}$$

Let $F_m(t)$ be the distribution function of the lifetime or running time, or the time between failures, of the machine, and let $\overline{F}_m(t) = 1 - F_m(t)$. Then the probability density function of the limiting forward recurrent time RT_m will be $\lambda_m.\overline{F}_m(t)$. Let $T_m(i)$ be the processing time for the operation i on machine m. Let $GT_m(i,t)$ be its distribution function. Intuitively, we can say that machine m has an availability on operation i when the processing time of this operation i on machine m is less than the time between failures. The *availability* should reflect the chance (or probability) that operation i can be performed on machine m (before it fails if it is not already failed), we call that kind of availability as machine availability $av(i_m)$ which can be computed based on result from the alternative renewal process:

$$av(i_m) = r_m.P\{T_m(i) \leq RT_m\} = r_m.\lambda_m.\int_t GT_m(i,t)\,\overline{F}_m(t)\,dt \tag{4}$$

Reachability is defined in the following context: suppose that we already set up for one particular layout, the criteria for arranging facilities in this layout are minimizing TCR and minimizing total travel time between facilities. Dynamically, when a particular routing fails due to machine breakdown, the layout should be able to be changed to another rout by selecting other machines which have enough availability (not down), not decreasing the TCR and not increasing total travel time very much. In other words, the parts should be able to move to their *neighbors*. The word *neighbor* means that the machine being used has the same characteristics and, of course, being available. On the other hand, the neighbor machines are the machines which can be reached by the current part in the system and not violating other constraints. The *reachability* of the machines, in this context, is understood as the characteristic showing that the machine can reach to its neighbor (s). The *reachability* of a part in a currently being down machine to it's neighbors can be viewed in Fig. 5.

If machine A is down then we select another "good" machine, machine B, for the current operation; we shall arrange machine B to be "close" to machine A at the beginning; the word "close" is understood as "*available* and *reachable*". This statement shows that we can establish a criterion for arranging machines in PLP so that it can obtain a high flexibility:

$$FX = \sum_{i_n \ne i_{As}} av(i_n) / d_{i_n i_{As}} \tag{5}$$

where:

$av(i_m)$: availability of machine n corresponding with operation i.

$d_{i_n i_{As}}$: distance between machine n and pre-assigned machine As for operation i.

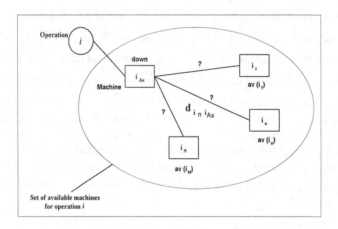

Fig. 5. Representation of *reachability*

3 Model Formulation for PLP

3.1 For the Minimization of Total Travel time

3.1.1 With Rectilinear Distance

The constraints can be separated into those in the same cell and those in different cells (modified from the work of Heragu and Kusiak (1990):
For machines which are in the same cell (Fig. 6)

$$|x_i - x_j| \geq 1/2\,(l_i + l_j) + c_{ij} \tag{6}$$
$$i = 1,...,\ n_A - 1$$
$$j = i + 1,\ ...,\ n_A$$
$$|y_i| > l_A\ ,\ i = 1,...,\ n_A \tag{7}$$

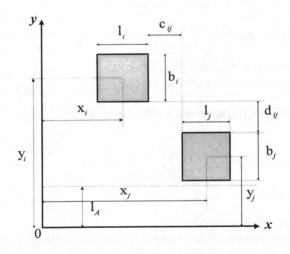

Fig. 1. Geometrical relationship between two machines which are in the same cell.

For machines which are lied in different cells:
Two machines i and j are not restricted by the clearance distances. Only y_i and y_j are constrained by the aisle wide l_A.

$$|y_i| > l_A \tag{8}$$
$$|y_j| > l_A \tag{9}$$
$$|y_i - y_j| > 1/2\,(b_i + b_j) + 2\,l_A \tag{10}$$

where:
x_i , y_i : co-ordinates of machine i;
l_i : length of machine i; b_i : width of machine i;

l_A : aisle length from x axis to base line;

n_A : number of machines to be arranged with AGVs

l_i, b_i : length and width of machine i respectively.

3.1.2 With Euclidean Distance

We consider Euclidean distance in case the material is moved by robot. Actually, the flow path of materials moving from C to B is the arc CB (see Fig. 1), however, to avoid clash between the robot 's arm and machine and for simplicity, we assign the flow path as direct line segment CB, where CB $= [(x_B - x_O)^2 + (x_C - x_O)^2]^{1/2}$. The constraints are geometrical constraints considering the clearance distance on x-axis between machines and the situation that the machines which are served by robot must be placed in the same cell.

$$|x_i - x_j| \geq 1/2 \, (l_i + l_j) + c_{ij} \tag{11}$$

$$(x_i - x_R)^2 + (y_i - y_R)^2 = (x_j - x_R)^2 + (y_i - y_R)^2 \tag{12}$$

$$|y_i| > l_A \tag{13}$$

$$j = 1, 2, ..., n_R$$

where:

n_R : number of machines to be served by robot

x_R, y_R : x- and y-coordinate of robot.

The first objective formulation can be arranged as follows:

Min.

$$\sum_{i=1}^{n_A-1} \sum_{j=i+1}^{n_A} f_{ij} \left(|x_i - x_j| + |y_i - y_j| \right) + \sum_{p=1}^{n_R-1} \sum_{q=p+1}^{n_R} f_{pq} \sqrt{(x_p - x_q)^2 + (y_p - y_q)^2} \tag{14}$$

where :

f_{ij} : flow cost from machines i to j.

x_i, y_i : x-coordinate and y-coordinate of machine i which is served by AGV.

x_j, y_j : x-coordinate and y-coordinate of machine j which is served by AGV.

x_p, y_p : x-coordinate and y-coordinate of machine p which is served by robot.

x_q, y_q : x-coordinate and y-coordinate of machine q which is served by robot.

n_A : number of machines served by AGVs.

n_R : number of machines served by Robot.

The first term of objective function shows the total travel time for machines which are served by AGVs while the second term represents for total travel time between machines served by robot. The above model is used with some assumption: there is only one robot serving the machines, those machines are placed in same cell and we do not take into account the plant size. In larger and real problems we can add more terms representing other robot services as well as add plant size constraints.

3.2. For the Maximization of Total Closeness Ratings (TCR)

Max.

$$\sum_{i=1}^{n_A-1} \sum_{j=i+1}^{n_A} r_{ij} / (|x_i - x_j| + |y_i - y_j|) + \sum_{p=1}^{n_R-1} \sum_{p=q+1}^{n_R} r_{pq} / \sqrt{(x_p - x_q)^2 + (y_p - y_q)^2} \quad (15)$$

where:

r_{ij} : closeness ratings from machines i to j, considering rectilinear distance (for machines which are served by AGVs).

r_{pq} : closeness ratings from machines p to q, considering Euclidean distance (for machines which are served by robot).

n_A : number of machines served by AGVs.

n_R : number of machines served by robot.

The first term of objective function shows the total closeness rating score for machines which are served by AGVs while the second term represents for total closeness rating score of machines served by robot.

3.3. For the Maximization of Layout Flexibility

$$Max. \sum_{t=1}^{T} \sum_{i=1}^{I_t} \sum_{i_n \neq i_{As}} av(i_n)/d_{i_n i_{As}} \quad (16)$$

where:

T : number of part type which are produced in the system

I_t : number of operations needed to produce part t.

i_{As} : machine which is pre-assigned for operation i.

i_n : the current machine in the set of available machine for operation i.

In the objective function, $d_{i_n i_{As}}$ represents the physical distance between machine i_{As} (machine which is pre-assigned to perform operation i) and machine i_n (another "available", in the set of available machines, which can be replaced i_{As} in case machine i_{As} fails). The objective function will achieve maximal value when machine i_n which having highest availability $av(i_n)$ is placed close to machine i_{As}.

An important remark should be made for equation (16) is the formulation is not a representation of *additive* characteristic but rather a lump sum of individual ratio of *availability* $av(i_n)$ to physical distance $d_{i_n i_{As}}$, this point also shows that we do not consider a *system availability* but rather *individual availability*.

4. Case Application

Consider a job shop with 9 Computer Numerically Controlled (CNC) machines including 2 lathes, 3 milling machines, 2 drilling machines and 2 inspection stations. The lathes are numbered and 2, milling machines 3 to 5, drilling machines 6 and 7 and the inspection stations 8 and 9. The shop has to produce 5 types of parts namely I, II, III, IV and V. In the job shop, machines 1, 3 and 6 are served by a robot while the rest are served by an AGV. The machines are pre-grouped into two cells: cell 1 includes machines 2, 4, 5 and 7 and the remaining machines are in cell 2 (Fig. 7). This cell formation procedure is adopted from Heragu and Kusiak (1990).

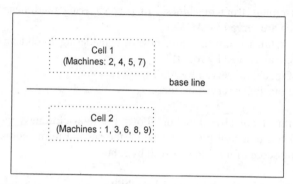

Fig. 7. The cell formation for the given job shop

The process plan, processing times and alternative machines for those plans are shown in Table 1.

Table 1. Process plans for the given job shop.

Parts	Process plans and alternative machines					
I	1 (.5)	3 (.35)	6 (.4)	9(1.4)		
Alt.mcs.	2 (.6)	4 (.3) 5(.4)		8(2.0)		
II	3 (.6)	2 (.45)	4 (.5)	7(.2)	8(.65)	
Alt. mcs.	4 (.65) 5(.60)	1 (.50)	5 (.5)	6(.3)	9(.75)	
III	6 (.15)	5 (.25)	2 (.2)	9(.3)	3(.25)	8 (.3)
Alt. mcs.	7 (.2)	3 (.25)	1(.25)		4 (.3)	
IV	2 (.5)	3 (.3)	1 (.4)	4(.35)	5 (.2)	9(.45)
Alt. mcs.	1 (.55)	4 (.35)		5(.35)	3(.25)	8 (.5)
V	5 (.4)	6 (.45)	2(.35)	8(.3)		
Alt. mcs.		7 (.5)	1 (.4)	9(.35)		

The alternative machines are placed under the pre-assigned machine, for instance, consider part I which has to be processed by machine 1 with a processing time of 0.5; this processing time is placed into a bracket (.5). The alternative machine is machine 2, processing time is 0.6 (0.6). The processing times are assumed to be deterministic. According to production plans, lot sizing specification as well as other aspects in layout designing, the closeness rating scores are established based on management experiences. These relationships normally reflect the qualitative justification in a real job shop (see Francis et. al., 1992). The following closeness rating scores for the above machines is presented as in Table2.

Life time and repair distribution of the machines in job shop are all exponentially distributed, the failure and repair rate are: $\lambda_1 = \lambda_2 = \lambda_4 = 0.45$, $\lambda_3 = 0.3$, $\lambda_6 = \lambda_7 = 0.4$, $\lambda_8 = \lambda_9 = 1$; and $\mu_1 = \mu_2 = \mu_3 = 1.25$, $\mu_4 = \mu_5 = 3.5$, $\mu_6 = \mu_7 = 2.5$, $\mu_8 = 4$, $\mu_9 = 4.5$.

Table 2. Closeness ratings matrix (r_{ij})

Mcs.	1	2	3	4	5	6	7	8
1								
2	E/250							
3	A/1250	E/250						
4	U/0	E/250	O/10					
5	E/250	O/10	E/250	A/1250				
6	X/-1250	E/250	O/10	U/0	I/50			
7	X/-1250	E/250	O/10	E/250	I/50	A/1250		
8	U/0	E/250	E/250	O/10	U/0	O/10	E/250	
9	X/-1250	X/-1250	E/250	U/0	E/250	E/250	O/10	A/1250

The model was formatted in Multiobjective Nonlinear Goal Programming (MNGP) and solved using Genetic Algorithms (GAs). The detailed application of GAs for MNGP has not discussed in this paper, herein we address only the results obtained.

5.Results

The results are presented both by individual objective functions (Table 3-5) and by considering them simultaneously as MCDM (Table 6). The obtained results are tabulated by three columns named: MCs, x co-ordinate and y co-ordinate.

Table 3. Machine coordinates for the first case, minimizing Total Travel Time

MCs.	x co-ordinate	y co-ordinate
1	3.21	-0.65
2	0.553	0.71
3	2.024	-1.60
4	4.35	0.923
5	1.33	1.560
6	0.612	-1.253
7	3.89	1.62
8	3.91	-0.85
9	4.834	-0.82

The objective function $F^*_1 = 4622.045$, the remaining objectives if we substitute value of decision variables will be $F_2 = 1823.419$ and $F_3 = 6.5946$.

Table 4. Machine coordinates for the case maximizing Total Closeness Rating Scores

MCs.	x co-ordinate	y co-ordinate
1	0.51	-1.021
2	0.553	1.16
3	1.522	-0.41
4	2.053	1.765
5	1.242	2.236
6	3.168	-1.235
7	3.156	0.712
8	4.745	-1.16
9	5.01	-1.22

The objective function value $F^*_2 = 5524.959$; the remaining objective functions are $F_1 = 5585.692$ and $F_3 = 9.7010$.

The results obtained when we consider only the objective of Maximizing Layout Flexibility is presented in Table 5.

Table 5. Machine coordinates for the third case, maximizing Layout Flexibility

MCs.	x co-ordinate	y co-ordinate
1	0.5	-1.6
2	0.553	0.71
3	0.965	-0.65
4	1.65	1.03
5	3.62	0.834
6	3.183	-1.376
7	3.324	0.67
8	5.01	-1.16
9	4.47	-1.22

The objective function value $F^*_3 = 15.2008$; $F_1 = 5380.63$ and $F_2 = 3017.461$.

When considering the three objectives simultaneously, the model is constructed by GP format with the initial priority structure as follows: the first priority P_1 is assigned to the objective of minimizing total travel cost, deviation variables are d_1^+ and d_1^-; second priority P_2 is for maximizing total closeness ratings with two deviation variables d_2^+ and d_2^-; the priority P_3 is for maximizing total flexibility corresponding with two deviation variables d_3^+ and d_3^-. All of deviation variables received the same weight.

The overall objective can be written:

$$Min. \; \{(P_1)d_1^+ + (P_2)d_2^- + (P_3)d_3^-\} \tag{18}$$

Equation (18) shows the intended objective is to minimize the sum of deviation variables. The first priority is given to the minimizing of over-achievement of total travel distance (d_1^+) showing that we do not want these distances going higher and this goal must be achieved firstly. The second priority is given to the minimizing of under-achievement of total closeness ratings scores (d_2^-) with an intention that we want these scores going high as much as possible. The third priority is given to the minimizing of under-achievement of total maximizing layout flexibility (d_3^-); this is what we want to against going down of layout flexibility. This priority setup in this paper is preliminarily, for fully investigating the problem, the sensitivity analysis by changing order of priorities should be done for further study.

The result obtained from the model is summarized in the Table 6 and the arrangement for the layout is shown in Fig. 8.

Table 6. Result summary - MCDM case.

Mcs.	X co-ordinate	Y co-ordinate	Objective Fct.	Dev. Variables
1	0.50	-1.25	$F_1 = 5371.13$	$d_1^+ = 0$
2	0.551	0.71		$d_1^- = 121.56$
3	2.02	-0.65		
4	3.05	1.353	$F_2 = 4086.21$	$d_2^+ = 0$
5	1.86	1.36		$d_2^- = 879.12$
6	3.186	-1.593		
7	4.35	1.62	$F_3 = 9.870$	$d_3^+ = 1.13$
8	5.037	-0.85		$d_3^- = 0$
9	5.602	-0.85		

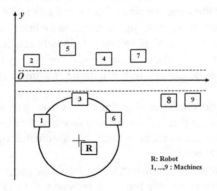

Fig. 8. The proposed layout for given problem.

Looking at Table 6 and Table 7, we might recognize that the deviation $d_1^+ = 0$ and the deviation is $d_3^- = 0$ showing that the first goal (for minimizing total travel distance) and the third goal (for maximizing total layout flexibility) are fully achieved. The second goal (for maximizing total closeness ratings scores) is not fully achieved and we have to accept a tradeoff value, $d_2^- = 879.12$.

A summarized result of the multiple objective case is shown in Table 7, which represents the payoff value of the three objectives. The values in the diagonal of the matrix are the ideal values of individual objective. The compromised values (values obtained from multiple objective case) are placed in the last column.

Table 7. Payoff Matrix for the three objectives

	F_1 (Min.)	F_2 (Max.)	F_3 (Max.)	Compromised Values
F_1 (Min.)	4622.045	5585.692	5380.63	5371.13
F_2 (Max.)	1823.419	6509.463	3017.461	4086.21
F_3 (Max.)	6.5946	9.7010	15.2008	9.87

6. Concluding Remarks

Summarizing the above preliminary results, we conclude that the objectives which were considered by some previous works such as minimization of total travel time and maximization of total closeness rating scores are necessary for layout design but they are not sufficient for a dynamic designing context. A new objective *maximizing layout flexibility* added to PLP would make the layout more realistic. Through this study, a new view of modeling the PLP is considered. It is more advantageous to follow this direction when there is more flexibility in the system.

In a MCDM environment, considering simultaneously three objectives with given priorities shows that the final result is somewhere in between the best and the worst values. Depend upon Decision Maker point-of-view, the priority and also the weights which are given to the deviations will determine the trade off needed for each objective.

Of course, there is no guarantee that the real system will work exactly as we proposed but we can assure that if there is some *machine failure* during the working phase, the system will be able to adapt so that it can handle the work load, at least, the system will not be totally down.

To do a better experiment, a simulation study should be carried out. In the context of simulation, all possible processing time distributions will be examined. Scheduling rules for operations will also be applied so that the "best" configuration of the given layout will be obtained.

REFERENCES

[1]   **Abdou, G. and Dutta S. P. (1990),** "An integrated approach to facilities layout using Expert Systems", *Int. J. Prod. Res.,* 28/4, pp. 685-708.

[2]   **Apple, J. M. (1977),** *Plant Layout and Material Handling, 3rd ed.,* John Wiley & Sons.

[3]   **Brill, P. H. and Mandelbaum, M. (1990),** "Measurement of adaptivity and flexibility in Production Systems.", *European Journal of Operational Research,* 49, pp. 325 – 332. Elsevier Science Publishers B. V.

[4]   **Davis, L. (1987),** *Genetic Algorithms and Simulated Annealing.* Pitman.

[5]   **Downlatshahi, S. (1992),** "Product Design in a Concurrent Engineering Environment: an Optimization Approach", *Int. J. Prod. Res.,* 30/8, pp. 1803-1818.

[6]   **Fortenberry, J. C. and Cox, J. F. (1985),** "Multiple citeria approach to the facilities layout problem". *Int. J. Prod. Res.,* 23/4, pp. 773-782.

[7]   **Francis, R. L. and White, J. A. (1974),** "Facility Layout and Location: An analytical approach", Prentice-Hall Inc., New Jersey, USA.

[8]   **Francis, R. L. et. al. (1992),** "Facility Layout and Location: An analytical approach", (2nd ed.), Prentice-Hall Inc., New Jersey, USA.

[9]   **Goldberg, D. E. (1989),** *Genetic Algorithms in search, optimization and machine learning.* Addison-Wesley.

[10] **Harmonosky, C. M., Tothero, G. K. (1992),** "A multi-factor plant layout methodology", *Int. J. Prod. Res.,* 30/8, pp. 1773-1789.

[11] **Heragu, S. S. and Dutta S. P. (1990),** "Machine layout: an optimization and knowledge-based approach", *Int. J. Prod. Res.,* 28/4, pp. 615-635.

[12] **Heragu, S. S. and Kusiak, A. (1988),** "Machine layout problem in Flexible Manufacturing Systems", *Operations Research,* 36/2, 258-268

[13] **Heragu, S. S. and Kusiak, A. (1990),** "Machine layout: an optimization and knowledge-based approach", *Int. J. Prod. Res.,* 28/2, 615-635.

[14] **Kumar, V. (1987),** "Entropic Measures of Manufacturing Flexibility.", *Int. J. Prod. Res.,* 25/7, pp. 957-966.

[15] **Kumar, V. (1988),** "Measurement of Loading and Operations Flexibility in Flexible Manufacturing Systems: An Information Theoretic Approach". *Annals of Oper. Res.,* Vol. 15, pp. 65 - 80.

[16] **Kusiak, A. (1990),** *Intelligent Manufacturing Systems,* Prentice Hall, Englewood Cliffs, New Jersey, USA.

[17] **Kusiak, A. (1993),** *Concurrent Engineering: Automation, Tools and Techniques,* John Willey and Sons, Inc. Pub., 1993.

[18] **Malakooti, B. (1989),** "Multiple objective facility layout: a heuristic to generate efficient alternatives", *Int. J. Prod. Res.,* 23/4, pp. 773-782

[19] **Malakooti, B. and D'Souza, G. I. (1985),** "Multiple objective programming for the quadratic assignment problem", *Int. J. Prod. Res.,* 25/2, pp. 285-300.

[20]   **Malakooti, B. and Tsurushima, A. (1989)**, "An expert system using priorities for solving multiple-criteria facility layout problems", *Int. J. Prod. Res.,* 27/5, pp. 793-808.

[21]   **Muther, R. (1973)**, *Systematic Layout Planning,* 2nd ed. Cahners Books, Boston.

[22]   **Partovi, F. Y. and Burton J. (1992)**, "An Analytical Hierarchy Approach to facility layout.", *Computers Ind. Eng.,* 22/4, pp. 447-457.

[23]   **Rosenblatt, M. J. (1979)**, "The facilities layout problem: A multigoal approach.", *Int. J. Prod. Res.,* 17/4, pp. 323-332.

[24]   **Ross, S. M. (1985)**, "Introduction to Probability Models", Academic Press Inc., USA.

[25]   **Saaty, T. L. (1980)**, *The Analytic Hierarchy Process*, McGraw-Hill, New York.

[26]   **Shang, J. S. (1993)**, "Multicriteria facility layout problem: An integrated approach", *European J. of Opl. Res.,* Vol. 66, pp. 291-304.

[27]   **Smith, G. F. and Browne, G. (1993)**, "Conceptual Foundations of Design Problem Solving", *IEEE Transactions on System, Manufacturing and Cybernetics,* 23/5, pp. 1209-1219.

[28]   **Spano, M. R, Sr., Ogrady, P. J. and Young, R. E. (1993)**, "The design of flexible manufacturing systems", *Comp. in Industry,* Vol. 21, pp. 185-198.

[29]   **Tabucanon, M.T. (1988)**, *Multiple Criteria Decision Making in Industry*, The Netherlands Elsevier.

[30]   **Thurston, D.L. (1990)**, "Multiattribute Uitlity Analysis in Design Management", *IEEE Trans. on Eng. Management,* 37/4, 296-301.

[31]   **Thurston, D. L. and Locascio, A. (1993)**, Chapter 11: Multiattribute Design Optimization and Concurrent Engineering, *Concurrent Engineering: Contemporary Issues and Modern Design Tools,* Chapman and Hall Pub., pp. 207-230.

[32]   **Urban, T. L. (1987)**, "A multiple criteria model for the facilities layout problem", *Intl. J. Prod. Res.,* Vol. 25, No.12, pp. 1805-1812

[33]   **Yao, D. D. (1985)**, "Material and Information Flows in Flexible Manufacturing Systems.", *Material Flow,* No. 2, pp.143 - 149. Elsevier Science Publishers B. V.

[34]   **Yao, D. D. and Pei, F. F. (1990)**, "Flexible Parts Routing in Manufacturing Systems.", *IIE Transactions,* Vol. 22, No. 1, pp.48 - 55.

Judgmental Modelling as a Benchmarking Tool: A Case Study in the Chemical Industry

Pete Naudé[1]

[1]Manchester Business School, Booth Street West, Manchester, M15 6PB, UK

Abstract. Throughout the world industries are facing increasing threats from a variety of forces such as globalisation, technological changes, and deregulation. One consequence of these changes has been the radical restructuring of organisations to be more responsive to market requirements, coupled with continual benchmarking against best-in-class competitors to ensure that companies retain a competitive position. Despite the recognition of the need for benchmarking, there is no consensus on the best approach. This paper shows how judgmental modelling can be used as a benchmarking tool to analyse both market requirements and competitive positioning. The information is based on 112 personal interviews carried out in a sector of the chemical industry across Europe. The results clearly indicate how the judgmental modelling approach can be used to generate meaningful insights in this complex area.

Keywords: Competitive Benchmarking, Judgmental Modelling

1 Introduction

1.1 Judgmental Modelling and Benchmarking

The process of benchmarking has been the focus of increasing management attention over the last decade (van de Vliet, 1996). Studies usually involve searching for some combination of assessing internal and external (often but not exclusively competitive) performance, and monitoring changes over time (Fiegenbaum et al, 1996). It is generally recognised (see for example, Watson, 1993) that the process of benchmarking consists of the four different stages of planning a study, searching for and collecting the data, analysing it, and finally adapting/improving before the process starts again. While the competitive needs to benchmark are obvious - increased competition from global sources, and new competitors from changes in technology or deregulation, what is not as clear is how to collect the data, with fairly simplistic research methodologies being the norm (see, for example, Andersen and Pettersen, 1996). Whatever method is adopted, it is obviously not sufficient to collect information that focuses just on whether *A* is better or worse than *B*. What is required is a process that allows the managers concerned to identify just how much better *A* is than *B*, and what the difference means in real terms.

It seemed that judgmental modelling offered an ideal way to approach this problem. Similar in concept to other MCDM approaches such as the AHP (Saaty, 1980), the approach is based on an underlying compensatory model (Wind, 1982) which involves decomposing the problem into three separate hierarchical issues:

> What attributes are important in making this decision?
> How do they vary in importance?
> How do the various competing suppliers perform on the same attributes?

In this instance, the company involved could call upon on a large amount of previously available market research that effectively meant that they already had a very good understanding of the attributes underlying the decision making process in the market place, and these are shown in Appendix 1.

The more traditional approaches to judgmental modelling suffer from the problem identified above. For example, standard application of the AHP model allows the user to conclude that A scores say 34% on an attribute, and B 50% - but these values have no real interpretation, other than to conclude that B is in some sense 'better' (see Korpela and Tuominen (1996) for a recent AHP-based application). For this reason, in the research undertaken, a modified form of judgmental modelling that relied on word-models was used (Islei et al, 1991; Naudé et al, 1993; Naudé, 1995). Examples of the word-models used are given in Appendix 2. As can be seen there, these are simply zero-anchored rating scales with standard reference frames that permit the user to draw more meaningful conclusions from differences between scores. For example, if one company were to score 100 on the attribute stable range and another to score 50, management could draw clear conclusions from what the differences meant.

In this study all interviews were carried out using a simple PC-based judgmental modelling package called JAS (Islei and Lockett, 1988; Lockett and Islei, 1989 & 1993). Like the AHP, the approach uses pairwise comparisons to collect the attribute weights, but then has the option of using a form of direct entry on word model scores to collect perceptions on the performance of different suppliers. When used in this way, we did not restrict respondents to just those scores shown in Appendix 2 (for example, a supplier could be scored 90, or 95). In addition, it is of course possible that all suppliers, or at least many of them, could be scored equally on an attribute indicating that the particular attribute tends not to be a discriminator in the market place.

1.2 Methodology

During late 1995 and early 1996, 112 personal interviews were carried out across Europe in a sector of the chemical industry. Table 1 gives an indication of how the sample was split by region and also by other characteristics. The work was sponsored by a large multi-national company, a large and active player in the market, in order to assess both the needs of the market place and the company's own competitive positioning within that market. The market itself is a mature one, with the product having been available for many years, with no practical substitute. A number of large multi-national companies supply the product, which has characteristics of being both a commodity and a speciality chemical, depending largely on the environment in which the buyer operates. Three factors tend to characterise the market: the product usually forms a relatively high percentage of the buyers' input costs, levels of demand and supply vary according to the stage of the economic cycle, and the production process involved has been the topic of much criticism from members of the environmental lobby. These three factors, plus the fact that there is no real substitute, meant that the product was always well known to the respondents interviewed, in spite of having the characteristics of being more of a straight rebuy than a modified rebuy (Robinson et al, 1967). In addition, it did mean that it was usually possible to identify just one individual in the company who was responsible for the purchase decision, rather than having to cope with the methodologically more complex issue of multiple participants in a decision making unit (see, for example, Håkansson, 1982).

Table 1. The respondent profile

Country			Commercial	87
France	19		Technical	25
Benelux	8			
Iberia	26		Segment A	77
Italy	19		Segment B	30
United Kingdom	18		Segment C	5
Germany	22			
Total	112		Customers	81
			Non-Customers	31

Given the prior knowledge concerning the attributes of potential importance, the field research focused on gaining insights into just two issues: the extent to which the different respondents varied in their assessment of the attribute importance, and how they perceived the different suppliers to be performing. In marketing terms, the first of these issues relates to how the market is segmented, and the second to the competitive positionings of the major players. The results of the segmentation analysis are presented elsewhere (Naudé and Hipkin, 1997), and this paper focuses solely on the use of the approach as a benchmarking tool.

2 Data Analysis

The results of the interviews yielded two inter-related datasets. In the more simple case, we had a 112-by-32 matrix reflecting the scores of the respondents on the attribute importance. However, the scores relating to the performance of the different suppliers on the same attribute set were more problematic. Clearly, it would have been naive to expect all respondents to have a view about all the different suppliers operating in the market place. In the interviews, therefore, the respondents were asked to comment about and evaluate only those suppliers with whom they were familiar. This meant that different respondents evaluated not only different suppliers, but also different numbers of suppliers. In France, for example, there were 19 respondents, but 84 data sets covering data on 9 different suppliers on each of the 32 attributes. Overall, the dataset of 112 respondents yielded just over 500 such data vectors.

Analysing such data is problematic on two accounts. Firstly, there are two very different ways of viewing the data: should the performance of the different suppliers be handled in its raw form (i.e. representing the raw word-model scores shown in Appendix 2), or should they be weighted according to the importance that each buyer puts onto each of the attributes? Secondly, it is clear that the sample was not in any sense a random sample: it consisted of a judgmentally determined list of customers and non-customers, pared down by those who refused to be interviewed. In addition, the size of the respondents, and indeed the size of the national markets, varies dramatically, and as such no confidence-interval type analysis is really acceptable.

The problems underlying the data analysis are shown in Figure 1, where we look just at the French data, although the results from the other countries display similar properties. In this case we have used Correspondence Analysis to view the data (Greenacre, 1984 & 1985; Hoffman and Franke, 1986; Greenacre and Hastie, 1987). While we do not wish to discuss in detail the supporting statistical output, we can see how the different buyers' perceptions on the attribute importance correspond to the different suppliers.

In this figure, a letter followed by a number relates to a particular attribute. With reference to Appendix 1, R2 (bottom left) relates to the second sub-attribute within people/relationships (i.e. the Rep's authority), while I2 and I3 (top right) are Customer Focus and Reputation. A capital letter followed by a small letter represents the different respondents. For example, Ch (far left) represents respondent C's views about supplier h (which we can see is closely related to attributes such as R2, R3, R4, and R6, but not to I2 and I3).

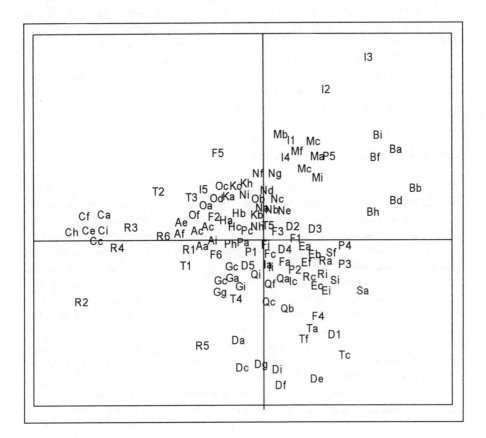

Figure 1 The French Respondents

The interesting aspect about figure 1 is the extent to which the individual respondent's perceptions all plot close to each other. For example, all respondent C's scores are to the left, all B's to the right, D's to the bottom, etc. The reason for this is firstly that the weighted scores were used. While using such weighted scores is appealing in the sense of being a more accurate reflection of buyer behaviour, it does not in fact help management in benchmarking itself. The reason for this is clear: if an attribute were to score say 20%, and the company scored a perfect wordmodel score (i.e. 100), the overall weighting would be reflected as 20. This score would also be achieved by an attribute weighted at

40% on which the company scored a relatively poor 50 on the wordmodels - can we really say that both scores are 'worth' 20?.

The second reason for the suppliers being plotted close together relates to the nature of the market itself. It is a mature market with some characteristics of being a commodity where all major players are well known, and sources of differentiation giving competitive advantage difficult to attain. In the minds of the buyers, therefore, the suppliers are all relatively similar. Once we have the compounded effect of weighting the word-model scores with those of the attribute importance, the differences between suppliers become even less marked. For this reason, in the rest of the paper only the unweighted scores are used.

2.1 Analysis of Current Performance

Figures 2 and 3 on subsequent pages give an indication of the level of analysis that is possible. For example, in Figure 2 the data is analysed for one particular supplier across the different countries. The advantage of this level of analysis is that it quickly becomes apparent where the different multi-national companies are concentrating their efforts, and what the focus of their strategies is. Because the same word models were used in all countries, direct comparisons can be made. It is clear from Figure 2 that Supplier d is focusing attention on Benelux and Iberia, and doing so through a product and delivery based approach, while it is adopting more of a skimming strategy in Italy and the UK. By analysing the data differently, it is easy to understand how competitors' strategies vary by country, as shown for the French market in Figure 3. Again, it is quick and easy for local management to develop an understanding of exactly what the different competitors are stressing in their local marketing and sales efforts: it is clear, for example, that Supplier f is the favoured supplier, with a, d, and i all being undercut on price and outperformed in most attributes other than delivery.

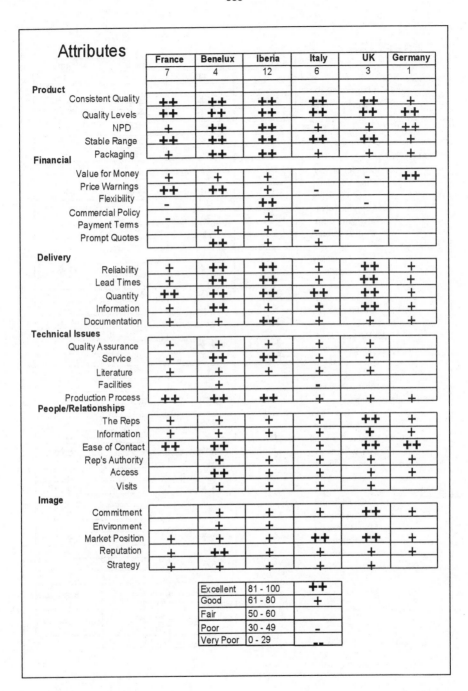

Attributes	France	Benelux	Iberia	Italy	UK	Germany
	7	4	12	6	3	1
Product						
Consistent Quality	++	++	++	++	++	+
Quality Levels	++	++	++	++	++	++
NPD	+	++	++	+	+	++
Stable Range	++	++	++	++	++	+
Packaging	+	++	++	+	+	+
Financial						
Value for Money	+	+	+		-	++
Price Warnings	++	++	+	-		
Flexibility	-		++		-	
Commercial Policy	-		+			
Payment Terms		+	+	-		
Prompt Quotes		++	+	+		
Delivery						
Reliability	+	++	++	+	++	+
Lead Times	+	++	++	+	++	+
Quantity	++	++	++	++	++	+
Information	+	++	+	+	++	+
Documentation	+	+	++	+	+	+
Technical Issues						
Quality Assurance	+	+	+	+	+	
Service	+	++	++	+	+	
Literature	+	+	+	+	+	
Facilities		+		-		
Production Process	++	++	++	+	+	+
People/Relationships						
The Reps	+	+	+	+	++	+
Information	+	+	+	+	+	+
Ease of Contact	++	++		+	++	++
Rep's Authority		+	+	+	+	+
Access		++	+	+	+	+
Visits		+	+	+	+	
Image						
Commitment		+	+	+	++	+
Environment		+	+			
Market Position	+	+	+	++	++	+
Reputation	+	++	+	+	+	+
Strategy	+	+	+	+	+	

Excellent	81 - 100	++	
Good	61 - 80	+	
Fair	50 - 60		
Poor	30 - 49	-	
Very Poor	0 - 29	--	

Figure 2 Supplier by Country

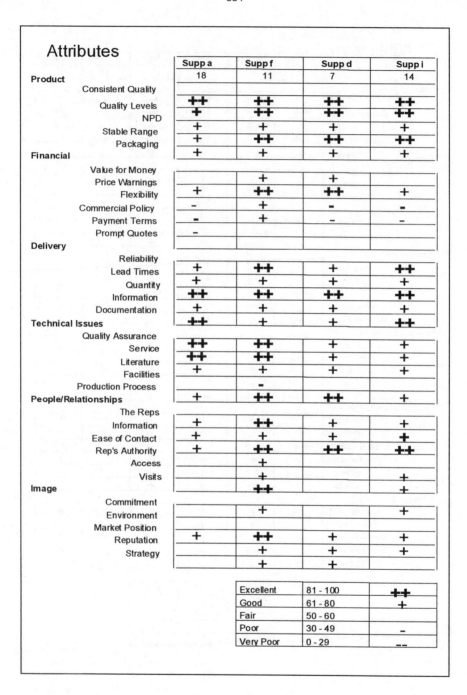

Figure 3 The Different Suppliers in France

2.2 Setting Future Objectives

It is clear that the whole process of benchmarking must be geared towards changing the internal operations of the organisation to a more efficient system in the near future. One of the advantages of the proposed approach based on judgmental modelling is the ease with which such a system is implemented. As shown in Figure 4 overleaf, it is relatively easy to use the same approach to do this. The figure combines the overall attribute importance (which can easily be handled separately at country, industry, or even individual level) with objectives on each of the word models. As shown here, the company decided that it had to excel at all product, delivery, and technical aspects, and would try to achieve an acceptable (i.e. good) score on the attributes of lesser importance (i.e. people/relationships and image). On price, it was decided that it should attempt to achieve a fair score on value for money, somewhere between fair and poor on price warnings and flexibility, and to be happy with achieving only a poor score on the last three attributes, on the assumption that the excellent scores on the other attributes would compensate.

3 Conclusion

The competitive pressures under which managers now have to operate in order to sustain and grow their companies mean that some form of benchmarking will increasingly have to be part of the normal management process. However, while there may be agreement on the need for benchmarking, there is little consensus as to how to collect the appropriate data. In this paper a process based on the use of judgmental modelling is proposed. One advantage is that the raw data is collected straight onto a PC through the use of a judgmental modelling package. Not only are interviewees still relatively underexposed to this method of collecting data (which resulted in a high level of interest and interaction from the sample), but the data is also quickly analysed in a number of ways to suit company needs on any specific attribute.

Attributes	Score	France	Benelux	Iberia	Italy	UK	Germany
Product	**0.239**	++	++	++	++	++	++
Consistent Quality	0.074	++	++	++	++	++	++
Quality Levels	0.067	++	++	++	++	++	++
NPD	0.039	++	++	++	++	++	++
Stable Range	0.038	++	++	++	++	++	++
Packaging	0.021						
Financial	**0.220**						
Value for Money	0.062						
Price Warnings	0.038		−		−		−
Flexibility	0.039	−		−		−	
Commercial Policy	0.032	−	−	−	−	−	−
Payment Terms	0.031	−	−	−	−	−	−
Prompt Quotes	0.018	−	−	−	−	−	−
Delivery	**0.187**						
Reliability	0.060	++	++	++	++	++	++
Lead Times	0.039	++	++	++	++	++	++
Quantity	0.036	++	++	++	++	++	++
Information	0.026	++	++	++	++	++	++
Documentation	0.025	++	++	++	++	++	++
Technical Issues	**0.154**						
Quality Assurance	0.042	++	++	++	++	++	++
Service	0.040	++	++	++	++	++	++
Literature	0.026	++	++	++	++	++	++
Facilities	0.024	++	++	++	++	++	++
Production Process	0.022	++	++	++	++	++	++
People/Relationships	**0.129**						
The Reps	0.025	+	+	+	+	+	+
Information	0.023	+	+	+	+	+	+
Ease of Contact	0.023	+	+	+	+	+	+
Rep's Authority	0.023	+	+	+	+	+	+
Access	0.021	+	+	+	+	+	+
Visits	0.014	+	+	+	+	+	+
Image	**0.071**						
Commitment	0.017	+	+	+	+	+	+
Environment	0.013	+	+	+	+	+	+
Market Position	0.015	+	+	+	+	+	+
Reputation	0.014	+	+	+	+	+	+
Strategy	0.013	+	+	+	+	+	+

Excellent	81 - 100	++
Good	61 - 80	+
Fair	50 - 60	
Poor	30 - 49	−
Very Poor	0 - 29	−−

Figure 4 Objectives by Country

Bibliography

Andersen, B. and Pettersen, P-G (1996) *The Benchmarking Handbook*, London: Chapman & Hall

Fiegenbaum, A., Hart, S., and Schendel, D. (1996) Strategic Reference Point Theory, *Strategic Management Journal*, 17:3, pp. 219-235

Greenacre, M. J. (1984) *Theory and Applications of Correspondence Analysis*, Academic Press, London

Greenacre, M. J. (1985) *Correspondence Analysis: Program SIMCA (Version 1.0) for the IBM Personal Computer*, University of South Africa, Pretoria, South Africa

Greenacre, M. J. and Hastie, T. (1987) The Geometric Interpretation of Correspondence Analysis, *Journal of the American Statistical Association*, (82:398), pp. 437-447

Hoffman, D. L. and Franke, G. R. (1986) Correspondence Analysis: Graphical Representation of Categorical Data in Marketing Research, *Journal of Marketing Research*, 23 (Aug), pp. 213-227

Håkansson, H. (1982) *International Marketing and Purchasing of Industrial Goods: An Interaction Approach, (A Study by the IMP Group)*, Chichester: John Wiley And Sons

Islei, G and Lockett, A. G. (1988) Judgmental Modelling Based on Geometric Least Squares, *European Journal of Operational Research*, (36:1), pp. 27-35

--------, --------, Cox, B., Gisbourne, S. and Stratford, M. (1991) Modelling Strategic Decision Making and Performance Measurements at ICI Pharmaceuticals, *Interfaces*, (21:6), pp. 4-22

Korpela, J. and Tuominen, M. (1996) Benchmarking Logistics Performance with an Application of the Analytic Hierarchy Process, *IEEE Transactions on Engineering Management*, 43:3, pp. 323-333

Lockett, A.G. and Islei, G. (1989) *JAS-Judgemental Analysis System, Version 1.2*, Manchester Business School, UK

-------- and -------- (1993) *JAS for Windows, Version 1.0*, Manchester Business School, UK

Naudé, P., Lockett, G. and Gisbourne, S. (1993) Market Analysis via Judgmental Modelling, An Application in the UK Chemical Industry, *European Journal of Marketing*, 27:3, pp. 5-22

--------, (1995) Judgmental Modelling as a Tool for Analysing Market Structure, *Industrial Marketing Management*, 24, pp. 227-238

--------, and Hipkin, I. (1997) Analysing MCDM Data from Multiple Respondents, *Paper presented at the 13th International Conference on Multiple Criteria Decision Making*, Cape Town, January

Robinson, P.J., Faris, C.W., and Wind, Y. (1967) *Industrial Buying and Creative Marketing*. Boston: Allyn and Bacon

Saaty, T. L. (1980) *The Analytic Hierarchy Process*, New York, McGraw-Hill

van de Vliet, A. (1996) To Beat the Best, *Management Today*, (Jan), pp. 56-60

Watson, G. H. (1993) *Strategic Benchmarking*, New York: John Wiley & Sons

Appendix 1 The Attributes Used

Product		
p1	Stable Range	*Extent to which the range is stable over time*
p2	Quality Levels	*Absolute quality levels of the product*
p3	Consistent Quality	*Consistent quality and to agreed specifications*
p4	Packaging	*The quality and suitability of the packaging used*
p5	New Prod. Dev.	*Extent of NPD, and willingness to work together with customers*
Delivery		
d1	Reliability	*Offering products on time, making and keeping the promise*
d2	Documentation	*Extent to which suppliers supply accurate documentation*
d3	Information	*The clarity and correctness of delivery information*
d4	Lead Times	*The time lapse between placing and receiving the order*
d5	Delivery Quantity	*Extent to which orders are delivered in full*
Technical Issues		
s1	Service	*Includes both rapidity and effectiveness of the Technical Service.*
s2	Literature	*The quality of the technical literature*
s3	Quality Assurance	*Extent to which the supplier offers a QA service*
s4	Production Process	*Degree of impact of the supplier's production process on our business*
s5	Facilities	*Extent to which supplier's facilities are available for our use*
Image		
i1	Strategy	*The clarity and consistency of strategy*
i2	Customer Focus	*Extent to which they put customers first*
i3	Reputation	*Covering both innovation and being a caring company*
i4	Market Position	*Extent to which they are a global but local company, committed to the future*
i5	Environment	*Environmental performance of both product and packaging*
Financial		
f1	Value for money	*Extent to which price and quality combine to give value*
f2	Flexibility	*Extent to which they are flexible on price negotiation*
f3	Commercial policy	*Extent to which their commercial policy is fair*
f4	Prompt Quotes	*Extent to which quotes are given on time*
f5	Payment terms	*Extent to which suppliers are flexible in determining payment terms*
f6	Price Warnings	*Extent to which warning is given of impending changes to price*
People/Relationships		
r1	The Reps	*Their knowledge, quality, professionalism*
r2	Rep's authority	*The extent of their decision authority*
r3	Ease of contact	*When information is needed or there are problems*
r4	Visits	*Regular, face-to-face visits from appropriate staff*
r5	Access	*Ready access to management when needed*
r6	Information	*Information flows about potential problems and market trends*

Appendix 2 A Selection of the Word-models Used

Product
Stable Range *Extent to which the range is stable over time*
100	I know of, and agree with, the way in which they manage their grade range
75	I sometimes find that they have changed their range without notification
50	About what I would expect from a supplier
25	An annoyingly high frequency of changes to their product range
0	They change so often that it stops me buying from them

Delivery
Reliability *Offering products on time, making and keeping the promise*
100	All deliveries are on time
75	The vast majority of deliveries are on time
50	Most deliveries are on time
25	An annoying amount of deliveries are not on time
0	Their deliveries are very inefficiently handled

Technical Issues
Production Process *Degree of impact of the supplier's production process on our business*
100	They are concerned about their production process and always try to improve it
75	They show concern and seem to be doing something to improve it
50	They show concern but don't seem to be doing anything about it
25	They show minimal concern
0	They just don't care what process they use

Facilities *Extent to which supplier's facilities are available for our use*
100	Their testing facilities are a major source of advantage to me
75	Using their testing facilities gives me a real competitive advantage
50	They sometimes give me help to test products
25	Their facilities are largely irrelevant to my success
0	I do not have access to their testing facilities

Image
Environment *Environmental performance of both product and packaging*
100	Dealing with them has a positive impact on my environmental performance
75	They are actively improving their environmental record, which helps me
50	Buying from them will not present me with environmental problems
25	I am not happy about the environmental impact of using some of their products
0	Using their product is a hazard to the environment

Spatial conflicts in transport policies: structuring the analysis for large infrastructure projects[1]

Euro Beinat

Institute for Environmental Studies, Free University, Amsterdam, The Netherlands

Abstract. Policy making in the European Union pursues the integration of national and local policies while maintaining national and regional identities. Spatial conflicts often occur in this framework. This paper discusses the nature and characteristics of spatial conflicts for transport policies and introduces a methodological framework for their study. The goal is to gain insight into horizontal conflicts (between policy units at the same level, like regions) and vertical conflicts (between nested policy units, such as the European Union, nations, regions, etc.). Spatial Analysis, Decision Analysis and Organisational Theory provide the methodologies for the analysis. The paper illustrates the approach and describes the problem structuring phase. A large transport infrastructure project is considered for this purpose. The spatial nature of actors is analysed and their objectives and concerns are made explicit. Together with the spatial information for the decision this leads to a spatial representation of decision context of each actor, which is at the basis of conflict analysis.

Keywords: spatial conflicts, decision analysis, spatial information, GIS

1 Introduction

The transport sector in the European Union (EU) accounts for a substantial share of the European GNP. It is a source of employment, a crucial factor for the economic growth of the EU and for its international competitiveness. The EU transport policy focus is manifold. However, transport infrastructures play an important role, both in

[1] This research is part of the DTCS project (Spatial decision support for negotiation and conflict resolution of environmental and economic effects of transport policies), financed by the European Commission DG12-D5 "Human dimension of environmental change" (contract ENV4-CT96-0199). The project is developed by the IVM (Amsterdam, NL), CERTE (Canterbury, UK), CESUR (Lisbon, P) and MIP (Milano, I). The author wishes to thanks all consortium members and in particular Carlos Bana e Costa, Michiel van Drunen and Margaret Jones for comments, ideas and GIS support. Part of the data used in this paper are provided by the Province of Gelderland, which is also gratefully acknowledged. The views expressed in this work reflect only the author's views.

terms of management of existing infrastructures (information systems, capacity management, road pricing, etc.) and in terms of provision of new infrastructures (Commission of the European Communities, 1993). New infrastructures are meant to increase the competitiveness of European regions and to achieve higher level of efficiency of the transport sector. However, infrastructures are a blunt-edged instrument. They address the competitiveness-efficiency nexus only indirectly and have several side effects (such as environmental and social impacts). On the other hand, they involve large investments and therefore are precious employment generators.

Large infrastructures like the Trans-European Networks (TEN) affect large spatial areas. Their costs and benefits affect several different regions, which usually share different identities, environmental conditions and priorities. Conflicts usually occur between the different spatial entities involved, resulting in complex decision negotiations or interruptions in the decision process.

This paper addresses the spatial conflicts of transport infrastructures and provides a methodological framework for their study. It describes spatial conflicts, the main factors which originate conflicts and provides a general scheme for their analysis. A large TEN project is used as an example (the Betuweroute project in the Netherlands) to show the spatial distribution of the decision actors, their different concerns and objectives and the way this drives the evaluation of the project. Due to space limitations, the paper focuses especially on the structuring phase, which leads to the identification of spatial conflicts.

2 A characterisation of spatial conflicts

Spatial planning and spatial conflicts analysis share many of the critical attributes of public decision processes (Allor, 1991). In particular, the need to consider: (1) a multiplicity of perceptions, interests and expectations, which are differently held by (2) multiple actors, who are organised and embedded in a (3) formal procedure or informal relationship which leads to a decision.

Actors may present themselves as people (Mark), as role occupants (the city Mayor), as groupings (farmers), as organisations (the ministry of environment), as occupational groupings (the engineers), as pressure groups (environmentalists) and more. An actor is any individual or grouping affected by or affecting the decision to which we can associate a relatively consistent point of view (Burgoyne, 1994).

Conflicts between actors can be classified into cognitive and interest conflicts. Cognitive conflict exists when actors base their decisions on different facts or when the same facts are used and interpreted differently. Interest conflict appear when the actors have different value systems and world views (Obermayer, 1994; cf. also Keeney, 1992). In the ideal situation in which all actors are perfectly informed and they are satisfied with their role in the decision process, then the cognitive-interest distinction is appropriate. However, in real cases the information for the decision is unevenly distributed and the role played by the actors is also a matter of dispute. These factors, linked to the way the decision process is organised, are also generators of conflicts (cf. Bogetoft and Pruzan, 1991, and Beinat, 1998, for a more detailed analysis of conflicts and of their origin).

Spatial decision problems have additional characterisations. This is due to the location and the scale dimensions. The location dimension affects both facts and values for the decision. The (expected) effects are spatially distributed usually over large area. This holds for direct impacts (for instance noise and pollution) and indirect impacts (for instance spill over economic benefits between regions). The nature, use and vocation of land is also spatially differentiated, implying the existence of different measurement criteria for the same effects in different places. Additionally, communities affected by the decision may share different value systems in different places.

The scale dimension indicates that different facts and values are appropriate at different scales. For instance, the national consequences of large transport infrastructures are addressed in terms of GNP contribution, economic growth, greenhouse gasses emissions etc. At the local level, the same infrastructure will cause noise pollution, housing relocation etc. The links between the macro and micro level are not always evident and usually imply that different decision actors evaluate effects at different scales.

There is also an intrinsic relationship between space and (a subset of) decision actors which coincide with spatial policy areas. (e.g. nations, provinces, districts etc..). However, it is important to point out that not all spatial actors play the same role in the process. Some actors are active players and pursue their specific objectives; others may have no stakes in the decision and be involved only to protect their interests.

The location and scale dimensions also suggest the distinction between vertical and horizontal conflicts. Vertical conflicts arise between policy units nested into each other. They are mainly due to the scale effects. Examples are the conflicts between the Dutch Transport Authorities (representing national interests) and the provincial and local authorities on the acceptability of the Betuweroute railway project (see below). Horizontal conflicts occur between policy units at the same spatial level, for instance between local authorities competing for the allocation of funds. Horizontal conflicts occur mainly due to location effects.

3 A methodological framework for spatial conflict analysis

At present, a theory which systematically addresses spatial conflicts is not available. Rather, there is a set of interdependent theories and methodologies which deal with single or few of the aspects of spatial conflicts (cf. among others, Densham, 1991; Armstrong, 1994; Douwen, 1996). Broadly speaking, one recognises three main approaches: spatial analysis and planning, decision analysis and organisational theory. The spatial analysis and spatial planning approach focuses on the spatial component of a decision problem (cf. Maguire et al, 1991). Geographical Information Systems (GIS) play an essential role in this context due to the amounts and complexity of the information needed for spatial problems. Decision analysis aims at an analytical and systematic exploration of the value systems of individuals involved in the decision. It provides methodologies for problem structuring, value judgement assessment, decision suggestions and sensitivity analysis (cf. among others Rosenhead, 1989; Keeney, 192; Vinke, 1992). Organisation theory focuses on the relationship and interaction between actors in a decision process, including the role of procedures, information flows and behavioural effects (Simon, 1991; Hersey and Blanchard, 1988; Huigen et al., 1993).

The common ground of these approaches is usually referred to as Collaborative Spatial Decision Support (CSDS; NCGIA, 1996). The components of this approach can be organised through the general scheme of Stakeholder Analysis (Burgoyne, 1994). Figure 1 graphically illustrates this scheme. It indicates the context in which each decision actor operates during the following stages: structuring and design; evaluation and negotiation; recommendation. Multiple actors are involved in all these stages, although the actor set may change along the process. Each actor is characterised by a role in the process, a value system and the set of facts which are relevant from his/her point of view. The system of role, values and facts of an actor at a certain stage of the process is the decision context of this actor.

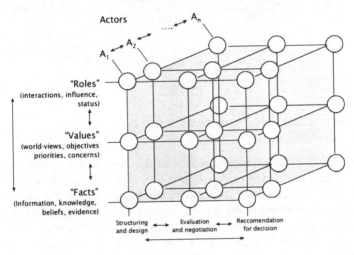

Figure 1. General framework for stakeholder analysis (adapted from Burgoyne, 1994).

The following sections briefly illustrate this approach applied to a large transport infrastructure problems. A full description of the case can be found in Beinat (1998).

4 A case study: the Betuweroute project

The Betuweroute is a new railway project which will connect the port of Rotterdam to Germany (cf. Figure 2). It is one of the priority Trans-European Network projects. The Betuweroute is meant to improve the hinterland connections of the Rotterdam harbour, to increase the quality and availability of goods transports to the rest of

Europe, to stress the role of The Netherlands as a distribution country and to meet the future demands of goods transport. The proposed railway will follow a route of about 130 km from the Port of Rotterdam to the city of Emmerich in Germany. Although some of the area, around Rotterdam, is urbanised, most of the area affected is agriculture, pasture land or natural area.

A number of administrative units and decision actors are involved in the policy making process: the European Union, the Dutch Government, the Dutch railway Company, the Dutch ministries of Environment and Transport, the Provinces of Gelderland and South Holland, 27 municipalities, several NGOs and interest groups.

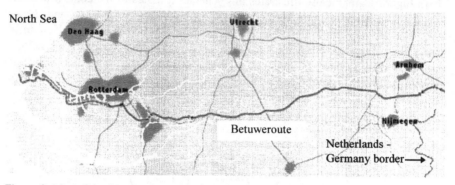

Figure 2. Map of the Betuweroute railway.

5 Structuring the value system of actors

Figure 3 shows a simplified representation of the main Dutch spatial policy units and actors, together with their objectives and concerns at the evaluation- negotiation stage. The figure distinguishes between the national perspective, the perspectives of the two provinces involved in the routing decision (South Holland and Gelderland) and the perspectives of the 27 municipalities involved (the third map illustrates only the perspective of two of them). As it can be seen, the explicit objectives and concerns of spatial policy actors are different and the emphasis changes by changing the scale and the location of the policy units.

Decision analysis provides the natural approach for addressing the value systems of spatial actors and their revealed preferences. For each actor, it is possible to structure a value tree and the set of attributes which serve to measure the performances of the decision according to the point of view of the actor. A synthetic, qualitative, overview of the result of this process is shown in Table 1.

The concerns shown are the collection of all concerns expressed by the spatial actors in Figure 3. Each item in the list aggregates several concerns which have a similar origin. As an example, "noise concerns" groups aspects such as "the number of houses within the 57dB noise contours"; "peak noise levels during the night", "cumulating of noise with other activities", "disruption of quietness in natural areas", which are relevant for different actors.

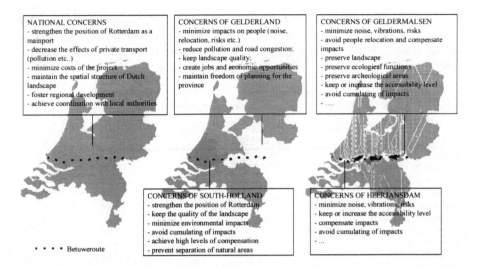

Figure 3. A simplified representation of the main spatial policy units and their objectives and concerns in the decision process of the Betuweroute railway project.

The √ marks in Table 1 are a qualitative indication of the relative psychological importance attached to each concerns. This information originates from surveys and public hearings held during the decision process (TB, 1996) and from a lexicographic analysis of the written complaints presented by the actors. The set of municipalities has also been re-grouped, separating the 8 municipalities closer to the Rotterdam harbour (TB, 1996: Group A), which perceive the economic benefits of the project, and the remaining 19 municipalities, more concerned with the social and environmental effects (Group B).

It is interesting to note that the relative, psychological, importance[2] attached to different performances of the project varies greatly across the actor set in line with scale considerations. While the national authorities stress the importance of decreasing air pollution, decreasing traffic congestion and fostering economic benefits, local authorities focus on the direct effects to the population (noise, risks, visual impacts etc.) to a much greater extent. The role of the provinces, as intermediate spatial actors, is somewhat in the middle. On some aspects, the emphasis is closer to that of the national actors (e.g. risks) while in others it is closer

---

[2] It is important to notice that there is no immediate and direct relationship between the psychological importance expressed by the actors and the concept of weights as used in multicriteria (MCA) methods. These psychological importance judgements are based on the value systems of the actors and on the information they hold on the expected effects of the project. However, they are not the result of an explicit weighting procedure which deliberately separates between preferences on different levels of impacts and the impacts themselves (cf. Keeney, 1992; Beinat, 1997).

to that of the local actors (e.g. noise). However, it is also possible to notice the relevance of the scale dimension. The effects on water resources, for instance, is stressed mainly by provincial authorities. This is consistent with the spatial scale of water problems, which is best addressed at a regional level.

Table 1. Concerns of the main policy units regarding the Betuweroute.

Concern	Ministries	Province of South Holl.	Province of Gelderland	Municip. Group A	Municip. Group B
People					
Community Effects	√	√	√	√√√	√√√
Noise, vibrations	√√	√√√	√√√	√√√	√√√
Cumulating effects	√	√√	√√	√√√	√√√
Increased risk	√√	√√	√√	√√√	√√√
Air pollution	√√√	√√	√√	√√√	√√√
Visual Impact	√√	√√√	√√√	√√√	√√√
Accessibility	√	√√	√√	√√√	√√√
Congestion	√√√	√√	√	√√	√
Resources					
Agriculture	√	√	√	√	√
Ecology	√√	√√	√√	√	√√√
Landscape	√√√	√√√	√√√	√	√√√
History and culture	√	√√	√√	√	√√√
Water management	√	√√√	√√√	√	√
Land-use	√	√	√√√	√√	√√
Economy					
Job Losses	√	√	√√	√	√√
Job Gains	√√√	√√√	√√√	√√√	√√
Cargo transport	√√√	√√	√	√	√
Costs	√√√	√	√	√	√
Damage compensation	√	√	√	√√√	√√√

Legend: √√√ = Major concern, √√ = Moderate concern, √= Minor concern

This information indicates potential sources of conflicts if the project does not satisfy the concerns of all actors. In this case, the project has been designed mainly to satisfy the national concerns, resulting into a performance profile which suites the goal of the national authorities to a large extent. However, this has resulted into poor performances on aspects relevant for the local decision actors and raised strong conflicts.

6 Spatial information and impacts for conflict analysis

Impact assessment is used to measure the effects of the railway in line with the demands of the decision actors. As shown above, the views and concerns differ spatially. This implies that a unique information basis for the whole project is not appropriate. Instead, it is necessary to tailor the provision of spatial information on the demands of individual actors at different spatial scales and locations. The use of Geographical Information Systems (GIS) is here very useful. GIS serve multiple purposes. They serve to store, organise and visualise spatial information and impacts of the project spatially. However, their also offer sophisticated methodologies to process data and to establish a link between spatial impacts and the information necessary to the actors to assess the project under they view point.

By means of spatial modelling it is possible to estimate the impacts of the project along the full length of the line. This calculations are typically carried out in Environmental Impact Assessment (EIA) studies. However, EIA is often unsuitable for conflict analysis as it implies a single-perspective evaluation of effects, substantially independent from spatially differentiated perspectives. In other words, the magnitude of the impacts does not necessarily reflect its acceptability and cannot be evaluated without considering the perspective of those who are affected. The following example illustrates the point.

The municipalities of Elst and Bemmel in the eastern part of the Betuweroute (Figure 4) are affected by significant direct impacts of the railway (noise, demolitions, landscape modifications, etc.). The impacts in these two areas are different, but compared to other municipalities, the differences are relatively small. For instance, the number of houses which will be demolished or that loose their function are 17 in Elst and 15 in Bemmel; the number of houses exposed to noise levels higher than 57dB are 34 in Elst and 28 in Bemmel; the lenght of the railway in characteristic landscape is 5.4Km in Elst and 8Km in Bemmel. Nevertheless, the reactions and the behavious of these two municipalities has been very different during the decision process (Drunen and Beinat, 1998). For instance, the inhabitants of Elst have filed nine noise specific complaints to the national authorities, compared to the two in Bemmel. In addition, when municipalities were required to declare their intention to co-operate with the national authorities in the planning process, the municipality of Bemmel declared its intention of doing so, while Elst formally opposed the choice.

The analysis of the value system of the local authorities based on multicriteria value functions (cf. Beinat 1997) served to highlight the relative importance of different factors. For instance the relative importance of noise, in terms of trade-off with other impacts, is rather similar in Bemmel and Elst. Significant differences, on the contrary, appear on the relative importance of the loss of agricultural area and on the number of houses to be demolished (cf. Beinat, 1998). The result for noise partly contraddicts the previous evidence, which associated to Bemmel a lower number of noise-related complaints. On the other hand, this is in line with the consideration that the concept of relative importance as a trade-off between impacts is not necessarily equivalent to the notion of psychological importance.

The explanation of the differences between these two local policy units is very complex and can hardly be based solely on the socio-economic features of the areas. It also important to stress that the impact indicators used in this case (and in almost all practical cases) are only proxies of the real effects which trigger the assessment (cf. the concept of proxy attribute in Beinat, 1997). For instance, the number of houses to be demolished in a community can capture only partially the meaning associated to the loss of some specific buildings, located in specific areas and inhabited by specific people. These aspects, in addition to the bare number of building demolished, determine the importance of loosing a certain number of specific houses in a given community. In some cases the inclusion of more specific impact indicators may help, but these indicators are usually very site specific and different indicators are necessary for different areas.

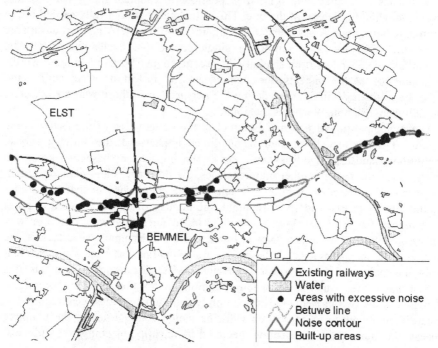

Figure 4. Noise contours and areas exposed to noise levels higher than 57dB in the municipalities of Bemmel and Els.

7 Conclusions

This paper illustrates an approach for the analysis of spatial conflicts in transport policies. The approach combines spatial analysis and decision analysis within the scheme of stakeholder analysis. This offers a general framework for the study of the perceptions and concerns of multiple actors and the possibility of linking the consequences of decision to the value systems of the actors involved in the process. The example of the Betuweroute project shows the spatial differentiation of the objectives and concerns of the actors and the differences in the priorities they attach to different concerns. This can be used as an indication of the spatial conflicts and offers indication both for the design of improved options and for the focus of the negotiation efforts. The analysis of the impacts of the project also demonstrated that the project acceptance does not simply depend on the magnitude of impact but also, and especially, on the evaluation strategies of the actors affected. This type of approach is still in its development stage, but these preliminary results indicate a large scope for the applications and discloses opportunities for addressing conflicts at an early stage of the decision process. It also help increasing the transparency of the process and the communicability of the results.

References

Allor, D.J. (1991) Options for dispute resolution in the public processes on urban land development, in S.S. Nagel, M.K. Mills (Eds.), Systematic analysis in dispute resolution, Quorum, New York.

Armstrong, M.P. (1994) Requirements for the development of GIS-based group decision support systems. Journal of the American Society for Information Science, vol. 45, no.9, pp. 669-677.

Beinat, E. (1997) Value functions for environmental management, Kluwer, Dordrecht.

Beinat, E. (Ed.) (1998) Transport policies and spatial conflicts, Institute For Environmental Studies, Amsterdam.

Burgoyne, J.G. (1994) Stakeholder analysis, in C. Cassell, G. Symon (Eds.) Qualitative methods in organisational research, Sage, London.

Bogetoft, P., Pruzan, P. (1991), Planning with Multiple Criteria. North-Holland, Amsterdam.

Commission of the European Communities (1993) Trans-European networks, Office for Official Publications of the European Communities, Brussels.

Densham, P. J. (1991) Spatial decision support systems, in D. Maguire, M.F. Goodchild, D. Rhind (Eds.), Geographical Information Systems: principles and applications, Wiley, New York.

Drunen, M. van, Beinat, E. (1998) Public participation in the Betuweroute project. Institute for Environmental Studies, Vrije Universiteit, Amsterdam.

Douwen, W. (1996) Improving the accessibility of spatial information for environmental management, PhD thesis, Free University of Amsterdam, Amsterdam, The Netherlands.

Hersey, P., K.H. Blanchard (1988) Management of organisational behaviour, Prentice-Hall, N.J.

Huigen, J., P.H.A. Frissen, P.W. Tops (1993) Het project Betuwelijn: spoorlijn of bestuurlijke co-productie? PhD thesis, KUB, Tilburg.

Keeney, R.L. (1992), Value-Focused Thinking, Harvard, Cambridge.

Maguire, D., M.F. Goodchild, D. Rhind (Eds.) (1991), Geographical Information Systems: Principles and Applications, Wiley, New York.

NCGIA: National Center for Geographical Information and Analysis (1996) http://www.ncgia.edu/research/i17/I-17_home.html

Obermayer, N.J. (1994) Spatial conflicts in the information age, http://wwwsgi.ursus.maine.edu/gisweb/spatb/urisa/ur94024.html

Rosenhead, J. (1989) Rational analysis for a problematic world, Wiley, Chichester.

Simon, H.A. (1991) Bounded rationality and organisational learning, Organisation Science 2, 125-139.

TB (1996) Tracebesluit Betuweroute, SDU uitgeverij, Den Haague.

Vincke. Ph. (1992), Multicriteria Decision-Aid, Wiley, Chichester.

CONGESTION MODELING AND MULTI-CRITERIA SMOOTHING HEURISTICS

Sydney C.K. Chu

Department of Mathematics, University of Hong Kong, Hong Kong

Abstract. The background of this work is the operation of a specialist out-patient clinic under the Hospital Authority of Hong Kong. Congestion is one main problem that needs to resolve. This in turn leads to long patient waiting for specialist doctors' consultation. We focus our analysis on congestion modeling to evaluate its multi-criteria performance measures and propose a smoothing heuristics as the solution approach. Numerical results are provided to illustrate the possibilities of notable improvement.

Keywords: Congestion modeling, multi-criteria heuristics

1 Introduction

The objective of this study is the operational improvement of a major specialist out-patient (SOP) clinic under the Hospital Authority of Hong Kong. We present in this paper our designed approach for its congestion analysis: the aspects on modeling, on multi-criteria performance measures and on a smoothing heuristics.

Medical consultations of different medical specialties are organized into morning and afternoon sessions on two different floors of the SOP clinic building. Patient demands are very heavy; while doctors' time and space (consultation rooms and waiting lobbies) are severely limited.

Such tightly constrained service resources have led to high average (and maximum experienced) waiting times for patients; as well as serious congestion problems of waiting patients in many of the sessions.

In previous studies of ours on clinic operations, different approaches have been adopted to incorporate service measures such as patients' waiting times into traditional multi-criteria mathematical programming formulations for medical resources [1]. In some cases, detailed simulations with process-tracing (or patient-following) technique [4] have been useful in estimating the model parameters.

For the emphasis on congestion modeling here, we propose an analytical model with linearly related input and output measures (more often used in inventory models [2]) to describe the predicted level of patient congestion over time.

This novel approach of congestion modeling is then extensively validated by simulation, a technique widely adopted in modeling the flows of patients [3,5,6]. The results show extremely good model consistency for various scenarios of operational policies. This congestion model subsequently becomes the performance evaluator (a subroutine) embedded in an overall local smoothing heuristics aimed at helping the congestion bottleneck situation.

For the heuristics, our experience with the system has suggested various prioritized goals, including the four measures of maximum/average numbers of waiting patients, maximum/average waiting times of patients, and their derived combinations.

The simple form of our congestion model has thus contributed to a feasible approach to ease the service- and session-specific level of congestion. Numerical results are provided to illustrate this otherwise (by simulation alone) difficult-to-achieve objective.

2 Congestion Modeling and Analysis

The stated goal of our project is the operational improvement of our benchmark specialist out-patient clinic, with outcomes of the study applicable to other such similar clinics in Hong Kong. To this end, we adopt a modeling approach flexible enough for and capable of approximating various scenarios of SOP clinic operations. The focused attention is on space congestion caused by over-crowding of patients awaiting eventual consultation from the medical specialists.

2.1 Mathematical Model

For a particular specialist session, the mathematical model takes the form of a simple (single-item) production-inventory model over its scheduled duration. Schematically, its graphical representation can be illustrated as in Figure 1 for a certain combination of input model parameters.

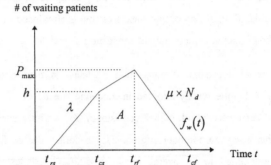

Figure 1. Congestion Model of the Approx. No. of Waiting Patients in a Session.

Notations

Parameters:

N_p = No. of patients served in the session

N_d = No. of doctors for the session

t_{rs} = Registration start-time

t_{rf} = Registration finish-time

t_{cs} = Consultation start-time

t_{cf} = Consultation finish-time

Variables:

λ = Patient arrival rate

μ = Consultation rate per doctor

A = Area under the curve

h = Slope changing point (at either time t_{rf} or t_{cs})

t = Time

Functions:

$f_w(t)$ = No. of waiting patients at time t during the session

Performance Measures:

P_{max} = Max. no. of waiting patients

W_a = Average waiting time for a patient

Figure 1 (Cont'd). Congestion Model.

N_a = Average no. of waiting patients

W_{max} = Max. waiting time for a patient

where $\lambda = \dfrac{N_p}{t_{rf} - t_{rs}}$; $\mu = \dfrac{N_p}{\left(t_{cf} - t_{cs}\right) \times N_d}$; $W_a = \dfrac{A}{N_p}$; $N_a = \dfrac{A}{t_{cf} - t_{rs}}$

Therefore

$$P_{max} = \left(t_{cf} - t_{rf}\right)\mu \times N_d$$

$$= N_p \frac{\left(t_{cf} - t_{rf}\right)}{\left(t_{cf} - t_{cs}\right)}$$

$$h = \left(t_{cs} - t_{rs}\right)\lambda$$

$$= N_p \frac{\left(t_{cs} - t_{rs}\right)}{\left(t_{rf} - t_{rs}\right)}$$

$$A = \frac{\left(t_{cs} - t_{rs}\right)h}{2} + \frac{\left(P_{max} + h\right)\left(t_{rf} - t_{cs}\right)}{2} + \frac{\left(t_{cf} - t_{rf}\right)P_{max}}{2}$$

$$= \frac{\left(t_{rf} - t_{rs}\right)h + \left(t_{cf} - t_{cs}\right)P_{max}}{2}$$

$$W_a = \frac{A}{N_p}$$

$$N_a = \frac{A}{t_{cf} - t_{rs}}$$

$$f_w(t) = \begin{cases} \lambda \times \left(t - t_{rs}\right) & \text{if } t \in \left[t_{rs}, t_{cs}\right] \\ h + \left(\lambda - \mu \times N_d\right)\left(t - t_{cs}\right) & \text{if } t \in \left[t_{cs}, t_{rf}\right] \\ \left(t_{cf} - t\right)\mu \times N_d & \text{if } t \in \left[t_{rf}, t_{cf}\right] \\ 0 & \text{otherwise} \end{cases}$$

$$W_{max} = \max\left[\left(t_{cs} - t_{rs}\right), \left(t_{rf} - t_{rs} - \frac{\left(t_{rf} - t_{cs}\right)\mu N_d}{\lambda}\right), \left(t_{cf} - t_{rf}\right)\right]$$

This standard OR model is used here to represent the number of waiting patients in a particular specialist session by the "inventory" level over time, with new arrivals interpreted as "production" and departing patients after consultation as "demand". This representation thus has a very strong intuitive appeal provided we can accept the basic model assumption that patient arrival and consultation service rate are approximately constant over their respective time intervals.

Figure 1 illustrates the case of consultation start-time before registration end-time and the arrival rate larger than total (over all specialists) service rate. There are a total of five cases arising from the different combinations of these four parameters. They lead to similar (but different in shape) congestion curves and formulae for calculating the four performance measures: maximum/average numbers of waiting patients as well as maximum/average waiting times of a patient. For conciseness, graphs of these other four cases are not shown here.

Finally, since waiting patients for different specialities scheduled in a same session "aggregate" in the same floor lobby, their numbers add up. The approximation of the total number of waiting patients (during a session) can therefore be obtained by summing their individual appropriate congestion curves as illustrated schematically in Figure 2.

Figure 2. Aggregating the Approx. Total Number of Waiting Patients in a Session.

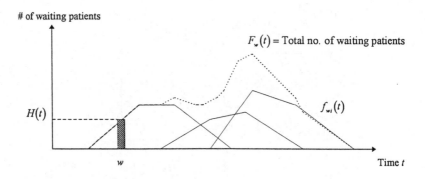

Notations
Variables:
w = Width of time interval (for discretizing the curve $F_w(t)$)
T = Total time under consideration
Functions:
$f_{wi}(t)$ = No. of waiting patients at time t for the i-th specialist session
$F_w(t)$ = Total no. of waiting patients at time t
$H(t)$ = Frequency count of waiting patients at time t
Performance Measures:
P_{max} = Max. no. of waiting patients

Figure 2 (Cont'd). Aggregating the Approx. Total Number of Waiting Patients.

W_a = Average waiting time for a patient

N_a = Average no. of waiting patients

W_{max} = Max. waiting time for a patient

where

$$F_w(t) = \sum_i f_{wi}(t)$$

$$H(t) \approx \frac{F_w\left(t + \frac{w}{2}\right) + F_w\left(t - \frac{w}{2}\right)}{2}$$

$$P_{max} = \max H(t) \qquad \forall t; \qquad A \approx \sum H(t) \times w$$

$$W_a = \frac{A}{\sum N_p} \quad ; \quad N_a = \frac{A}{T}$$

$$W_{max} = \max\{w_{max} | w_{max} = \text{max. waiting time for the given speciality}\}$$

2.2 Model Validation

The simple congestion model requires approximating actual arrival as well as service rates by their constant counterparts. Such validity is established by rather extensive simulations. It is compared against various modes of postulated patient arrival behavioral patterns (of a same set of collected data), namely "Fix Arrival", "Uniform Arrival" (within a half-hour interval), "Exponential Arrival" and under a "Planned Booking System". The comparisons are based on the performance measures of maximum number of patients observed, average waiting time per patient and average number of patients in a session.

These three measures are first tabulated and then subject to the non-parametric pair-wise comparison Kolmogorov-Smirnov tests. Very high model consistency and validation are indicated by the tests' p-values all being less than 10^{-6}. As an illustration, we show the one case of differences in average waiting times between the "inventory" model and the exponential arrival scenario (for all 58 specialty-session pairs) in Figure 3.

Figure 3. Validation by Simulation with Exponential Arrival Patients.

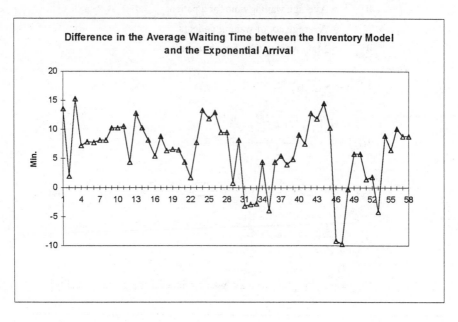

2.3 Multiple Performance Measures

For performance evaluation, our focused attention suggests multiple prioritized goals, consisting of measures of maximum/average numbers of waiting patients, maximum/average waiting times of patients, and their derived combinations. This multi-criteria objective for each session is represented by its lexicographically ordered vector given by the following.

2.4 Lexicographical Vector Ordering

Session congestion vector v is defined as:
$$v_i = [T_i * AvgP_i, MaxP_i, AvgW_i, AvgP_i, MaxW_i] \qquad \text{for } i = 1, 2, \ldots, N_s$$
where

Ns	= Total no. of sessions
Ti	= Duration of session i (i.e. time from the first arrival to the last departure).
$AvgP_i$	= Average no. of patient in session i
$MaxP_i$	= Max. no. of patient in session i
$AvgW_i$	= Average waiting time of session i
$MaxW_i$	= Max. waiting time of session i

Lexicographical order is then defined as:

$$v_i = [a_1, a_2, \ldots, a_l] \succ v_j = [b_1, b_2, \ldots, b_l]$$
$$\text{if } a_k - b_k \geq \varepsilon \text{ and } |a_r - b_r| < \varepsilon \ \forall r < k$$

Each session's congestion can thus be evaluated. The overall (multi-criteria) objective is to "minimize" various sessions' congestions taken as a whole.

2.5 Peak Smoothing Heuristics

Congestion has different meanings for different specialties, since different amount of space is considered as necessary for different types of patients. (For example, wheel chair orthopedic patients occupy more space than general medical patients.) Nevertheless, it is possible to characterize a congestion index for each specialty. For a session, it will be classified as "congested" if the aggregated patient-peak (i.e. maximum number of session-patients) derived from these indices exceed a specified limit. Such sessions then require help in relieving their "congestions" under the current schedule. It is therefore appropriate to devise a patient-peak removal procedure to primarily ease the congestion by reducing the total "work" of that session; and hopefully improves on the patients' average and/or maximum waiting times. Our proposed peak smoothing heuristics is specifically designed for this purpose.

Operational rules also impose natural limits on the number of times of thus "splitting" and/or "merging" of individual specialties; and such moves are considered only for more than 10% immediate improvement on congestion vectors (of the multiple performance measures). A general description of the algorithm is given as follows.

2.6 Patient-peak Smoothing Heuristics Algorithm

Step 1. Rank all the individual sessions' congestion vectors by their lexicographical ordering. Label sessions as "congested" or "non-congested" relative to their congestion indices. If there exists no congested sessions, go to Step 5.

Step 2. Select the most congested session as the candidate to "split"; and the least congested session as the candidate to "merge".

Step 3a. Check for compatibility of the two candidate specialties; as well the constraints on splitting/merging limits and improvement thresholds. In case of failure to move, select the second least congested session. Repeat this try until a compatible session is identified or no non-congested sessions remain.

Step 3b. Perform the move if the candidates are confirmed in Step 3a; and then go to Step 4. Otherwise select the second most congested session (to "split") and the least congested session (to "merge"); and return to Step 3a. In the second case, if there exists no further choice for a congested session, go to step 5.

Step 4. Recalculate and rank the sessions' congestion vectors. Go back to Step 2.

Step 5. Stop.

We note that the move (in Step 3b above) involves also determining as many patients and doctors as possible to reschedule to another receiving (and compatible) session, without overcrowding the latter. That is, after the move, the identified "congested" session will still be more congested than the (receiving) "non-congested" session. (This requires a simple line search for the optimal number of patients to be moved.) An illustration of the before-and-after performance based on this multi-criteria smoothing heuristics is given in Figures 4 - 7 below.

Figure 4. Performance of Smoothing Heuristics on Max. Patient Number.

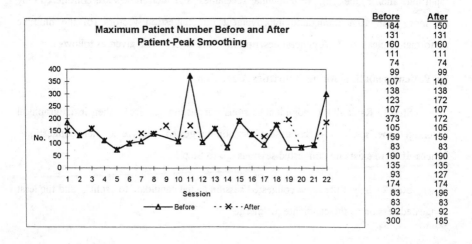

Before	After
184	150
131	131
160	160
111	111
74	74
99	99
107	140
138	138
123	172
107	107
373	172
105	105
159	159
83	83
190	190
135	135
93	127
174	174
83	196
83	83
92	92
300	185

Figure 5. Performance of Smoothing Heuristics on Avg. Patient Number.

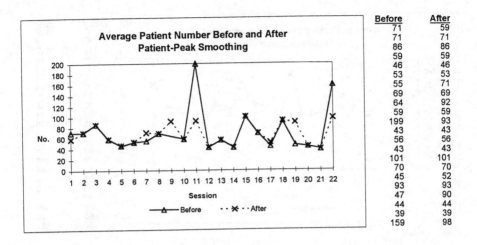

Before	After
71	59
71	71
86	86
59	59
46	46
53	53
55	71
69	69
64	92
59	59
199	93
43	43
56	56
43	43
101	101
70	70
45	52
93	93
47	90
44	44
39	39
159	98

While Figures 4 and 5 above indicate the (improved) performance measures in terms of patient numbers, Figures 6 and 7 below concern the measures of waiting times.

Figure 6. Performance of Smoothing Heuristics on Max. Waiting Time.

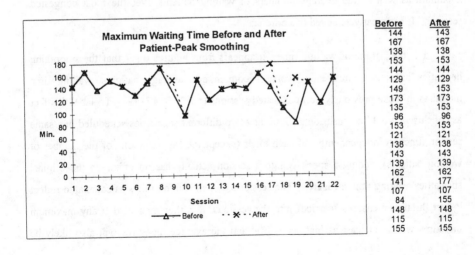

Before	After
144	143
167	167
138	138
153	153
144	144
129	129
149	153
173	173
135	153
96	96
153	153
121	121
138	138
143	143
139	139
162	162
141	177
107	107
84	155
148	148
115	115
155	155

Figure 7. Performance of Smoothing Heuristics on Avg. Waiting Time.

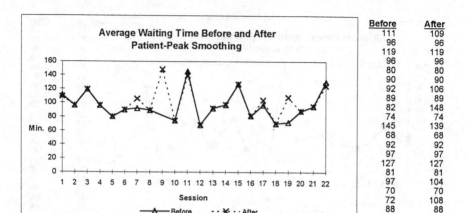

Before	After
111	109
96	96
119	119
96	96
80	80
90	90
92	106
89	89
82	148
74	74
145	139
68	68
92	92
97	97
127	127
81	81
97	104
70	70
72	108
88	88
95	95
130	125

3 Conclusion

As the specific design of the smoothing heuristics is to remove (or lower) the patient-peak (the maximum number of patients experienced during a session), the rescheduling of a number of patients from congested sessions (such as no. 11 and 22 in Figures 4 - 5) to other non-congested sessions (such as no. 7, 9, 17 and 19) has profound impact on the maximum as well as the average numbers of waiting patients over the most congested sessions. It has clearly achieved this criteria.

For the criteria of waiting times, it appears from Figures 6 - 7 that the smoothing heuristics has led to noticeably worse maximum and average patients' waiting time measures in those previously non-congested sessions (no. 7, 9, 17, 19). To label this fact as conclusive would be misleading. Since any two different specialties scheduled in a same session function independently of each other (except the "aggregation" of the number of waiting patients), adding a specialty into a session actually has no effect on the original specialties' waiting time measures. The session's overall waiting times as calculated reflect in fact the time measures "carried" into the sessions. It is then evident that any maximum measures will be at least as high as before; and the average measures will also likely be averaged out to be higher since the added specialty (into the session) was necessarily from a congested, hence worse-performing setting.

The actual waiting times as experienced by the patients after such patient-peak smoothing should remain largely comparable as before. The less overall congested situation also implies further improvement possibilities on waiting times by addressing the start-end times of patients' arrival and registration, relative to doctors' consultation.

While this reconfiguration of registration/consultation does not fall into the premise of our current study, a planned (patients') booking system is being actively pursued. Our work here will very likely form the basis to establish that system as the norm of operation of many of the SOP clinics in Hong Kong in the near future.

4 References

[1] S.C.K. Chu and M.P.P. Ho, "A Modelling Framework for the Doctor Allocation Problem of a Multi-Branch Medical Practice", Proceedings of the Third Conference of the Operational Research Society of Hong Kong, (1993), 159-166.

[2] S.C.K. Chu, "Inventory Minimization and Production Scheduling: the case of a large Hong Kong company", International Transactions in Operational Research, Vol. 1, No. 3, (1994), 271-283.

[3] M.P.P. Ho, "A Case Study on a General Practice in Hong Kong", Proceedings of the First Conference of the Operational Research Society of Hong Kong, (1991), 31-39.

[4] M.P.P. Ho, S.C.K. Chu, and Y.L. Wong, "Simulation Modelling for the Operation of a Comprehensive Clinic", Proceedings of the Second Conference of the Operational Research Society of Hong Kong, (1992), 130-140.

[5] G. Romanin-Jacur and P. Facchin, "Optimal Planning of a Paediatric Semi-intensive Care Unit via simulation", European Journal of Operational Research, Vol. 29, (1985), 192-198.

[6] M. Wright, "The Application of a Surgical Bed Simulation Model", European Journal of Operational Research Society, Vol. 36, No. 6, (1985), 517-523.

A Real-World MCDA Application in Cellular Telephony Systems

Carlos A. Bana e Costa[1], Leonardo Ensslin[2], Italo J. Zanella[2]

[1] Technical University of Lisbon, IST, Av. Rovisco Pais, 1000, Lisbon, PORTUGAL
[2] Federal University of Santa Catarina - EPS, 88.010-970, Florianópolis/SC, BRAZIL

Abstract. This paper addresses important issues which were present in the Structuring and Evaluation phases of the constructive learning process of municipality selection for expanding the cellular phone system in the State of Santa Catarina, in southern Brazil. In particular, we discuss questions such as "what is the problem?" and "how to approach it?", "how to identify and make operational the fundamental points of view for municipality evaluation?", and "how to deal with context and judgemental dependencies?".

Keywords. MCDA, structuring, cognitive mapping, successive choice

1 Introduction

Telesc Cellular is a business unit of TELESC S/A, the state company responsible for phone services in the State of Santa Catarina, in southern Brazil. As a part of the on-going privatization process launched recently by the Brazilian Government, the Telecommunication Ministry will next year, sell to private operators concessions for a new frequency zone, called "b" band. Besides that, in the short-term, Telesc Cellular will be transformed into a new independent company and privatized.

Not surprisingly, given this increased pace of change, Telesc Cellular managers have become increasingly concerned about the competitive capacity of the new company in the open-market.

It was decided that it is strategically very important to accelerate the expansion of Telesc Cellular services throughout the State of Santa Catarina. This will of course increase the selling value of the (future) company. Thirty municipalities with expansion potential have been identified based on screening criteria. A budget has been fixed, and the costs of expanding the system in (or to) each municipality have been estimated.

The problem was to identify priorities for expansion in the near future. The process by which this selection was made using multicriteria decision-aids is the subject of this paper.

We start by discussing in section 2 the "decision-making" and "decision-aiding" *problematiques* of the case - see (Bana e Costa, 1996). Sections 3 and 4 describe the structuring phase of the process, and section 5 approaches the evaluation and selection of the municipalities recommended for integration into the expansion program. A final note is presented in section 6.

2 Decision-making and decision-aiding problematiques

The first problem that confronted the analyst was to understand "what is the problem?" from the decision-maker's or client's perspective (the *decision-making problematique*). This was essential in order to get a wise answer to the question "how shall the study approach the problem?" (the *decision-aiding problematique*).

For the Telesc Cellular managers - our (collective) decision-maker - the problem was how to select (for investing in) the municipalities which will best increase the market position of the company. This selection is important to create a "sustainable competitive advantage" (cf. Aaker, 1995). The implementation of a good expansion plan will anticipate and therefore restrict the market opportunities for future competitors.

The initial budget allocated for expanding the system was $30 million, and the estimated expansion costs in the thirty selected municipalities range from a maximum of $10 million in Florianópolis to $150 thousand in Maravilha (see Table 1).

Table 1. Municipalities and their respective expansion costs

Municipality	Cost (x $1,000)	Municipality	Cost (x $1,000)
Florianópolis	10,000	Ibirama	500
Joinville	8,000	Itapiranga	400
Blumenau	3,500	Timbó	400
Baln. Camboriú	2,500	Corupá	400
Criciúma	2,000	Capinzal	400
Itajaí	1,700	Gov. Celso Ramos	400
Chapecó	1,200	Ilhota	400
Brusque	1,200	Turvo	350
Jaraguá do Sul	1,000	Guabiruba	350
Araranguá	900	Paulo Lopez	300
Lages	800	Campos Novos	250
Tubarão	800	S. Miguel do Oeste	200
Concórdia	700	Taio	200
S. Bento do Sul	700	Ituporanga	150
Joaçaba	500	Maravilha	150

What then is the set A of potential actions? One possibility would be to adopt a "comprehensive conception" for A (Roy, 1996) and compare all the sub-sets of municipalities with a total cost close to $30 million. The problem would be formulated so as to choose the "best" combination (a technical choice-problematique). However, the search for the "best" would technically require the difficult definition of criteria to compare sub-sets of municipalities (and not to compare the municipalities with each other). More importantly, this formulation would not fit the decision-maker's desire for ranking the municipalities by their individual attractiveness for investment (a decision-making ranking-problematique).

A "fragmented conception" of A (in the sense that investing in one municipality does not preclude investment in another - Roy, 1996) was considered adequate for the Telesc case. But, should a conventional ranking procedure be adopted? In fact, no, because each action is not independent, given that the expansion of the cellular phone service in (or to) a particular municipality may also benefit (serve) adjacent zones within other municipalities.

How then have we dealt with this dependence phenomenon? Actually, we adopted what Bana e Costa (1996) called a technical problematique of "successive choice", formulating the problem in terms of a sequence of choices of the "best" action (see Fig. 1). A municipality selected at a choice stage was taken out of the set for the next stage, and before going ahead the problem was re-framed to incorporate the repercussions of the previous choice over the performances of the remaining municipalities. For example, Telesc managers said, the eventual expansion of the cellular phone service in Joinville will cause an improvement in the technical conditions in Florianópolis, and vice-versa. Consequently, the selection of one of the two would decrease the expansion priority of the other. (Note that the municipalities are "alternatives" at each choice stage.)

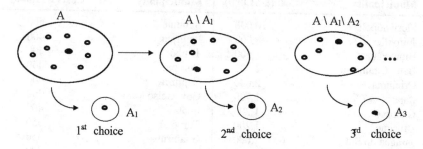

Fig. 1. Successive choice problematique

3 Fundamental points of view (FPsV)

Having defined the technical problematique of the study and the type of set of potential actions, the next decision-aiding activity was the identification of the "fundamental points of view" (FPsV) (*cf.* Bana e Costa, 1992) in terms of which the municipalities should be compared. This has been an interactive learning process which started with individual interviews with the five managers of Telesc Cellular. Each manager had been invited to express his own thinking on the objectives to be attained in expanding the system and the characteristics of the municipalities that should be taken into account. Moreover, cause-effect or means-ends relations among stated aspects were identified. After the initial interviews, we developed a "cognitive map" - CM (*cf.* Eden, 1988; Eden *et al.*, 1992) integrating all the relevant points of view (PsV) of the managers in the form of a "means-ends network-like structure" (Belton *et al.*, 1997). The process then evolved to a group discussion of the CM, as a basis for identifying common PsV. During the workshop some new PsV emerged, while other ones were eliminated or re-defined, and new interrelations were identified.

Fig. 2 shows the central part of the final CM representing the problem as the five managers collectively perceived it. Note that the final CM was constructed live and interactively during the workshop, with the support of the Graphics COPE software (Banxia Software Ltd., 1995), evolving incrementally from the initial CM that we had proposed.

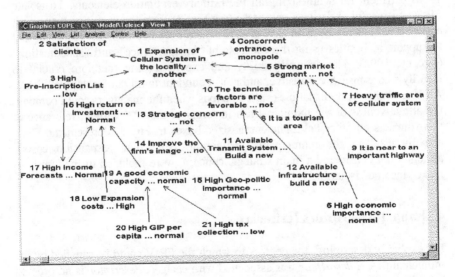

Fig. 2. Cognitive map for the Telesc Cellular case

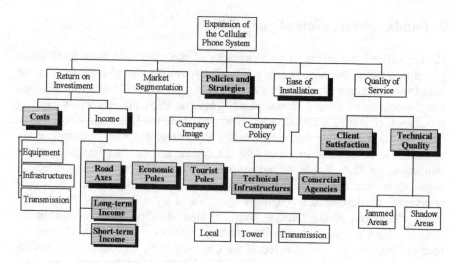

Fig. 3. Telesc Cellular tree of FPsV

As detailed elsewhere (*cf.* Bana e Costa *et al.*, forthcoming), our way of approaching an initially ill-structured problem, by cognitive mapping, is intentionally oriented towards the progressive appearance of the FPsV, in this case for municipality evaluation, which are isolatible "significant axes" (Roy, 1996) pertinent for synthesizing all the primary evaluation elements. To isolate FPsV, it is useful to derive a value tree (see Fig. 3) from the cognitive maps built earlier. This has not been an easy task, for there does not exist any technical tool to support it, besides using the concepts of "independence" and "specification" (Keeney, 1992). The analyst's ability and experience are therefore crucial to identify a consensual, exhaustive, and non-redundant family of "isolatible" FPsV. A PV tended to be an isolatible evaluation axis when the managers could compare the attractiveness of the municipalities in terms of that PV without reference to their impacts on other PsV. The emphasized boxes in Fig. 3 represent the FPsV actually used in the municipality evaluation. Note that "expansion costs", although not considered a main strategic concern, were included in the family to make "income" isolatible.

4 Impact descriptors (criteria)

For explicitly describing the degrees to which the FPsV could be satisfied by the municipalities, a *descriptor* was associated with each. A *descriptor* is an <u>ordered</u> set of plausible *impact levels* intended to serve as a basis to describe, as objectively and unambiguously as possible, the *impacts* of the actions in terms of a given PV (Bana e Costa, 1992). This is similar to Keeney's (1992) notion of an *attribute* and fits with Bouyssou's (1990) broad definition of a *criterion* as "a tool allowing comparison of alternatives according to a particular PV".

Note that lower-level PsV are shown in the value tree of Fig. 3. They are "elementary" PsV specifying some FPsV, and they *helped define* the criteria actually used (*de facto*, some of them only emerged at this stage of the process).

The FPsV "costs", "long-term income", "short-term income", and "client satisfaction" were made operational for municipality evaluation by quantitative descriptors: respectively, "total expansion costs" (see Table 1), "average annual municipal taxes" (both measured in dollars), "number of registered requests for new cellular phones", and "number of complaints over system disruptions (reported by Telesc Client Assistance Service)". The first attribute is a "direct" or "natural" descriptor (where the lower the cost the better it is), and the other three indicators are "indirect" or "proxy" descriptors (for all three, the attractiveness for expansion directly increases with the level of impact). (For a discussion of different types of descriptors see Bouyssou, 1990, Keeney, 1992, Bana e Costa, 1992.)

Qualitative descriptors were associated with all the other FPsV. Some of them were easily constructed, like that for "existence of a commercial infrastructures", which can simply be described by the dichotomous scale "yes (a Telesc agency already exists in the municipality)" or "no, ...".

For "road axes", which natural description would require the unobtainable data for clients in transit on highways in the State, a rough and ready appraisal in terms of the importance of highways crossing the municipalities was accepted (see Table 2).

The remaining FPsV are significant axes synthesizing several inter-related concerns. In these cases, a "decomposition-recombination" technique was used for overcoming the difficulties associated with the presence of several dimensions. To avoid lengthening the paper unnecessarily, we will only detail the application of this construction technique in the two illustrative and interesting cases of rendering operational the FPsV "existence of local infrastructures" and "technical quality", which involve respectively three and two dimensions.

Table 2. Descriptor of "road axes"

Impact levels	Description
N5	Municipality crossed by the highway BR-101.
N4	Municipality crossed by the highway SC-470.
N3	Municipality crossed by the highway BR-282.
N2	Municipality crossed by the highway BR 116.
N1	No important highway crosses the municipality.

Table 3. Descriptor of "existence of local infrastructures"

Levels	Description
N4	There exists a place to install the cellular system, a tower for the antenna and available transmission lines.
N3	There exists a place to install the cellular system, a tower for the antenna but transmission lines are not available, or there exists a place to install the cellular system and available transmission lines but a tower to install the antenna does not exist, or there exists a tower to install the antenna and available transmission lines but a place to install the cellular system does not exist.
N2	There exists a place to install the cellular system but neither a tower to install the antenna nor transmission lines are available, or there exists a tower to install the antenna but neither a place to install the cellular system nor transmission lines are available, or Transmission lines are available but a place to install the cellular system and a tower for the antenna do not exist.
N1	A place to install the cellular system and a tower for the antenna do not exist and transmission lines are not available.

Table 3 shows the constructed descriptor of the FPV "existence of local infrastructures", specified by the elementary PsV "existence of a place for installing the system", "existence of a tower to install the antenna", and "availability of transmission lines". First, each of these PsV was defined by a binary yes-no descriptor. After they were recombined, and the eight possible combinations of elementary states "yes" and "no" were appropriately ordered by the Telesc managers. This gave rise to the four impact levels in Table 3. Note that the combined profiles involving two elementary states "yes" and one "no" (that is, "yes-yes-no", "yes-no-yes" and "no-yes-yes") were judged equally attractive, thus defining only one comprehensive impact level (the same applies for the combinations with only one "yes").

A more subtle case is that of "technical quality". We tended to consider the PsV "existence of shadow areas" and "existence of jammed areas" (see Fig. 3) as isolatible (note that a shadow area is never jammed), and separate descriptors have been associated with them. The former have been described by two levels: "shadow areas are critical (where communication is very difficult to establish)" and "shadow areas are not critical". For the latter, the managers viewed it necessary to distinguish three levels: "critically jammed areas (where congestion goes over a given threshold)", "not critically jammed areas (low congestion)", and "without jammed areas" (no congestion).

When asked if these descriptions were not too ambiguous or vague, the managers argued that every expert in Telesc Cellular would understand their meaning. So, our alternative suggestion for constructing some precise impact indexes has not been followed.

Because, where critical shadow areas existed the priority for expanding the system was judged high whatever the level of congestion, and *vice-versa*, the two PsV were apparently mutually preferentially independent. Nevertheless, some hidden synergetic value-effects were later revealed when we confronted the managers with the comparison of vectors of impact levels of the two PsV. For example, consider four dummy municipalities **m1**, **m2**, **m3**, **m4** with equal impacts in all the other PsV but with the following impacts in the two PsV in question:

m1 - with *critical shadow areas* and with <u>critically jammed areas</u>;

m2 - with *critical shadow areas* and with <u>jammed area **not** critical</u>;

m3 - with *no critical shadow area* but with <u>critically jammed areas</u>;

m4 - with *no critical shadow area* and with <u>jammed area **not** critical</u>.

When asked to semantically judge the difference of attractiveness between **m1** with **m2**, the managers considered it to be *moderate*. Nevertheless, they judged the difference of attractiveness between **m3** and **m4** to be only *weak*. These judgements reveal the existence of value "difference dependencies" between the two PsV. When confronted with this interesting dependence phenomenon, the managers explained it very easily and maintained their initial judgements.

It was then decided to compose the two PsV in a synthesized FPV (labeled "technical quality" in Fig. 3). It is worthwhile noting that this was not absolutely necessary for municipality evaluation, but it was dictated by the *constructive* wish to maintain the validity of the working hypothesis of *additivity* in the construction of the evaluation model (see next section). Accordingly, the descriptor of "technical quality" shown in Table 4 was constructed by ordering the six binary impact level combinations of "existence of shadow areas" and "existence of jammed areas". Note the *lexicographical* structure of the descriptor, revealing that the managers always gave a higher priority to lowering the (more, or less) negative effects caused by the shadow areas (which always exist and cannot be totally eliminated), regardless of the level of congestion existing in the municipalities.

Once descriptors for all FPsV were defined, the "initial" impact-profile of each municipality **m** $(g_1(m), ..., g_j(m), ..., g_{11}(m))$ was identified by associating **m** with one impact level of each descriptor ("initial" in the sense that, as explained in section 2, each impact is not necessarily constant throughout the successive choice stages).

Table 4. Descriptor of "technical quality"

Levels	Description: (municipality ...)
N6	... with critical shadow areas and with critically jammed areas.
N5	... with critical shadow areas and with jammed areas not critical.
N4	... with critical shadow areas but without jammed areas.
N3	... without critical shadow areas but with critically jammed areas.
N2	... without critical shadow areas but with jammed areas not critical.
N1	... without critical shadow areas and without jammed areas.

5 Evaluation and selection of municipalities

Evaluating the municipalities under conditions of certainty was a reasonable working hypothesis. Over and above this, it was decided to construct an additive value model to measure the overall attractiveness of the municipalities. Accordingly, the next phase after structuring the problem was devoted to the construction of (cardinal) partial value-functions to measure the relative attractiveness of the municipalities in terms of each FPV, and to the assessment of scaling constants ("weights") to harmonize the partial values. In both activities we used MACBETH ("Measuring Attractiveness by a Categorical Based Evaluation Technique), the evaluation procedure developed by Bana e Costa and Vansnick (1994 and forthcoming). Although in this paper the evaluation process will not be described, we refer interested readers to (Bana e Costa and Vansnick, 1997) and (Bana e Costa et al., 1998), where similar applications of MACBETH are presented in detail.

In the framework of a descriptive or prescriptive approach, the adoption of a riskless model could be seen as unrealistic, since the municipalities selected do involve risk, at least with regards to the FPsV "long-term income" and "short-term income". Faced with similar circumstances, Dyer and Lorber (1982, p. 674) give the following (prescriptive) argument in favour of the "assumption" of certainty:

> "There are very few real-world problems where the consequences of a decision are unambiguously determined with complete certainty. On the other hand, the assumption of certainty may simplify the problem sufficiently to offset a slight loss of realism in the analysis."

On the other hand, a (descriptive) argument is given by von Winterfeldt and Edwards (1986, p. 308): "In many practical situations, utility and value functions elicited with risky versus riskless methods are hardly distinguishable because of judgement error and because of stimulus and response mode effects ...". And they conclude: "Our position is this: since there is little *experimental* evidence for drastic differences between risky and riskless utility functions and since there are good *theoretical* arguments that they are identical, the choice between risky and riskless models should be left to the analyst."

We agree with both the above arguments. However, as we adopt a constructive approach in decision-aiding, "loss of realism" and "judgement error" are expressions with little substantive meaning in such an operational framework. Therefore, we constructively prefer to say that the family of FPsV would only be "exhaustive" (*cf.* Roy, 1996, §10.3.3) if the Telesc managers were to be indifferent to two municipalities with equal impacts (in all the FPsV) regardless of different "income" uncertainties. This has actually been the case.

Fig. 4 shows the scheme of the successive-choice model adopted for selecting the most attractive municipalities for expanding the cellular phone system in the State of Santa Catarina. At a particular choice stage, the overall value of each municipality still under consideration was calculated by the additive value model, and the one with the highest overall value was selected. Of course, two or more municipalities could be selected at the same choice stage, if their overall values were close enough to provoke rank reversals for slight weight variations. Here, sensitivity analysis played a crucial role. We note that, between two choice-stages, not only were the impact-profiles (and consequently the partial values) of the municipalities up-dated to incorporate the repercussions of the last choice, as explained in section 2, but also that the weights of the criteria must be eventually re-scaled (if, for at least one FPV, the best or worst impacts changed).

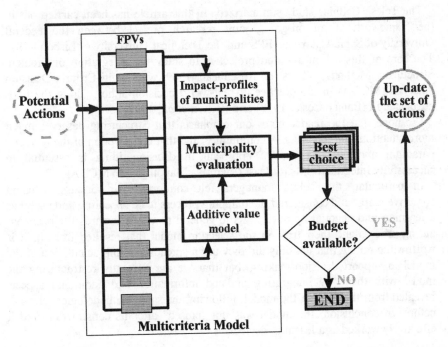

Fig. 4. Scheme of the successive-choice model for the Telesc Cellular case

Table 5. Selected municipalities

Evaluation	Municipality	Cost (x $1,000)	Σ
1st choice	Joinville	8,000	8,000
2nd choice	Florianópolis	10,000	18,000
3rd choice	Blumemau	3,500	21,500
4th choice	Itajaí	1,700	23,200
5th choice	Criciúma	2,000	25,200
6th choice	Baln. Camboriú	2,500	27,700
7th choice	Chapecó	1,200	28,900
8th choice	Jaraguá do Sul	1,000	**29,900**

Finally, by applying the successive-choice model while taking into account the budget limit of $30 million, eight municipalities have been recommended for integration into the Telesc Cellular expansion program (see Table 5).

6 Final note

The Telesc Cellular study summarized in this article has been carried out in the framework of an on-going joint research project between the Federal University of Santa Catarina - EPS and the Technical University of Lisbon - IST. The core of this action-oriented project is to study the integration of Problem Structuring Methods - PSMs (*cf.* Rosenhead, 1989) with Multi-Criteria Decision Analysis (MCDA) in the practice of real-world decision-aiding, along the lines advocated in (Bana e Costa, 1992) and (Belton *et al.*, 1997).

We hope that the reader notes our emphasis that Structuring deserves much more attention from the MCDA community, and that the adoption of the *learning* paradigm and the *constructive* approach in decision-aiding is essential to successfully integrate PSMs (at least Cognitive Mapping) into MCDA.

In particular, the Telesc managers felt comfortable in discussion using cognitive maps, and considered the mapping process to be appealing and useful as a learning and negotiating device. As to general reactions towards the decision-aiding process and their trust in the selection model that resulted from it, it is worthwhile noting that not only did they use a spreadsheet implementation of the model to support decision-making, but that we are currently restructuring the model with them, by integrating updated information and additional aspects revealed from their use of the model, following the same methodological steps as before. In conclusion, the model was not "closed" after its construction and it effectively worked as a learning device.

7 References

Aaker, D.A. (1992), *Strategic Market Management* (fourth edition), John Wiley & Sons, New York.

Bana e Costa, C.A. (1992), "Structuration, construction et exploitation d'un modèle multicritère d'aide à la décision", PhD Thesis, Universidade Técnica de Lisboa.

Bana e Costa, C.A. (1996), "Les problématiques de l'aide à la décision: vers l'enrichissement de la trilogie choix-tri-rangement", *Recherche Opération-nelle/Operations Research*, 30, 2 (191-216).

Bana e Costa, C.A., Ensslin, L., Corrêa, E.C., Vansnick J.C. (forthcoming), "Decision Support Systems in action: Integrated application in a multicriteria decision aid process", *European Journal of Operational Research*.

Bana e Costa, C.A., Vansnick J.C. (1994), "MACBETH - An interactive path towards the construction of cardinal value functions", *International Transactions in Operations Research*, 1, 4 (489-500).

Bana e Costa, C.A., Vansnick J.C. (1997), "Applications of the MACBETH approach in the framework of an additive aggregation model", *Journal of Multi-Criteria Decision Analysis*, 6, 2 (107-114).

Bana e Costa, C.A., Vansnick J.C. (forthcoming), "The MACBETH approach: basic ideas, software, and an application", in N. Meskens and M. Roubens (eds.), *Advance in Decision Analysis*, Kluwer Academic Publishers.

Banxia Software Ltd. (1995), *Graphics COPE User Guide*, Glasgow.

Belton, V., Ackermann, F., Shepherd, I. (1997), "Integrated support from problem structuring through to alternative evaluation using COPE and V.I.S.A" (Wiley-prize winning paper, 1995), *Journal of Multiple Criteria Decision Analysis*, 6, 3 (115-130).

Bouyssou, D. (1990), "Building criteria: A prerequisite for MCDA", in C.A. Bana e Costa (ed.), *Readings in Multiple Criteria Decision Aid*, Springer-Verlag, Berlin (58–80).

Dyer, J., Lorber, H.W. (1982), "The multiattribute evaluation of program-planning contractors", *OMEGA*, 10, 6 (673-678)

Eden, C. (1988), "Cognitive Mapping: A review", *European Journal of Operational Research*, 36 (1-13).

Eden, C., Ackermann, F., Croppers (1992), "The analysis of cause maps", *Journal of Management Studies*, 29, 3 (309-324).

Keeney, R.L. (1992), *Value-Focused Thinking: A Path to Creative Decision-making*, Harvard University Press, Cambridge, MA.

Rosenhead, J. (ed.) (1989), *Rational Analysis for a Problematic World*, John Wiley & Sons, Chichester.

Roy, B. (1996), *Multicriteria Methodology for Decision Aiding*, Kluwer Academic Publishers, Boston.

von Winterfeldt, D., Edwards, W. (1986), *Decision Analysis and Behavioral Research*, Cambridge University Press, Cambridge, MA.

Interactive MCDM Support System in the Internet

Kaisa Miettinen and Marko M. Mäkelä

Department of Mathematics, University of Jyväskylä, P.O. Box 35,
FIN-40351 Jyväskylä, Finland

Abstract. NIMBUS is an interactive multiobjective optimization system. Among other things, it is capable of solving complicated real-world applications involving nondifferentiable and nonconvex functions. We describe an implementation of NIMBUS operating in the Internet, where the World-Wide Web (WWW) provides a graphical user interface. Different kind of visualizations of alternatives produced are available for aiding in the solution process.

WWW-NIMBUS is a pioneering interactive optimization system in the Internet. The main principles of the implementation have been centralized computing and distributed interface. Typically, the delivery and update of any software is problematic. Limited computer capacity may also restrict the range of applications. Via the Internet, the calculations can always be performed with the latest version and a modest microcomputer may have capabilities of a mainframe computer. No special tools, compilers or software besides a WWW browser are needed in order to utilize WWW-NIMBUS.

1 Introduction

Even though there exists a wide collection of algorithms for continuous multiobjective optimization problems (see, for example, [1], [8], [9] and references therein), their computer implementations are difficult to find. And even if one succeeds in finding some software, it is hard to keep up with updates. Another restricting fact is that the functionality of the program depends on the computing capacity of the user's computer. An alternative is to utilize the capacity of larger computers via computer networks. In this approach, the user interface is usually clumsy and a new user environment may be unfamiliar. From the viewpoint of the software producer, the delivery of the updates and the preparation of separate versions to different operating systems is laborious.

A novel way of avoiding the problems above is to employ the Internet. Internet itself enables calculation in different computers, and the World-Wide Web (WWW) provides a required interface. We can centralize the computing part to one server computer and distribute the interface to the computers of each individual user. When the user has a WWW browser, (s)he has access to the computing capacity and programs of the server computer. It is always the latest version of optimization software that is available. The user does not need high computing capacity or specific compilers in her or his computer. The browser is the only requirement. The graphical interface via the WWW is flexible and independent of the computer and operating system used.

NIMBUS (Nondifferentiable Interactive Multiobjective BUndle-based optimization System) with a WWW interface is the realization of the ideas of centralized computing and distributed interface. NIMBUS is an interactive multiobjective optimization method. It is capable of solving nonlinear and even complicated real-world applications involving nondifferentiable and nonconvex functions. The NIMBUS algorithm has been introduced in [3], [4] and [7].

NIMBUS offers flexible ways of interactive evaluation of the problem and the preferences of the decision maker during the solution process. The interaction phase is aimed at being comparatively simple and easy to understand for the decision maker. While the system now operates in the Internet, the WWW provides a graphical user interface with supporting illustrations. Both the optimization and the production of graphical visualizations are carried out in a server computer (located at the University of Jyväskylä).

The continuation of the paper is organized as follows. Chapter 2 is devoted to a brief description of the NIMBUS algorithm. The interface and its realization in the WWW are introduced in Chapter 3. The paper is concluded in Chapter 4.

2 NIMBUS Algorithm

2.1 Concepts and Notations

We study a multiobjective optimization problem of the form

$$(2.1) \quad \begin{aligned} \text{minimize} \quad & \{f_1(\mathbf{x}), f_2(\mathbf{x}), \ldots, f_k(\mathbf{x})\} \\ \text{subject to} \quad & \mathbf{x} \in S, \end{aligned}$$

where we have k *objective functions* $f_i \colon \mathbf{R}^n \to \mathbf{R}$. They are assumed to be locally Lipschitz continuous (see, for example, [4]). The *decision vector* $\mathbf{x}$ belongs to the convex and compact (nonempty) feasible set S.

In the following, we denote the image of the feasible set by $Z \subset \mathbf{R}^k$. The elements of Z are called *criterion vectors* and denoted by $\mathbf{f}(\mathbf{x}) = (f_1(\mathbf{x}), f_2(\mathbf{x}), \ldots, f_k(\mathbf{x}))^T$. The components of an *ideal criterion vector* $\mathbf{z}^\star \in Z$ are obtained by minimizing each objective function separately. To avoid trivial cases, we suppose that the ideal criterion vector is infeasible.

As the concepts of optimality we employ Pareto optimality and weak Pareto optimality. A decision vector $\mathbf{x}^\star \in S$ and the corresponding criterion vector $\mathbf{f}(\mathbf{x}^\star) \in Z$ are *Pareto optimal* if there does not exist another decision vector $\mathbf{x} \in S$ such that $f_i(\mathbf{x}) \leq f_i(\mathbf{x}^\star)$ for all $i = 1, \ldots, k$ and $f_j(\mathbf{x}) < f_j(\mathbf{x}^\star)$

for at least one objective function f_j. *Weak Pareto optimality* is defined by employing k strict inequalities.

Because of the possible nonconvexity of the problem (2.1), (weak) Pareto optimality is to be understood in a local sense in what follows. We assume that we have a *single decision maker* (DM) and that less is preferred to more to her or him.

2.2 Classification

The NIMBUS method is based on the classification of the objective functions. The DM examines the values of the objective functions calculated at a point $\mathbf{x}^h$ at the iteration h and divides the objective functions into up to five classes. The classes are functions f_i whose values should be decreased ($i \in I^<$), functions whose values should be decreased down till some aspiration level ($i \in I^\leq$), functions whose values are satisfactory at the moment ($i \in I^=$), functions whose values are allowed to increase up till some upper bound ($i \in I^>$), and functions whose values are allowed to change freely ($i \in I^\circ$).

The DM is asked to specify the aspiration levels $\bar{z}_i^h < f_i(\mathbf{x}^h)$ for $i \in I^\leq$ and the upper bounds $\varepsilon_i^h > f_i(\mathbf{x}^h)$ for $i \in I^>$. The difference between the classes $I^<$ and $I^\leq$ is that the functions in $I^<$ are to be minimized as far as possible but the functions in $I^\leq$ only till the aspiration level.

The classification is the core of NIMBUS. However, the DM can tune the order of importance in the classes $I^<$ and $I^\leq$ with optional positive weighting coefficients w_i^h summing up to one. Notice that the weighting coefficients do not change the primary orientation specified in the classification phase. If no weights are specified, we set them equal to one.

According to the classification and the connected information, we form a new single objective optimization problem

$$
\begin{aligned}
\text{minimize} \quad & \max_{\substack{i \in I^< \\ j \in I^\leq}} \left[w_i^h (f_i(\mathbf{x}) - z_i^\star), w_j^h \max \left[f_j(\mathbf{x}) - \bar{z}_j^h, \, 0 \right] \right] \\
\text{(2.2)} \quad \text{subject to} \quad & f_i(\mathbf{x}) \leq f_i(\mathbf{x}^h), \ i \in I^< \cup I^\leq \cup I^= \\
& f_i(\mathbf{x}) \leq \varepsilon_i^h, \ i \in I^> \\
& \mathbf{x} \in S,
\end{aligned}
$$

where $z_i^\star$ for $i \in I^<$ are components of the ideal criterion vector.

When the problem (2.2) is solved by any nondifferentiable optimizer (see, for example [2]), a solution, denoted by $\hat{\mathbf{x}}^h$, is obtained. It is weakly Pareto optimal under minor conditions (see [7]). If the DM dislikes the new solution for some reason, no dead end is at hand. (S)he can explore and compare a desired amount of alternative solutions between $\mathbf{x}^h$ and $\hat{\mathbf{x}}^h$.

2.3 Algorithm

The NIMBUS algorithm is depicted in the form of a flowchart in Fig. 2.1. The detailed algorithm can be found in [7].

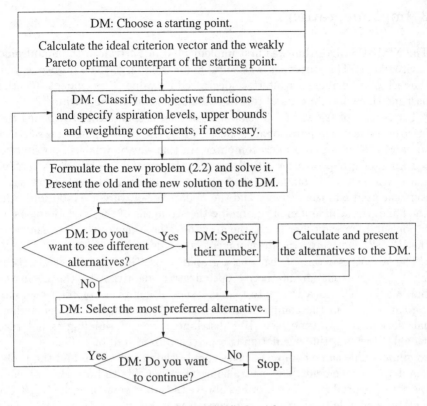

Fig. 2.1. NIMBUS algorithm

The Pareto optimality of generated solution vectors can be checked and guaranteed by employing standard auxiliary optimization problems (see [3]).

Weakly Pareto optimal counterparts of the starting point and the intermediate solutions can be calculated by using the classification $I^< = \{1, 2, , \ldots, k\}$ and solving the problem (2.2).

Notice that the DM must be ready to give up something in order to attain improvement for some other objective functions when we are in the weakly Pareto optimal set. The search procedure stops if the DM does not want to improve any criterion value.

As far as the history of NIMBUS is concerned, the first NIMBUS version was developed in [3], [4] and [5] and it was applied in a structural design problem in [6]. The first version differs mainly from the second one (presented here) in the form of the auxiliary problem (2.2). In the first version, a multiobjective auxiliary problem was formed according to the classification information. The second version was derived and applied in an optimal control problem of continuous casting in [7]. The two versions differ in performance. The second version is more controllable to its users while the first one is computationally more efficient.

3 Implementation

The NIMBUS algorithm was first implemented in Fortran in the mainframe environment. This approach was suitable for even large-scale problems, but the lack of a flexible user interface decreased its value. It is evident that the user interface plays a critical role in realizing interactive algorithms.

The second phase was to develop a functional user interface by paying the price of reduced computational capacity. Thus, the environment was selected to be MS-Windows in a microcomputer. In this way, it was also possible to reach larger groups of users. Both the mainframe and the Windows version share weaknesses in common. They set requirements on the computer environment used and the delivery and the update of software are laborious. On one hand, the willingness of combining the strengths of these two implementations and, on the other hand, the growth of the usage of the Internet (and the WWW) motivated the development of the described version.

The WWW is a way of realizing a graphical user interface. It is also possible to implement an interactive optimization algorithm by the means of the WWW. The idea is not to demand special compilers or high computing capacity from the computer of the user. When considering demanding calculation, a Java-type (see [10]) distributed way of thinking is not reasonable. Instead, the calculation can be centralized into one efficient server computer. The same computer takes care of the visualization and the problem data management. The centralizing offers benefits to both the user and the software producer. The user has always the latest version of the NIMBUS method available, and exactly one version is to be updated and developed, respectively. In addition, the user saves the trouble of the installation of the software. Naturally, the WWW environment sets limits of its own to the flexibility of the user interface. However, its benefits are much more significant.

3.1 Technical Realization

The implementation has been carried out by retaining into standard programming facilities and uncommercial tools. The system has been divided into several independent main modules. Each of the modules has been realized using most appropriate tools. The interface is the master module, being responsible for the functioning of the other modules: parser, code generator, solver and visualization. The interface has been programmed in the object-oriented Python language (see [11]). Reasons for selecting this language are the dynamical handling of strings and the existing Common Gateway Interface (CGI) modules for employing environmental variables. The interface contains submodules, namely, widgets and problem data management. The widgets are tools for efficient HTML document generation for the WWW environment. The problem data management means saving, loading and deleting the user-specified problems.

The parser has been implemented in C language. A compiled language is more efficient for parsing purposes than interpreting languages (such as Python). The parser checks the correctness of the inputs of the user. It also performs the symbolic differentiation of the objective and the constraint functions.

The code generator (in Python) modifies the parsed problem information into the form needed in the solver. The solver is basically the same in all the NIMBUS implementations, and it has been written in Fortran 77. This language was originally selected because the nondifferentiable optimizers based on the efficient proximal bundle method (see [2]) was coded in Fortran 77.

The different modules and their relationships have been depicted in Fig. 3.1. The arrows represent how the data is transferred, basically with the help of external files.

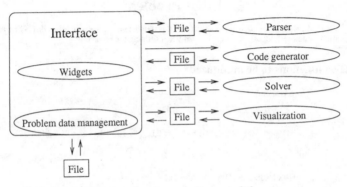

Fig. 3.1. Implementation modules

3.2 User Interface

The main features of the WWW-NIMBUS system from the point of view of its user are illustrated in Fig. 3.2. All the problem information is saved in the server computer. In order to guarantee the data security of the user's problems, (s)he has to specify a user name and a password for WWW-NIMBUS. Every user has a directory of her or his own. It is also possible to visit WWW-NIMBUS as a guest. Guests cannot save any problems to the system.

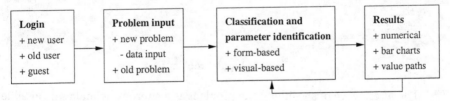

Fig. 3.2. Main features of WWW-NIMBUS in the eyes of its user

A new problem can be given to the system either by filling a form (as represented in Fig. 3.3.) or as a subroutine (of a specified type). The user has to indicate the dimensions of the problem (variables, objective functions, nonlinear and linear constraints) and optional names of the variables and the objective functions. The alternative of user-specified gradients of the objective and the constraint functions is to employ symbolic differentiation. Forthcoming are also options for automatic differentiation and difference approximations.

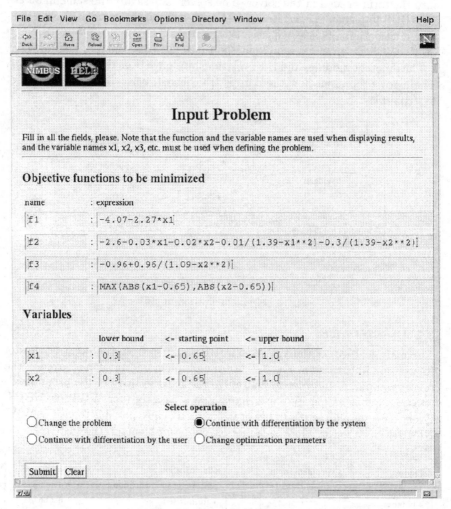

Fig. 3.3. Input Problem page

The alternative of specifying the problem as a subroutine makes it possible to solve complicated and/or large-scale problems that do not have explicit formulas of functions and derivatives. Such cases are, for example, optimal control problems where the function value evaluations involve solving iteratively some partial differential equation systems.

Classification can be done either graphically or using a form page. Graphical classification means that the user examines the current visualized solution and specifies desired levels for each objective function with a mouse. Based on this information, the system automatically classifies the functions into up to five classes and the required parameter information is obtained from the levels. In a form-based classification, the classes are first explicitly specified and then the connected parameter information is given. The feasibility of the classification is checked in both cases.

After the new problem (2.2) has been generated and solved, both the old and the new solution are presented numerically. If desired, the user can consider intermediate alternatives or take a look at graphical visualizations. There are bar charts and value paths available in different scales. The user can choose a desired combination of the figures. An example of value paths in the relative range of values is shown in Fig. 3.4. To support the decision making process, the user has a possibility of dropping undesirable alternatives from consideration and, thus, decreasing the amount of solution candidates. After the user has examined the results, (s)he selects the most preferred solution for continuation. It is either a basis for a new classification or a final solution whose Pareto optimality is checked.

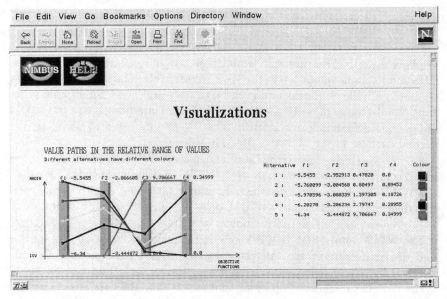

Fig. 3.4. Value paths on the Visualizations page

3.3 New Features

It was mentioned in Section 2.3 that the presented method is a second version of NIMBUS. The WWW implementation is also new. In addition to the fact that the underlying algorithm has changed, the new implementation differs in its interface.

Symbolic differentiation is a new optional possibility. Some simple nondifferentiable functions can also be subdifferentiated. Such functions are, for example, absolute and maximum values.

The graphical classification has several benefits. The user does not have to master the idea of different classes and their symbols. The classification takes place in a natural way. The user considers the Pareto optimal ranges of each objective function and specifies desires and hopes by moving levels within the ranges. The system takes care of the correctness of the classification and the specified levels can always be tuned numerically afterwards.

The new implementation contains many-sided features in the visualization of the solution alternatives. The user can choose a desired combination of six illustrations, which are bar charts and value paths in relative and absolute ranges. The loading of the images takes some time and, thus, the user can decide how much time (s)he can spend on visualization.

The efficiency of the implementation has been an important target. The system has been divided into small parts. The user's browser needs to load only those parts that are being used at a time. This realization aims at making good use of the interpreting language used, Python.

4 Conclusions

We have described an interactive multiobjective optimization method NIMBUS and its implementation in the World-Wide Web. The introduced system is able to handle nonlinear and even large-scale and nondifferentiable problems. Unlike many other interactive methods, the questions posed to the DM are not too demanding in NIMBUS. The DM is only asked to designate the desired changes in the objective function values.

Anyone who has access to the Internet can use WWW-NIMBUS. No special tools, compilers or software besides a WWW browser are needed. Further, the WWW environment assists user-friendly realization. A part of the user-friendliness is the fact that the DM does not have to specify complicated (sub)gradient information for nonlinear or nondifferentiable functions.

The WWW implementation is based on the ideas of centralized computing and distributed interface. All the calculations and data management are handled by a particular server computer. Having only one version of the system implies easiness of updating and means that the user always has the latest version available. The interface is formed with the WWW browser on the computer of the user. In this way, a modest microcomputer has the computing power of an efficient mainframe computer.

WWW-NIMBUS has been locally used successfully for research and teaching purposes from small to large-scale problems. Usage statistics indicate also active users in foreign companies and universities. So far, no inquiries about the type of problems solved or usage experiences have been made.

WWW-NIMBUS can be used free of charge. The URL of the system is http://nimbus.math.jyu.fi/.

Acknowledgements

The authors thank Mr. Jari Huikari, Mr. Joose Niemistö, Mr. Janne Mäkinen and Mr. Heikki Toivonen for helping in the implementation. This research was supported by the grants 8583 and 22346 of the Academy of Finland.

References

1. Hwang C.-L. and Masud A.S.M., "Multiple Objective Decision Making – Methods and Applications," Springer-Verlag, Berlin, Heidelberg, 1979.
2. Mäkelä M.M. and Neittaanmäki P., "Nonsmooth Optimization: Analysis and Algorithms with Applications to Optimal Control," World Scientific Publishing Co., 1992.
3. Miettinen K., "On the Methodology of Multiobjective Optimization with Applications," Doctoral Thesis, University of Jyväskylä, Department of Mathematics, Report 60, 1994.
4. Miettinen K. and Mäkelä M.M., *Interactive Bundle-Based Method for Nondifferentiable Multiobjective Optimization: NIMBUS*, Optimization **34** (1995), 231–246.
5. Miettinen K. and Mäkelä M.M, *Interactive Method NIMBUS for Nondifferentiable Multiobjective Optimization Problems*, in "Multicriteria Analysis," Proceedings of the XIth International Conference on MCDM, Edited by J. Clímaco, Springer-Verlag, Berlin, Heidelberg, 1997, 310–319.
6. Miettinen K., Mäkelä M.M and Mäkinen R.A.E., *Interactive Multiobjective Optimization System NIMBUS Applied to Nonsmooth Structural Design Problems*, in "System Modelling and Optimization," Proceedings of the 17th IFIP Conference on System Modelling and Optimization, Edited by J. Doležal and J. Fidler, Chapman & Hall, 1996, 379–385.
7. Miettinen K., Mäkelä M.M. and Männikkö T., *Optimal Control of Continuous Casting by Nondifferentiable Multiobjective Optimization*, Computational Optimization and Applications (to appear).
8. Steuer R.E., "Multiple Criteria Optimization: Theory, Computation, and Applications," J. Wiley & Sons, 1986.
9. Steuer R.E., Gardiner L.R. and Gray J., *A Bibliographic Survey of the Activities and International Nature of Multiple Criteria Decision Making*, Journal of Multi-Criteria Decision Analysis **5** (1996), 195–217.
10. Java Home Page URL http://java.sun.com.
11. Python Language Home Page URL http://www.python.org.

Using AHP for Text Book Selection

H. Roland Weistroffer and Lynda Hodgson

Department of Information Systems, Virginia Commonwealth University,
Richmond, VA 23284-4000, USA

Abstract. The selection of an appropriate textbook for a university level course
is an important determinant for the success and quality of that course.
Unfortunately, the selection process is often perfunctory, with last minute
decisions made based on quickly paging through available textbook alternatives.
This paper describes an attempt at using a formal procedure for textbook
selection, based on the analytic hierarchy process and utilizing the software
package Expert Choice. The process described includes the identification of
relevant selection criteria, the identification and evaluation of candidate
textbooks, and the quest for consensus among multiple decision makers in cases
where sections of a course are taught by different instructors. The course used in
this case study is a first course in systems analysis and design for third year
undergraduate information systems students and involves three decision makers.
However, many aspects of the model described apply to other courses in other
settings. The hierarchical model was designed in such a way that the course
specific part is separable from a more generally applicable part. The experiences
in implementing and using the model are discussed.

Key Words: AHP, application, textbook selection

1 Introduction

1.1 The Textbook Selection Problem

Textbooks are generally considered to be an integral part of teaching, and thus
the selection of an appropriate textbook is an important decision process, the
outcome of which can strongly affect learning [6]. Various approaches to
choosing a particular book at the university level have been identified. These
approaches include keeping the currently used textbook, choosing a text authored
by a colleague, selecting a book based on relevant reviews and comparisons, and
voting within the department or among the faculty teaching a certain course [7].
Clearly, these approaches are not all equally desirable.

Since faculty in university business schools (among others) is charged with teaching students how to make good decisions, they should be expected to use good decision techniques themselves, especially when making decisions that affect the learning process. Thus, a rational approach (or at least a structured approach) based on published decision models seems appropriate.

The textbook selection process can be considered to have three major components:

1) Identifying relevant selection criteria and prioritizing these criteria.
2) Identifying and evaluating candidate books using these criteria.
3) Reaching consensus among the involved parties.

In this paper we develop a process for the general textbook selection problem based on an existing decision model, the analytic hierarchy process (AHP), and try to validate the feasibility of this approach through a specific application.

1.2 Choice of Decision Model and Software

The first two decision-making components in the textbook selection problem, as listed above, are addressed directly by AHP. Furthermore, AHP also provides assistance for the third component, group agreement. As described by Saaty [11] [12], AHP provides a framework that supports multi-criteria decisions. It has been used in a variety of different applications, including strategic planning, facilities planning, vendor or distributor selection, and local tax planning (see for example [1] [2] [3] [4] [14] [16] and [17]). Some of the applications using AHP involve decisions made in a university setting, such as evaluating research papers [10] and selecting faculty [5].

AHP is able to structure complex problems that involve subjective judgments, while maintaining simplicity. The ability to structure a complex problem and then focus attention on individual components amplifies decision-making capability [9]. A 1982 study comparing five conceptually different approaches for multi-criteria weighting [13] found that AHP was perceived as the easiest method to use, and the one whose results were considered most trustworthy. A later study [18] in which twenty-four university business professors compared five decision-support packages to a spreadsheet package by solving six decision problems found that Expert Choice[1], an AHP-based package, was one of the highest rated. In that study, the spreadsheet tool was provided without a pre-designed template. Other authors have shown that spreadsheet packages can in

---

[1] Expert Choice, Inc., 4922 Ellsworth Avenue, Pittsburgh, PA 15213.

fact be used to implement decision models such as AHP (see for example [15]). Nevertheless, in the current paper we chose to use Expert Choice.

1.3 The Analytic Hierarchy Process

AHP allows decision-makers to structure a complex problem as a decision hierarchy, with the ultimate decision goal at the apex. At the next level are the decision criteria that contribute to the goal. AHP allows multiple levels of criteria. Thus, general criteria of the same relative magnitude or scope can be identified at one level, then some of these criteria can be broken down into sub-criteria at the next level. Decomposition into further levels of sub-criteria can continue until all contributing factors have been identified and assigned an appropriate place in the hierarchy. At the bottom level of the hierarchy are the actual decision alternatives.

Once the criteria have been identified, they must be evaluated for their relative contribution to the goal or the next higher-level criterion. These subjective evaluations are turned into quantitative weights so that more important criteria can be given more consideration in making the final decision. Expert Choice provides for a variety of user interfaces to turn subjective judgments into weights. The user has the choice of directly entering numerical weights for each criterion, or have Expert Choice derive these weights by requesting judgments in the form of pair-wise comparisons between criteria.

In pair-wise comparisons, every criterion is evaluated by comparing it to every other criterion that has the same parent node in the hierarchy. The user states, verbally, graphically, or by numerical scale whether a criterion is judged to be very strongly more important, strongly more important, moderately more important or equally important. Expert Choice turns the evaluations into a quantitative scale. Equal importance is rated low (1) on the scale, and high relative importance is rated high (9) on the scale. The result is a matrix as shown in the example below:

	Criterion A	Criterion B	Criterion C
Criterion A	1	2	6
Criterion B	1/2	1	3
Criterion C	1/6	1/3	1

Fig. 1. Example Comparison Matrix

In this example, Criterion A is equally to moderately more important than Criterion B, but is strongly more important than C. Criterion B is moderately more important than C. If the relationship is reversed, reciprocal numbers are used, i.e. since the relative importance of Criterion A to Criterion C is 6, the relative importance of C to A is 1/6. The diagonal of the comparison matrix always consists of 1s, and the bottom left of the matrix is always the reciprocal of the top right.

The relative scores in the table should be consistent with each other. Consistency has two meanings. The first is that similar ideas or objects are grouped according to homogeneity or relevance, and the second is that the intensities of relations among ideas based on a particular criterion justify each other in some logical way [12]. Consistent judgments allow the software to derive weights so that the sum of the weights for the contributing criteria (or sub-criteria) sum to 1. However, in realistic situations, subjective judgments can be expected to be somewhat inconsistent. Expert Choice determines weights by calculating eigenvalues of the comparison matrices. The package also provides an inconsistency measure. Inconsistency values of .1 or less are considered acceptable [9]. With Expert Choice, the decision-maker can simply revise the comparisons, or can request suggestions for improving consistency. The process of improving consistency helps the decision-maker think through and clarify the reasons for making a particular judgment or set of judgments.

Similarly to evaluating the criteria (or attributes) with respect to their relative impact on the overall objective by pair-wise comparisons, each alternative at the bottom of the hierarchy is evaluated according to its relative contribution to each attribute (criterion), also by pair-wise comparisons. Figure 2 shows an example of a three level hierarchy (i.e. no sub-criteria).

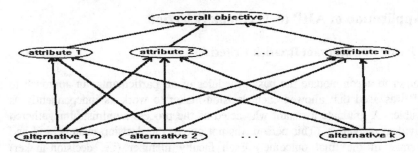

Fig. 2. Example Hierarchy

The evaluations of the alternatives with respect to each attribute and the evaluation of the attributes are then combined to get a utility measure for each alternative, i.e. a measure that indicates the degree to which each alternative satisfies or contributes to the overall objective. Expert Choice allows the user to view the results graphically and numerically.

1.4 Group Decision Making with AHP

At the university level, the choice of a textbook often involves multiple faculty members or a selection committee, complicating the decision process. Liberatore, et al. [10] note that "academic committees can rarely achieve consensus on anything expeditiously." The problems they encounter include the difficulty of agreeing on criteria, excessively time-consuming discussions, and the challenge of maintaining congeniality among committee members. AHP has been shown to save faculty time when used for the process of evaluating research papers. It has also been successfully used to gain committee consensus on the criteria for allocating funds to improve effectiveness in a university business school. A faculty selection committee used AHP to turn a meandering discussion of potential faculty members into a focused analysis that quickly identified the leading candidate [5].

Two types of group consensus building using AHP have been proposed. Saaty [12] suggests that the tool be used in group sessions, where all opinions are recorded and discussed until a consensus judgment is determined. When there is disagreement, this can be resolved by selecting the judgments that are most consistent with the judgments on which there is general agreement. The other approach is simply to average the judgments of all participants [10]. The first method requires more time, but hopefully allows all participants to feel comfortable with the final results. The second method probably is more expeditious.

2 Application of AHP to Textbook Selection

2.1 Identifying Relevant Decision Criteria

In order to accommodate the busy schedules of all participants, an approach to AHP was used that allowed each decision-maker to work as independently as possible. A graduate assistant who acted as the process administrator gathered individual responses. (This person was not one of the decision-makers, and had no stake in the final outcome.) Each faculty member (i.e. decision-maker) identified decision criteria that he considered important. These lists of criteria from all parties were consolidated into a single list, and shown to all participants for possible modifications (in a way, similar to a Delphi process, but less formal). A likewise procedure was adopted to arrive at a list of candidate textbooks.

The decision process began with a short interview session with each of the three faculty members involved in choosing the new text. They were asked to specify both, general textbook selection criteria, and criteria that would be relevant specifically to the choice of a systems analysis and design text. The responses were compared and integrated into a decision hierarchy, with general criteria at the top level (just below the overall objective), and specific criteria at the lower levels. The intention was to clearly separate course specific criteria from those that would be part of a general textbook selection model.

Expert Choice allows for only eight sub-criteria for any criterion or goal, and therefore the criteria had to be organized to accommodate this limitation. This organization process proved beneficial to the overall model because it forced further clarification of the terms used by the three decision-makers, and some duplication of concepts was removed. Grouping related criteria into higher levels of abstraction also resulted in improvements. Thus the criteria "instructor's manual", "example test questions", and "presentation materials" were grouped to be sub-criteria under the umbrella criterion "instructional support". The result was two levels of general textbook criteria.

Criteria that are specific to systems analysis texts were added at the third level. The resulting hierarchy is shown in Figure 3. Note that the same specific criteria needed to be evaluated for both the breadth and the depth general criteria. This was done because the decision-makers felt that it was important to cover certain subjects (breadth), but at the same time felt that detailed coverage of some of those subjects was less important (depth).

For the weighting process, the model was used to create three independent copies of the model – one for each decision-maker. The three decision-makers then worked completely independently using their personal model copies. All of them were allowed to choose the weighting process with which they felt most comfortable. One of the decision-makers chose pair-wise comparison, while the other two preferred to directly supply numeric weights.

Both weighting techniques resulted in modification of the original set of criteria, because the weighting process provided new insights into the model. It was discovered that certain criteria were actually elimination criteria. Books that did not meet these readily identified criteria were automatically rejected by all of the decision-makers. It was agreed by all three that no texts over three years old or texts based on any proprietary vendor tool or methodology would be considered. One of the specific criteria, data-flow diagramming, was renamed process modeling to reflect its intent more accurately. All three copies of the model were updated to reflect these changes. One of the decision-makers then decided to restructure his model to contain two levels of specific criteria instead of one. This person also chose to eliminate a few criteria that he felt were not relevant for his personal decision process. These differences were not significant enough to force a reworking of the other two model copies. Instead, the eliminated criteria were given weights of zero and the other items were rearranged making sure the number and names of all final leaves were consistent across the three model copies. The ordering and number of levels should not affect the outcome, as long the leaves are given the appropriate weights.

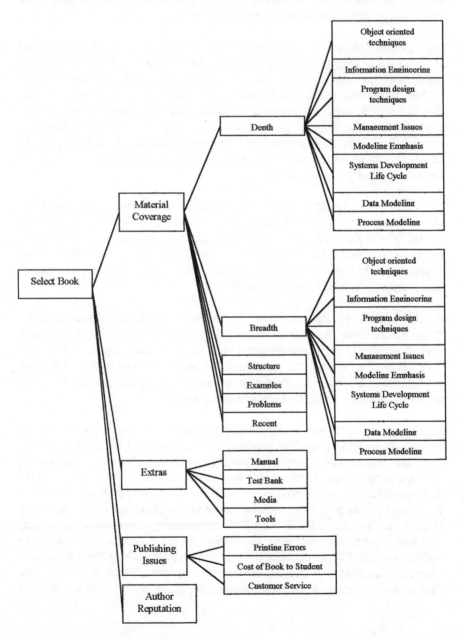

Fig. 3. Textbook Evaluation Criteria Hierarchy

As can be seen in Figure 3, the final list includes fourteen general criteria, of which six relate to material coverage, four relate to supplements (extras) provided by the publisher, three relate to the publisher directly (publishing issues), and one relates to the reputation of the author.

Depth of Material Coverage. Listed as "Depth" in Figure 3, this criterion relates to the detail of coverage of specific topics identified as being important for the course for which the textbook is being considered. For the systems analysis and design course, eight topics, listed as sub-criteria in Figure 3, were included: (1) object oriented techniques, (2) information engineering, (3) program design techniques, (4) management issues, (5) modeling emphasis, (6) systems development life cycles, (7) data modeling, and (8) process modeling.

Breadth of Material Coverage. Listed as "Breadth" in Figure 3, this criterion relates to the spectrum of topics covered that were identified as being important for the course for which the textbook is being considered. Generally, this would be the same topics specified for depth of coverage.

Structure. This relates to the structure in which the material is presented in the textbook, i.e. the sequence in which topics are introduced, the breakdown into chapters, and also the structure of chapters with regards to review questions, practice problems, etc.

Examples. This criterion relates to the quality and quantity of examples used in the textbook to help students understand the material.

Problems. This relates to the quantity and quality of practice problems that are provided in the textbook to help students assess their understanding of the material and to support the instructor in giving homework assignments to students.

Recent. This relates to the currency of the material provided, i.e. how recent the information in the textbook is. This is of major importance to courses in information systems, as technology changes so quickly.

Manual. Most textbooks nowadays tend to come with an instructor's manual that helps professors use the textbook to maximum advantage in their course. The quality of the instructor's manual may be an important consideration when adopting a textbook.

Test Bank. Most textbooks nowadays tend to come with a test bank, either in printed form, or increasingly more commonly in electronic form. This helps instructors designing their own tests that they may give to their students. Some test banks come with sophisticated software that allows instructor's to print different versions of tests, or even to give on-line tests. The quality of the test bank and the associated software may be an important consideration when adopting a textbook.

Media. Increasingly, instructors use multimedia in their class rooms, including transparencies, video clips, and slide presentations. Many textbooks come with some of these presentation materials. The type of presentation materials provided, and the quality thereof may be an important consideration when adopting a textbook.

Tools. Some courses require the use of certain software packages, and increasingly, textbooks provide this software free when a book is adopted for classroom use. The type and quality of the software provided may be an important consideration when choosing a textbook.

Printing Errors. Though usually of minor importance, some textbooks are published under great deadline pressure (in time for fall classroom adoption), sometimes resulting in a product with more than acceptable errors.

Cost of Book to Students. The price of textbooks has gone up quite a bit over the years. Instructors may not want to adopt a textbook that costs way above average, as students are often tight for money.

Customer Service. Instructors increasingly expect good customer service from textbook publishers (and their representatives), like providing free desk copies, presentation materials, software, etc. Some book representatives are faster in responding to the specific needs of university instructors, and rewarding these representatives by adopting their books may be a consideration when choosing a text.

Author Reputation. Though a textbook may speak for itself, some instructors are more comfortable in adopting a text written by a well known person in the field.

As both depth and breadth of material coverage include eight sub-criteria (specific to the analysis and design course), there are a total 28 criteria overall that were considered in the application example described in the remainder of this paper.

2.2 Evaluating Candidate Books

Candidate texts were nominated after the weighting process was completed. Each decision-maker was encouraged to nominate two or three texts. Because of the eight sub-level limitation of Expert Choice, as well as the time required to read through and rate each text, it was necessary to keep the number of nominations from each participant relatively small. Six texts, including the one currently in use, were nominated.

It was decided that because the number of final criteria was relatively high, pair-wise comparison to rate the texts was not practical. Using pair-wise comparison on six books (B) for each of 28 criteria (C) would have required $C*((B*(B - 1)/2)$ = 420 independent evaluations. Instead, a rating sheet was developed which allowed the books to be rated on a simple one-to-nine scale for each of the final criteria. This reduced the number of evaluations to $B*C = 168$. Each decision-maker was provided with a rating sheet, and copies of the nominated texts. When the sheets were filled out, each one was used to complete the rater's copy of the model. Figure 4 shows an example of the rating sheet for the specific criterion of breadth of coverage with respect to data modeling.

Criterion:
Material > Breadth > Data Modeling

	Does not Meet Criterion			Meets Criterion			Exceeds Criterion		
Current text	1	2	3	4	5	6	7	8	9
Hoffer, George, Valacich: Modern SA & D	1	2	3	4	5	6	7	8	9
Shelly, Cashman, Adanski: SA & D	1	2	3	4	5	6	7	8	9
Fertuch: SA & D with Modern Methods	1	2	3	4	5	6	7	8	9
Norman: Object-Oriented SA & D	1	2	3	4	5	6	7	8	9
DeWitt: SA & D and the Transition to Objects	1	2	3	4	5	6	7	8	9

Fig. 4. Rating Sheet Example

2.3 Group Agreement

The initial results showed that there was minimal overall agreement on most of the texts. No books were in the top two across all decision-makers. One book appeared in the top 50% for all decision-makers. Other than this there was no commonality in the rankings.

Further inspection of the responses shed some doubts on the reliability of one set of responses. It appeared that the rater simply rated the current text as above average in essentially all categories and rated all other texts average or below average in all categories. The bias was made clear by two particular responses. In these, the current text was rated above average in the "object oriented" material even though it mentioned object orientation only briefly in the text. The two competing books which were exclusively devoted to object orientation were rated low in this category!

At this point, it was decided that another set of weights and rates should be solicited, so that there would be at least three unbiased viewpoints contributing to the decision. A fourth decision-maker was identified and agreed to participate. Unfortunately, the results were only marginally better. This decision-maker made a good-faith effort at providing criteria weights, and provided book ratings for the two books with which he was familiar. However, this decision-maker did not rate any of the other books. There were two reasons cited for this. First, he did not feel he had time to adequately review the unfamiliar texts. Second, he felt that the selection process was unnecessary because he had already chosen a new book (one of the candidate books in this evaluation), and was currently using it in his classes.

In spite of these problems, the responses were examined to determine if some conclusions could be derived. Each decision-makers final outcome is shown in Figure 5 below.

	DM1		DM2		DM3		DM4	
	Book	*Score*	*Book*	*Score*	*Book*	*Score*	*Book*	*Score*
Top Choice	Current	0.264	Norman	0.295	DeWitt	0.183	Hoffer	0.184
	Hoffer	0.164	DeWitt	0.293	Norman	0.174	Norman	0.178
	Norman	0.155	Current	0.103	Current	0.173	Fertuck	0.167
	DeWitt	0.155	Hoffer	0.103	Hoffer	0.163	DeWitt	0.163
	Fertuck	0.146	Shelly	0.103	Fertuck	0.160	Shelly	0.160
Last Choice	Shelly	0.117	Fertuck	0.103	Shelly	0.146	Current	0.147

Fig. 5. Comparison of Individual Ratings

Figure 6 below shows the result of combining the four individual model copies into a single hierarchy (giving each decision-maker equal weight).

Book	Score
Norman	0.802
DeWitt	0.794
Current	0.687
Hoffer	0.614
Fertuck	0.576
Shelly	0.526

Fig.6. Outcome of Ratings for Combined Model

Further analysis was conducted to determine where disagreement existed. To determine whether there was agreement on the text ratings, the criteria weights were averaged and used to recalculate each decision-maker's model. This was accomplished by exporting each of the rater's models into a spreadsheet (a function built into Expert Choice). From this, an average was computed for each criterion. A new Expert Choice model was built by setting up a simple two-level hierarchy that contained all of the criteria with values corresponding to the Excel averages. Then each of the decision-maker's textbook ratings were applied to the model separately. In this way, if the outcomes were similar, it could be assumed that the major differences were in the criteria weights, rather than the textbook ratings. However, if the outcomes were still significantly different, then it could be assumed that there was little agreement on the textbooks. A similar procedure was performed with average textbook ratings applied to each of the decision-maker's set of criteria weights. This was done to determine if major differences existed in the criteria ratings instead of, or in addition to the differences in the textbook ratings.

The results seemed to indicate that the disagreement was mainly in the ratings of the books, not in the ratings of the criteria. Using average textbook ratings, but individual criteria ratings resulted in almost identical rankings of the textbooks among all four decision-makers.

3. Conclusions

In this paper we described a structured approach to selecting a textbook, making use of the analytic hierarchy process and the software package Expert Choice. As stated in the introduction, the selection process usually consists of three parts, i.e. determining the selection criteria, evaluating the books with respect to the selection criteria, and reaching consensus among multiple decision makers. The first part was accomplished successfully in this project, resulting in the model summarized in Figure 3. This by itself can in general be very helpful in any textbook selection process.

The second part was only partially successful. Reasons for this may be the busy schedules of the faculty members involved, and perhaps a relatively low priority assigned to this activity by these faculty members. Out of town commitments and teaching workloads resulted in lengthy delays in receiving individual responses. Further delays were encountered as questions arose over definitions. A graduate assistant acted as an intermediary to relay information between the decision makers until a clear definition was reached, however, using an intermediary proved to be very time consuming, and probably much less effective than person-to-person discussion. The importance of direct, personal interaction, of-course, should not come as a surprise, as we routinely teach that to our students in the systems analysis course. However, just because we preach it doesn't always imply we practice it. An entire semester elapsed before all three of the original decision-makers provided a complete set of criteria ratings. Another semester elapsed before all three sets of textbook ratings were provided.

Furthermore, there seemed to be reluctance (almost fear) by some of the decision-makers to commit themselves in such a formal process. The comment was made that to fairly evaluate the text books in this way would require using each of the textbooks for teaching an actual class first.

The third part, reaching consensus, really had not much of a chance to succeed, as it depended on the success of the other two parts. Despite these limitations, the authors feel that the approach is useful, at least in cases of a single decision-maker. Success in the case of multiple decision-makers would likely depend on the degree of commitment by all members of the group to the application of this approach.

References

[1] Arbel, A. and Orgler, Y.E. An application of the AHP to bank strategic planning: the mergers and acquisitions process. *European Journal of Operational Research*, 48(1), 27-37, 1990.
[2] Benjamin, C.O., Ehrie, I.C., and Omurtag, Y. Planning facilities at the University of Missouri-Rolla. *Interfaces* 22(4), 99-105, 1992.
[3] Carlsson, W., and Walden, P. AHP in political group decisions: a study in the art of possibilities. *Interfaces* 25(4), 14-29, 1995.
[4] Davies, M. A multicriteria decisions model application for managing group decisions. *Journal of Operational Research Society*, 45(1), 47-55, 1994.
[5] Dyer, R.F. and Forman, E.H. Group decision support with the Analytic Hierarchy Process. *Decision Support Systems* 8(2), 99-124, 1991.

[6] Farr, R. Textbook selection and curriculum change. *The Journal of State Government* **6**(2), 86-91, 1987.

[7] Let's ask the students what textbooks they want to use. *Marketing News* **21**(14), 24, 1987.

[8] Liberatore, M. J. An extension of the analytic hierarchy process for industrial R&D project selection and resource allocations. *IEEE Transactions on Engineering Management* **34**(l), 12-18, 1987.

[9] Liberatore, M.J. A decision support system linking research and development project selection with business strategy. *Project Management Journal* **19**(5), 14-21,1988.

[10] Liberatore, M.J., Nydick, R.L., and Sanchez, P.M. The evaluation of research papers (or how to get an academic committee to agree on something). *Interfaces* **22**(2), 1992.

[11] Saaty, T.L. *The Analytic Hierarchy Process.* McGraw-Hill, New York 1980.

[12] Saaty. T.L. *The Analytic Hierarchy Process – Planning, Priority Setting, Resource Allocation.* RWS Publications, Pittsburgh, Pennsylvania 1990.

[13] Schoemaker, P.J.H., and Waid, C.C. An experimental comparison of different approaches to determining weights in additive utility models. *Management Science* **28**(2), 182-196, 1982.

[14] Weistroffer, H.R., Wooldridge, B.E., and Singh, R. Local tax planning with AHP and Delphi. In G. Fandel, T. Gal, and T. Hanne (editors), *Multiple Criteria Decision Making*, Springer, Berlin 1997.

[15] Winston, W.L., and Albright, S.C. *Practical Management Science: Spreadsheet Modeling and Applications*, Duxbury Press, Belmont, California 1997.

[16] Yeoh, P.-L., and Calantone, R.J. *Journal of Global Marketing* **8**(3), 39-65, 1995.

[17] Zahedi, F. The analytic hierarchy process and its applications. *Interfaces* **16**(4), 96-108, 1986.

[18] Zapatero, E.G., Smith, C.H. and Weistroffer, H.R. Evaluating multiple attribute decision support systems. *Journal of Multi-criteria Decision Analysis* **6**, 201-214, 1997.

Acknowledgments to Referees: The assistance of the following referees in compiling this volume is gratefully acknowledged.

Per Agrell
Yasmin Aksoy
Elizabeth Atherton
Carlos Bana e Costa
Claude Banville
Euro Beinat
Valerie Belton
Slim Ben-Khalifa
Malin Brännback
Carl Brønn
Cathal Brugha
John Buchanan
Maria Captivo
João Clímaco
Matthias Ehrgott
Love Ekenberg
Paul Fatti
Abe Feinberg
Petr Fiala
Sarah Fores
Simon French
Xavier Gandibleaux
Luiz Gomes
Carlos Gomes
Jacinto González-Pachón
Manfred Grauer
Salvatore Greco
Jürgen Guddat
Linda Haines
Raimo Hämäläinen
Thomas Hanne
Mordecai Henig
Larry Jenkins
Dylan Jones
Janusz Kacprzyk
Nikos Karacapilidis
Birsen Karpak
Gregory Kersten
Murat Köksalan

Pekka Korhonen
Markku Kuula
Risto Lahdelma
Freerk Lootsma
Simo Makkonen
Thierry Marchant
Wotjek Michaelowski
Kaisa Miettinen
Pete Naude
Alan Pearman
Don Petkov
Victor Podinovski
Mari Pöyhönen
Sixto Rios-Insua
Maria Victoria Rodríguez-Uría
Marc Roubens
Pekka Salminen
Leanne Scott
Lisa Simpson
Yannis Siskos
Roman Slowinski
Moshe Sniedovich
David Soloveitchik
Jaap Spronk
Antonie Stam
Ralph Steuer
Jacques Teghem
Jozsef Temesi
Jorgen Tind
Tadeusz Trzaskalik
Alex Tsoukias
Frank Twilt
Marjan van Herwijnen
Jyrki Wallenius
Hannele Wallenius
Roland Weistroffer
Margaret Wiecek
Kobus Wolvaardt

Springer
and the
environment

At Springer we firmly believe that an international science publisher has a special obligation to the environment, and our corporate policies consistently reflect this conviction.

We also expect our business partners – paper mills, printers, packaging manufacturers, etc. – to commit themselves to using materials and production processes that do not harm the environment. The paper in this book is made from low- or no-chlorine pulp and is acid free, in conformance with international standards for paper permanency.

Druck: Strauss Offsetdruck, Mörlenbach
Verarbeitung: Schäffer, Grünstadt

Lecture Notes in Economics and Mathematical Systems

For information about Vols. 1–274
please contact your bookseller or Springer-Verlag